Monitoring the Comprehensive Nuclear-Test-Ban Treaty: Seismic Event Discrimination and Identification

Edited by
William R. Walter
Hans E. Hartse

SPRINGER BASEL AG

Reprint from Pure and Applied Geophysics
(PAGEOPH), Volume 159 (2002), No. 4

Editors:

William R. Walter
Lawrence Livermore National Laboratory
L-205 P.O. Box 808
Livermore, CA 94551
USA

Hans E. Hartse
Geophysics Group, EES-11
MS D408
Los Alamos National Lab
Los Alamos, New Mexico 87545
USA

A CIP catalogue record for this book is available from the Library of Congress, Washington D.C., USA

Deutsche Bibliothek Cataloging-in-Publication Data

Comprehensive nuclear-test-ban treaty: seismic event discrimination and monitoring identification / ed. by William R. Walter ; Hans E. Hartse. - Basel ; Boston ; Berlin : Birkhäuser, 2002
    (Pageoph topical volumes)
ISBN 978-3-7643-6675-9

ISBN 978-3-7643-6675-9   ISBN 978-3-0348-8169-2 (eBook)
DOI 10.1007/978-3-0348-8169-2

Originally published by Birkhäuser Verlag, Basel, Switzerland 2002

Printed on acid-free paper produced from chlorine-free pulp

9 8 7 6 5 4 3 2 1

Contents

Pure appl. geophys. 159 (2002) 619–620
0033 – 4553/02/040619 – 02 $ 1.50 + 0.20/0

© Birkhäuser Verlag, Basel, 2002

❘ Pure and Applied Geophysics

Monitoring the Comprehensive Nuclear-Test-Ban Treaty

Preface

The first nuclear bomb was detonated in 1945, thus ushering in the nuclear age. A few political leaders quickly saw a need to limit nuclear weapons through international cooperation and the first proposals to do so were made later that same year. The issue of nuclear testing, however, was not formally addressed until 1958 when the United States, the United Kingdom, and the Soviet Union, initiated talks intended to establish a total ban on that testing (a Comprehensive Test-Ban Treaty or CTBT). Those talks ended unsuccessfully, ostensibly because the participants could not agree on the issue of on-site verification.

Less comprehensive treaties did, however, place constraints on nuclear testing. The United States, the United Kingdom, and the Soviet Union, in 1963, negotiated the Limited Test-Ban Treaty (LTBT) which prohibited nuclear explosions in the atmosphere, outer space and under water. The Threshold Test-Ban Treaty (TTBT), signed by the United States and the Soviet Union in 1947, limited the size, or yield, of explosions permitted in nuclear tests to 150 kilotons.

Seismological observations played an important role in monitoring compliance with those treaties. Many of the world's seismologists set aside other research projects and contributed to that effort. They devised new techniques and made important discoveries about the Earth's properties that affect our ability to detect nuclear events, to determine their yield, and to distinguish them from earthquakes. Seismologists are rightfully proud of their success in developing methods for monitoring compliance with the LTBT and TTBT.

Although seismologists have also worked for many years on research related to CTBT monitoring, events of recent years have caused them to redouble their efforts in that area. Between 1992 and 1996 Russia, France and the United States all placed moratoria on their nuclear testing, though France did carry out a few tests at the end of that period. In addition, the United States decided to use means other than testing to ensure the safety and reliability of its nuclear arsenal, and all three countries, joined by the United Kingdom, agreed to continue moratoria as long as no other country tested. Those developments, as well as diplomatic efforts by many nations, led to the renewal of multilateral talks on a CTBT that began in January 1994. The talks led to the Comprehensive Nuclear-Test-Ban Treaty. It was adopted by the United Nations General Assembly on 10 September, 1996 and has since been signed by 161 nations. Entry of the treaty into force, however, is still uncertain since it requires ratification by all 44 nations that have some nuclear capability and, as of 1 November 2001, only 31 of those nations have done so.

Although entry of the CTBT into force is still uncertain, seismologists and scientists in related fields, such as radionuclides, have proceeded with new research on issues relevant to monitoring compliance with it. Results of much of that research may be used by the International Monitoring System, headquartered in Vienna, and by several national centers and individual institutions, to monitor compliance with the CTBT. New issues associated with CTBT monitoring in the 21st century have presented scientists with many new challenges. They must be able to effectively monitor compliance by several countries that have not previously been nuclear powers. Effective monitoring requires that we be able to detect and locate considerably smaller nuclear events than ever before and to distinguish them from small earthquakes and other types of explosions. We must have those capabilities in regions that are seismically active and geologically complex, and where seismic waves might not propagate efficiently.

Major research issues that have emerged for monitoring a CTBT are the precise location of events, and discrimination between nuclear explosions, earthquakes, and chemical explosions, even when those events are relatively small. These issues further require that we understand how seismic waves propagate in the solid Earth, the oceans and atmosphere, especially in regions that are structurally complex, where waves undergo scattering and, perhaps, a high degree of absorption. In addition, we must understand how processes occurring at explosion and earthquake sources manifest themselves in recordings of ground motion.

Monitoring a CTBT has required, and will continue to require, the best efforts of some of the world's best seismologists. They, with few exceptions, believe that methods and facilities that are currently in place will provide an effective means for monitoring a CTBT. Moreover, they expect that continuing improvements in those methods and facilities will make verification even more effective in the future. This topical series on several aspects of CTBT monitoring is intended to inform readers of the breadth of the CTBT research program, and of the significant progress that has been made toward effectively monitoring compliance with the CTBT.

The following set of papers, edited by Drs. William Walter and Hans Hartse, presents research results on Seismic Event Discrimination and Identification that are applicable for monitoring a CTBT. It is the sixth of eight topics addressed by this important series on *Monitoring the Comprehensive Nuclear-Test-Ban Treaty*. Previously published topics are Source Location, Hydroacoustics, Regional Wave Propagation and Crustal Structure, Surface Waves, and Source Processes and Explosion Yield Estimation. The final two topics, Data Processing and Infrasound, will appear in an ensuing issue.

Brian J. Mitchell
Saint Louis University
Series Editor

Pure appl. geophys. 159 (2002) 621–622
0033–4553/02/040621–02 $ 1.50 + 0.20/0

❚ Pure and Applied Geophysics

Introduction

WILLIAM R. WALTER[1] and HANS E. HARTSE[2]

For over fourty years verification seismologists have been reporting on advances in event identification research at Professional Society Meetings and more recently at the Seismic Research Symposium held annually in the United States. This research has evolved over time as international treaties banning testing in certain environments or limiting the yield of nuclear weapons tests have been negotiated and signed. In recent years this research has targeted the challenges of verifying a Comprehensive Nuclear-Test-Ban Treaty (CTBT). In September 1996, the United Nations General Assembly adopted the CTBT, prohibiting nuclear explosions worldwide, in all environments. The treaty calls for a global verification system, including a network of 321 monitoring stations distributed around the globe, a data communications network, an international data center, and on-site inspections, to verify compliance. The problem of identifying small-magnitude banned nuclear tests and discriminating between such tests and the background of earthquakes and mining-related seismic events, is a challenging research problem. The problem becomes particularly challenging as the magnitude of the event decreases, limiting the number of stations that see signal above the noise and increasing the number of background events that must be considered. This special volume is intended to report on seismic discrimination and identification research that has been ongoing since countries began to sign and ratify the CTBT in 1996.

A number of trends can be seen within these papers. First as the 170 seismic stations established by the CTBT for the International Monitoring Systems (IMS) are installed and updated there is an endeavor to ascertain how the IMS will perform for event identification and screening. Thus many papers collect data from these IMS sites to calibrate them and analyze their discrimination performance. Second, the emphasis is now on smaller events. This means that these papers now emphasize regional data. Hence, most event-station distances addressed in these papers are between 200 and 2000 km. Regional data also mean that a third trend, more seismic

---

[1] Geophysics and Global Security Division, Livermore National Laboratory, USA.
E-mail:bwalter@llnl.gov
[2] Los Almos Seismic Research Center, Los Alamos National Laboratory, USA.
E-mail:hartse@lanl.gov

stations and arrays from more locations worldwide, including many stations that are not part of the IMS, are now being used for discrimination studies. Within this volume a wide variety of traditional large arrays, smaller regional arrays, broadband Global Seismic Network (GSN) stations, other regional and local network stations, and even temporarily deployed stations, have all provided data for these investigations. The number of seismic stations that now provide data brings up a fourth important trend in verification seismology (and all other areas of seismological research) that has evolved over the last decade or so. Centralized digital data archives (such as the IRIS Data Management Center) and the advent of the autodrm (cf., KRADOLFER, 2000) have made it considerably easier and faster for researchers to acquire the large data sets used in discrimination studies.

The papers reveal a strong emphasis on short-period body-wave discriminants that can be applied to small events at regional distances. TAYLOR *et al.*, FAN *et al.*, and RODGERS and WALTER all report on source and path correction research applied to short-period P/S ratios. Other researchers address specific regions in their studies. RINGDAL *et al.* address P/S discrimination in the European Arctic, KOCH and KOCH and FÄH report on short-period discrimination research in Germany, and GITTERMAN *et al.* report on teleseismic detection and discrimination of explosions recorded within Israel. Not all papers here emphasize short-period body waves, however. KREMENETSKAYA *et al.* remind us that $m_b:M_s$ discriminants may be applicable to regional events in a second study of the European Arctic. Broadband and long-period data from mining-related events are studied by BOWERS and WALTER and HEDLIN *et al.* to separately identify these types of events. FISK *et al.* show experimental performance of the partial IMS network at screening out earthquakes using primarily $m_b:M_s$ and depth estimation while taking care not to screen out any explosions. Finally WOODWARD *et al.* emphasize the statistical considerations of event identification, particularly with incomplete training data.

REFERENCE

KRADOLFER, URS (2000), *Waves4U: Waveform Availability through Autodrms*, Seismol. Res. Lett. *71*, 79–82.

Pure appl. geophys. 159 (2002) 623–650
0033–4553/02/040623–28 $ 1.50 + 0.20/0

❘Pure and Applied Geophysics

Amplitude Corrections for Regional Seismic Discriminants

STEVEN R. TAYLOR,[1] AARON A. VELASCO,[1] HANS E. HARTSE,[1]
W. SCOTT PHILLIPS,[1] WILLIAM R. WALTER[2]
and ARTHUR J. RODGERS[2]

Abstract — A fundamental problem associated with event identification lies in deriving corrections that remove path and earthquake source effects on regional phase amplitudes used to construct discriminants. Our goal is to derive a set of physically based corrections that are independent of magnitude and distance, and amenable to multivariate discrimination by extending the technique described in TAYLOR and HARTSE (1998). For a given station and source region, a number of well-recorded earthquakes is used to estimate source and path corrections. The source model assumes a simple BRUNE (1970) earthquake source that has been extended to handle non-constant stress drop. The discrimination power in using corrected amplitudes lies in the assumption that the earthquake model will provide a poor fit to the signals from an explosion. The propagation model consists of a frequency-independent geometrical spreading and frequency-dependent power law Q. A grid search is performed simultaneously at each station for all recorded regional phases over stress-drop, geometrical spreading, and frequency-dependent Q to find a suite of good-fitting models that remove the dependence on m_b and distance. Seismic moments can either be set to pre-determined values or estimated through inversion and are tied to m_b through two additional coefficients. We also solve for frequency-dependent site/phase excitation terms. Once a set of corrections is derived, effects of source scaling and distance as a function of frequency are applied to amplitudes from new events prior to forming discrimination ratios. Thus, all the corrections are tied to just m_b (or M_0) and distance and can be applied very rapidly in an operational setting. Moreover, phase amplitude residuals as a function of frequency can be spatially interpolated (e.g., using kriging) and used to construct a correction surface for each phase and frequency. The spatial corrections from the correction surfaces can then be applied to the corrected amplitudes based only on the event location. The correction parameters and correction surfaces can be developed offline and entered into an online database for pipeline processing providing multivariate-normal corrected amplitudes for event identification. Examples are shown using events from western China recorded at the station MAKZ.

Key words: Seismic, discrimination, amplitude, magnitude.

Introduction

Corrections applied to seismic waves for source identification problems in nuclear test monitoring remains an important issue. For multivariate regional event identification, it is important to derive source and path corrections that make

---

[1] Los Alamos National Laboratory, Los Alamos, NM 87545, U.S.A. E-mail: taylor@lanl.gov
[2] Lawrence Livermore National Laboratory, Livermore, CA 94551, U.S.A.

discriminants independent of size (m_b) and range. To date, most of the research in regional seismic event detection has focused mainly on deriving propagation (e.g., distance) corrections to various combinations of amplitude ratios (Distance Corrected Ratios (DCR); e.g., TAYLOR et al., 1989; HARTSE et al., 1997). However, for discriminants that involve measurements in different frequency bands, source scaling can be a significant factor causing trends with source size (e.g., TAYLOR and DENNY, 1991; TAYLOR and HARTSE, 1998). If multivariate techniques are to be used for feature selection and event classification, it is important that biases caused by both propagation and source effects are removed.

In this paper, we outline a procedure for simultaneously correcting both propagation and source effects to seismic amplitudes prior to forming amplitude ratios. Correcting amplitudes offers several advantages over correcting ratios. First, correcting amplitudes using simple physical models offers much flexibility in that all information is retained as opposed to partial censoring of data by preselecting ratios. Also, because of hidden correlation structures of different ratios, it is difficult to *a priori* select a set of discriminants that will give optimum performance. The best discrimination performance is not necessarily achieved by preselecting the best amplitude ratios. In fact, combining a discriminant that by itself does not work well with one that does, may lead to improved overall performance (e.g., TAYLOR and HARTSE, 1997). The problem is exacerbated if little or no explosion calibration information is available, which is the case for most stations that will be used in monitoring the CTBT. Additionally, preselected ratios may be highly interdependent leading to a false impression of discrimination capability. It has been observed that seismic discriminants can show significant regional variability that is a function of factors such as crustal velocity structure and explosion-source material properties. Thus, in areas where little or no calibration information from explosions is available, it may not be prudent to preselect ratios.

We will illustrate that the amplitude-correction parameters can be based on simple physical models which can be difficult to do with ratios because of tradeoffs between different parameters. Thus, in uncalibrated or poorly calibrated regions, reasonable corrections can be obtained using expert opinion or extrapolation from geophysically similar regions. Further, we will show that for a given station and source region, the residuals from the one-dimensional model can be spatially interpolated for each phase and frequency band (e.g., using kriging) to account for lateral variations in propagation. It has been shown that geographical interpolation can be used to create correction surfaces that significantly reduce scatter in discriminants. The corrections appear to be particularly effective at low (1 Hz) frequencies (e.g., TAYLOR and HARTSE, 1998; PHILLIPS et al., 1998).

We have developed a methodolgy for removing m_b and distance trends from regional seismic amplitudes (MDAC – Magnitude and Distance Amplitude Corrections). MDAC is an improved version of the SPAC (Source and Path Amplitude Corrections) discussed by TAYLOR and HARTSE (1998) and TAYLOR and

VELASCO (1998). Improvements over SPAC include the inclusion of a grid search over trial values of stress drop, geometrical speading, and attenuation parameters. We have also established a tie between m_b and $\log M_0$ using the earthquake source model (rather than a linear relationship as assumed in TAYLOR and HARTSE, 1998).

The MDAC method combined with spatial interpolation of amplitudes (e.g., kriging) represents a station-specific approach for computing discriminants that are multivariate normal for a given source region. For a given station and source region, six MDAC parameters $(Q_0, \eta, \gamma, \sigma_b, F_{mb}, A_{mb})$ are derived for each regional phase. The first three terms are the Q values at 1 Hz (Q_0), the geometric spreading coefficient (η) and the power-law coefficient (γ) for frequency-dependent attenuation and control the path corrections. The fourth term is the Brune stress drop (σ_b). The last two terms (F_{mb}, A_{mb}) control the scaling between the seismic moment and magnitude (m_b) and control the source scaling. The F_{mb} and A_{mb} terms are unnecessary if seismic moments are routinely available (e.g., through an M_w coda magnitude scale; MAYEDA and WALTER, 1996). The MDAC-corrected amplitudes can be spatially interpolated using kriging (e.g., SCHULTZ et al., 1998) to derive a correction surface for each phase as a function of frequency. These surfaces can be used to reduce scatter in the corrected amplitudes and improve discrimination performance. The MDAC parameters and kriged surfaces can be developed offline and entered into an online database for rapid pipeline processing. Once a new event is located and a m_b or M_0 is calculated, the regional phases can be windowed and Fourier-transformed. The MDAC-corrected amplitudes for each phase and frequency can be computed using only the distance and magnitude. The spatial corrections from the kriged surfaces can then be applied to the corrected amplitudes based only on the event location. At this point, a set of multivariate-normal corrected amplitudes is available for event identification.

In this paper, we first describe the theory behind the MDAC method followed by an example using data from western China recorded at the IRIS GSN (Incorporated Research Institutions in Seismology Global Seismic Network) station MAKZ.

Estimation of Source and Path Parameters Using MDAC

In this section we describe the method we use for inverting source and path parameters and forming a set of frequency-dependent amplitude corrections for each phase recorded at a given station. In theory, the results could be extended to handle data from a network of stations. In our formulation, we follow a modification of the techniques of SERENO et al. (1988), TAYLOR and HARTSE (1998) and TAYLOR and VELASCO (1998). The advantage of using simple physical source and propagation models (as opposed to curve fitting) is that a priori information is often available for many of the parameters.

At a particular station, we assume the instrument-corrected amplitude spectrum for a given phase, $A_i(f)$, for source i, is given by

$$A_i(f) = G(r, r_0) S_i(f) \exp\left(-\frac{\pi f}{Q(f)v} r_i\right) P(f) , \tag{1}$$

where $S_i(f)$ is the source spectrum, r is the epicentral distance, f is the frequency, $Q^{-1}(f)$ is the frequency-dependent attenuation, v is the group velocity, $P(f)$ is a unitless phase site/excitation factor, and $G(r, r_0)$ is the frequency-independent geometrical spreading (STREET *et al.*, 1975) given by

$$\begin{aligned} G(r, r_0) &= r^{-1} & r < r_0 \\ G(r, r_0) &= r_0^{-1}(r_0/r)^\eta & r \geq r_0 \end{aligned} \tag{2}$$

where r_0 is a transition distance from spherical spreading to spreading rate η. We linearize equation (1) by taking logarithms

$$\log A_i(f) = \log G(r_i, r_0) + \log S_i(f) - \frac{\pi f \log e}{Q(f)v} r_i + \log P(f) . \tag{3}$$

In equation (3) we assume the following model for the source-time function (AKI and RICHARDS, 1980; BRUNE, 1970)

$$S_i(f) = \frac{M_0^{(i)} R_{\theta\phi}}{4\pi(\rho_s \rho_r V_s(P, S)^5 V_r(P, S))^{1/2} \left(1 + \left(\frac{f}{f_c^{(i)}}\right)^2\right)} , \tag{4}$$

where for event i, $M_0^{(i)}$ is the seismic moment and $f_c^{(i)}$ is the source corner frequency. The source and receiver density values are given by ρ_s, ρ_r, respectively. The parameters $V_s(P, S)$ and $V_r(P, S)$ correspond to P or S velocity in the source and receiver region, respectively (e.g., for shear phases use shear velocities). The radiation pattern coefficient is given by $R_{\theta\phi}$. Because radiation patterns are rarely known, we typically set $R_{\theta\phi}$ to 0.44 and 0.60 for P and S waves, respectively (BOORE and BOATWRIGHT, 1984). XIE and PATTON (1999) suggest that Pn amplitudes from earthquakes show considerable scatter due to a combination of radiation pattern and path effects. XIE (1998) found little evidence for radiation pattern effects on Lg for a well-studied earthquake in western Texas.

In equation (4), we explicitly assume a high-frequency decay of -2 which is commonly observed for earthquakes (e.g., HOUGH, 1996). For $f = 0$, the low-frequency level S_0 becomes

$$S_i(0) = S_0^{(i)} = \frac{M_0^{(i)} R_{\theta\phi}}{4\pi(\rho_s \rho_r V_s(P, S)^5 V_r(P, S))^{1/2}} = K M_0^{(i)} . \tag{5}$$

We further assume a simple scaling between event corner frequency and low frequency spectral level of the form (using equation (5))

$$f_c = cS_0^{-\kappa} = c(KM_0)^{-\kappa} \ . \tag{6}$$

For a BRUNE (1970) dislocation source model, $\kappa = 1/3$ and c is proportional to the cube root of the Brune stress drop, σ_b. The parameter κ can be used to allow for non-constant stress drop. The scaling of corner frequency and moment (or low-frequency level) has been observed to deviate from 1/3. For example, CONG *et al.* (1996) and NUTTLI (1983) have observed $\kappa = 1/4$ from central Asian earthquakes and mid-plate earthquakes, respectively. Additionally, MAYEDA and WALTER (1996) found evidence for 1/4 root scaling from western U.S. earthquakes. Although some studies have argued for an increase of stress drop with moment, in general there has been no definitive scaling observed (e.g., HOUGH, 1996). Using

$$f_c = 0.49 V_s(P,S) \left(\frac{\sigma_b}{M_0} \right)^{1/3} \tag{7}$$

from BRUNE (1970), the parameter c is given by

$$c = 0.49 V_s(P,S)(K\sigma_b)^{1/3} \tag{8}$$

or for non-constant stress drop ($\kappa \neq 1/3$)

$$c = 0.49 V_s(P,S) \left(\frac{\sigma_b}{M_0} \right)^{1/3} (KM_0)^{\kappa} \tag{9}$$

and

$$\sigma_b = \left(\frac{c}{0.49 V_s(P,S)} \right)^3 K^{-1} S_0^{1-3\kappa} = \left(\frac{c}{0.49 V_s(P,S)} \right)^3 K^{-3\kappa} M_0^{1-3\kappa} \ . \tag{10}$$

From equation (10), we can show that $\partial \log \sigma_b / \partial \log M_0 = 1 - 3\kappa$ showing that for $\kappa = 1/3$ the stress drop is independent of moment.

In practice, we select a stress drop common for all events and approximate c using equation (8). For $\kappa \neq 1/3$ we compute the stress drop for each event using equation (10) and individual corner frequencies using equation (6). We do this because if equation (9) is inserted into equation (6) (the fundamental equation relating corner frequency and moment), the scaling with κ drops out we are left with cube-root scaling. In this way, the units of c vary with κ as $m^{2\kappa} s^{\kappa-1}$ reducing to $(m/s)^{2/3}$ for cube-root scaling. The reasons for non-constant stress-drop scaling are unclear, but the main point is that we have an option to account for it through the non-physical parameter c in equation (6).

We assume a power-law frequency-dependent Q of the form

$$Q(f) = Q_0 f^{\gamma} \ . \tag{11}$$

Inserting equations (5), (6), and (11) into (3) gives the final equation to be used in the grid search

$$\log A_i(f)$$

$$= \log G(r_i, r_0) + \log S_0^{(i)} - \log \left[1 + \left(\frac{f(S_0^{(i)})^\kappa}{c} \right)^2 \right] - \frac{\pi \log e}{Q_0 v} f^{1-\gamma} r_i + \log P(f) \ . \ (12)$$

In practice, we select a frequently recorded single phase to be used to estimate the seismic moment (typically Lg). Using that phase at a given station, we grid-search through a range of σ_b, η, Q_0 and γ values looking for a combination of parameters that produces a good fit to the data using the residual sum of squares (RSS) as a guide. The parameter κ can also be adjusted as part of this process, but is not actually part of the grid search. For each combination of parameters, we solve for the seismic moment for each event, $M_0^{(i)}$ through linearized inversion (Appendix A). Moments for any or all events can be fixed if desired. As will be discussed further below, the final parameters to be used depend on the overall fit to the data, the physical reasonableness of the final parameters (based on prior information if possible), and the ability of the derived parameters to remove trends of corrected amplitudes as a function of both m_b and distance for each frequency. We then perform a grid search for η, Q_0 and γ values for the remaining phases using the same stress drop and moments for the phase used to estimate the source parameters. The unitless phase/site excitation factor, $P(f)$, is computed at each step from the mean residual at each frequency.

Once the moments are computed for each event, it is necessary to derive a scaling relationship between $\log M_0^{(i)}$ and m_b. Then to correct a new event for source scaling, the only parameter that is then needed is the magnitude. Equation (12) is then used to derive a predicted phase amplitude and the logarithm of the corrected amplitude, $A_i^{(c)}(f)$, is defined to be the difference between the logarithm of the observed, $A_i^{(o)}(f)$, and predicted, $A_i^{(p)}(f)$, spectral values. We will refer to these as the MDAC-corrected amplitudes. Explicitly

$$\log A_i^{(c)}(f) = \log A_i^{(o)}(f) - \log A_i^{(p)}(f) \tag{13}$$

where the predicted spectral amplitude is given by equation (12). Note that the corrected amplitudes given in equation (13) are basically residuals to the fit given by equation (12) where M_0 has been estimated from the m_b. In this way, an event having large phase amplitudes at a particular frequency will be characterized by a large positive residual.

In TAYLOR and HARSTE (1998) we tied $\log S_0^{(i)}$ to m_b using a linear relationship. To account for corner frequency scaling, we now use the functional form of equation (12) to derive a relationship between $\log M_0^{(i)}$ and m_b. We define the theoretical m_b which is a function of M_0 to be

$$m_b^{th} = \log M_0 - \log \left[1 + \left(\frac{F_{mb}(KM_0)^\kappa}{c} \right)^2 \right] + A_{mb} \tag{14}$$

where F_{mb} is defined to be the frequency at which m_b was measured. For a given $\log M_0^{(i)}$ we solve for the values of F_{mb} and A_{mb} that minimize the difference between the observed and calculated m_b. To do this, we perform a linearized inversion for F_{mb} and A_{mb} as described in Appendix A. This is an important step in the MDAC processing because F_{mb} and A_{mb} have a strong control on the MDAC corrections and their ability to remove magnitude and distance effects. Note that this step is unnecessary if seismic moments are routinely available through a calibrated M_w scale. Once the MDAC parameters $(Q_0, \eta, \gamma, \sigma_b, F_{mb}, A_{mb})$ have been computed for a particular phase, the only input parameters needed to correct the amplitudes from a new event are the magnitude (m_b) and epicentral distance.

The next step involves examining whether a particular set of parameters removes the trends of the corrected amplitudes (residuals) with both m_b and distance. It turns out that the best-fit solutions do not necessarily remove the trends at all frequencies. Thus, we generally identify the top 5% solutions based on RSS values by using the F-statistic and perform linear regressions of the corrected amplitudes versus m_b and distance at each frequency for each combination of parameters. For each regression, we compute the F and p values of the regression at a given significance level, α (typically $\alpha = 0.05$). If the p value is larger than α (i.e., the observed value of the F statistic for the regression is smaller than the tabulated value), then we conclude that the linear trend has been removed. In this way, different combinations of search parameters can be examined to see if they successfully remove trends with m_b and distance. Information regarding the interpretation of p values can be obtained in any standard textbook on regression (e.g., WEISBERG, 1980).

In practice, it is often very difficult to remove all trends with m_b and distance at all frequencies. Typically, we will select models that successfully remove trends with m_b and then let the spatial surface corrections (discussed below) complete removing the distance trends. We have added an option to include a set of non-physical parameters to finish removing all trends with m_b and distance if desired by removing a surface fit at each frequency given by

$$\log A_i^{(c)}(f) = a + bm_b^{(i)} + cr_i \tag{15}$$

resulting in three additional correction parameters at each frequency.

The F_{mb} and A_{mb} terms are computed for each phase and are added to aid in the elimination of trends in the corrected amplitudes with m_b and distance. These two parameters and the occasional use of equation (15) is necessary due to differences in apparent source scaling of the different phases. For example, PRIESTLEY and PATTON (1997) observed different scaling of Pn and Lg magnitudes with seismic moment from earthquakes. XIE and LAY (1995) noted that Lg scaling slopes can show a strong regional dependence. They suggested that source scaling, combined with propagation effects, can account for the differences in apparent source scaling.

As discussed in TAYLOR and HARTSE (1998) there are many tradeoffs associated with determining the parameters in equations (12) and (14). Our philosophy in deriving source and path corrections is to constrain as many parameters as possible based on prior information. The remaining parameters are checked to see if they are geophysically reasonable. The key is treating the problem as a nonunique curve-fitting exercise to derive reasonable corrections to be applied to seismic amplitudes used as discriminants that effectively remove dependencies on m_b and r.

Application of MDAC to Western China Data Recorded at MAKZ

As an illustration of the MDAC methodology, we present results from the IRIS GSN station MAKZ (Makanchi, Kazakhstan). All seismograms were obtained from the IRIS Data Management Center (DMC). The data are from four Lop Nor nuclear explosions and 412 earthquakes located in a region around Lop Nor extending between 35° and 50°N latitude and 80° and 100°E longitude (the Lop Nor box; Fig. 1). Distances range from 111 to 1931 km and we have restricted our analysis in

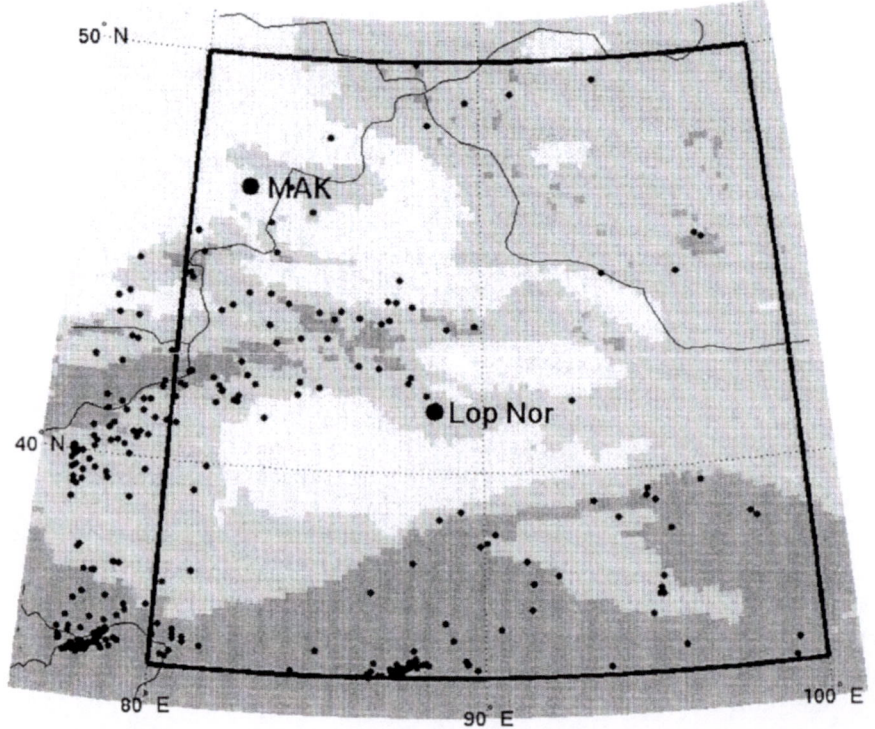

Figure 1
Map showing events used in this study within box surrounding Lop Nor.

the paper to PDE magnitudes greater than 3.5. Event locations and magnitudes were obtained from both the United States Geological Survey Preliminary Determination of Epicenters (USGS/PDE) catalogs maintained at the IRIS DMC and the Reviewed Event Bulletin (REB) from the Prototype International Data Centre (PIDC). As discussed by HARTSE *et al.* (1997), we measured RMS amplitudes taken in eight different 1-octave frequency bands ranging from 0.5 to 8 Hz. Because the methodology discussed above is for a spectral inversion, we convert the RMS amplitudes to pseudo-spectral amplitudes using Parseval's Theorem for the discrete Fourier transform (see Appendix B).

We first computed 6 MDAC parameters including moment for the Lg amplitudes. Using a signal-to-noise cutoff of 1 using prephase noise adjusted for window length (see Appendix B), we selected crustal events having at least 6 amplitude measurements out of a possible 7 ranging from 0.5 to 8 Hz. We use prephase noise because secondary phases may erroneously pass signal-to-noise tests using background noise when they are actually buried in signal-generated coda or physically blocked. As discussed by DEWEY (1999) a magnitude discrepancy exists between PIDC and USGS magnitudes. Typically, the USGS magnitudes exceed those of the PIDC by about 0.3 magnitude units between USGS m_b of 4.5 to 5.0. The relationship becomes more complicated at lower magnitudes. We performed an analysis of 170 events within the Lop Nor box that had both USGS and REB magnitudes and found that the USGS magnitudes on average to be 0.29 magnitude units greater than those of the PIDC. Thus, we adjusted the events having REB magnitudes upward by 0.29. Tests show that this correction had only a minor effect on the final results and that only 28 events of the 167 used for the Lg grid search had REB magnitudes.

A grid search was performed for each phase for six logarithmically spaced stress drops between 0.1 and 30 MPa, 50 points each for η, Q_0, and γ, between 0.5 and 1, 300 and 800, and 0 to 1, respectively resulting in 750,000 combinations of parameters. The parameter κ was set to 1/3 and the elastic parameters used to define the parameter K (equation (5)) and the group velocity (equation (1)) are given in Table 1.

Figure 2 shows an example of a Lg grid search for a stress drop of 0.3 MPa. The upper right corner shows orthogonal slices taken through the selected values of

Table 1

Elastic parameters (MKS)

Phase	$R_{\theta\phi}$	ρ_s	ρ_r	V_s	V_r	v
Pn	0.44	2700	2500	6100	5000	8100
Pg	0.44	2700	2500	6100	5000	6100
Sn	0.6	2700	2500	3526	2890	4500
Lg	0.6	2700	2500	3526	2890	3600

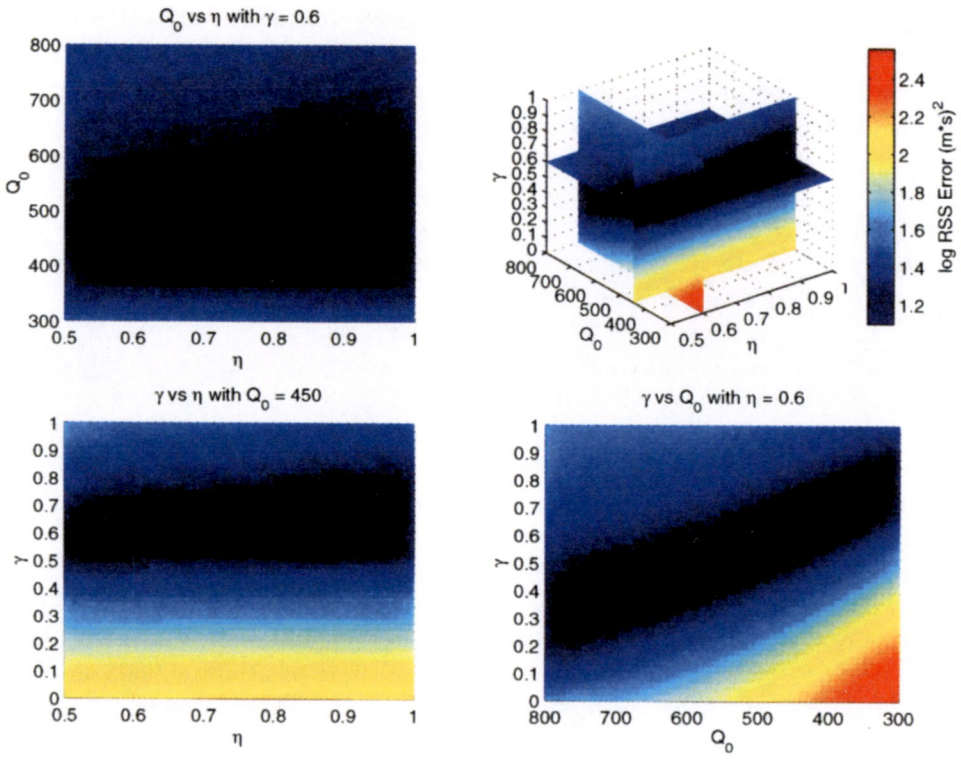

Figure 2

Representation of grid search results for Lg with stress drop of 0.3 MPa. Upper right shows orthogonal slices taken through the selected values of the three search parameters η, Q_0 and γ. Remaining three plots show projections along the major axes. Note that the final solution used is not necessarily the minimum solution.

the three search parameters η, Q_0, and γ. The remaining three plots show corresponding projections along the principal axes. Significant tradeoffs can be observed between the different parameters. The best-fit solutions actually appear to be located outside of the search space and note that the final solution used is not necessarily the minimum solution. However, these solutions were judged to be physically unreasonable based on information from previous studies (e.g., CONG *et al.*, 1996) and resulted in unstable moment estimates. Figure 3 shows a slice of the RSS volume for $\sigma_b = 0.3$ MPa and $\eta = 0.6$ where the tradeoff between Q_0 and γ can be observed.

From Q tomographic maps for Lg of MITCHELL *et al.* (1997) and CONG *et al.* (1996), Q_0 varies from approximately 350 in the southern portions of our study area to 800 to 1000 in the northern portions. The corresponding values of γ range from approximately 0.5 to 1. Most of our events and the Lop Nor test site are located in the zone having lower values of Q_0, and γ.

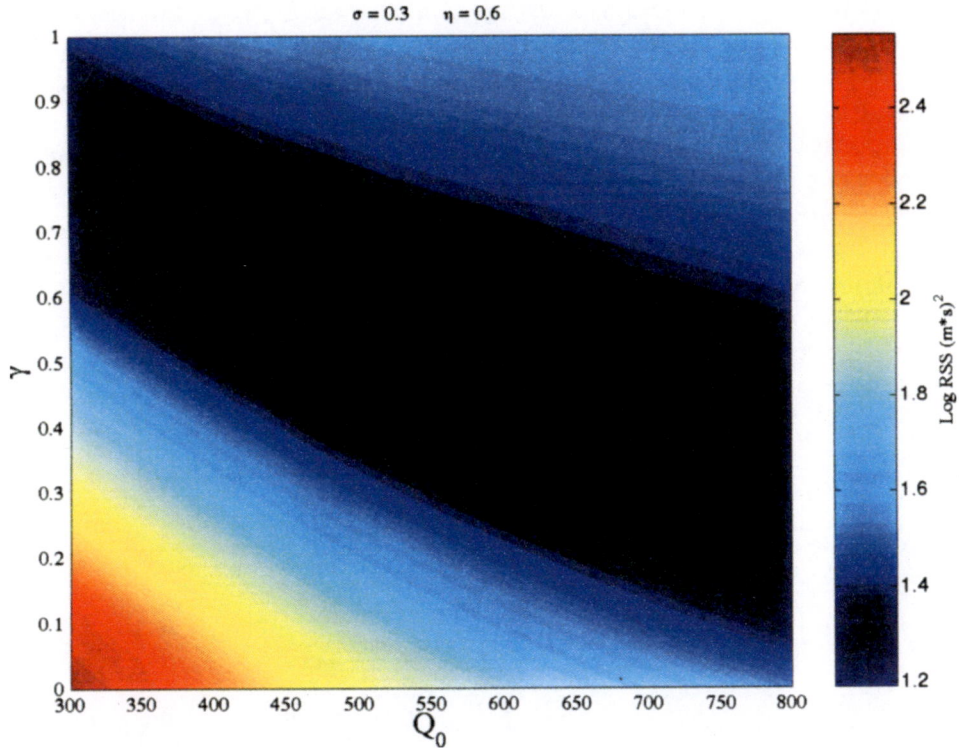

Figure 3
Representation of grid search results for Lg with stress drop of 0.3 MPa and $\eta = 0.6$ showing tradeoff between Q_0 and γ.

The next step involves examining the best-fit solutions to see if they adequately remove trends with m_b and distance. To do this we select a specified number of the best-fit solutions (typically the top 5% RSS values based on the F-statistic) and regress the corrected amplitudes (equation (13)) versus both m_b and linear distance. At this stage, we go though and recorrect the amplitudes based on the $m_b - \log M_0$ scaling relationship (equation (14)). The main parameter controlling the magnitude-moment relationship is F_{mb}. Because a least-squares algorithm is used (Appendix A), the estimated parameters can be biased by a few points at large magnitude. Thus, we find that it is often better to set F_{mb} to different values and examine the p values from the regressions. Because of the simple physical model used to represent the amplitudes, it is often difficult to completely remove trends with m_b and distance at all frequencies. We find that for given values of F_{mb} we can often completely remove trends with m_b but not distance or *vice versa*. Because additional correction surfaces discussed below can be used to remove unmodeled distance effects, we typically select parameters that are more effective in removing trends with m_b.

Figure 4 shows an example of Lg parameters and fits to our selected model. Because we used Lg to invert for log M_0, the site/excitation factors are small (upper left portion of Fig. 4). The numerous tradeoffs between the different parameters make it not difficult to get a good fit to the observed spectrum as seen by the upper right portion of Figure 4. As can be seen by the tradeoffs shown in Figure 3, a number of models can fit the data. Based on the prior information on Lg attenuation discussed above and analysis of the p values of regressions of corrected amplitudes versus m_b and distance (discussed below), we selected the model shown in Figure 4.

Figure 5 shows results of the linear regressions versus m_b and distance for selected frequencies. We found that a value of $F_{mb} = 0.5$ Hz was effective in removing most of

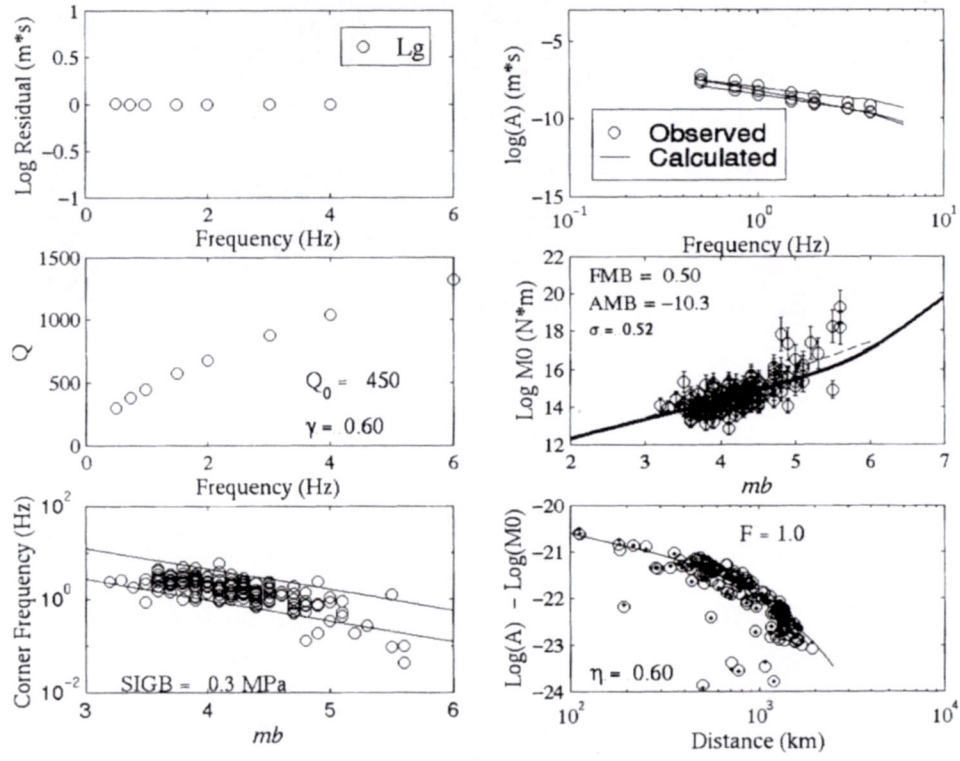

Figure 4

Plots illustrating various parameters for Lg MDAC model. Moving clockwise from upper left (UL): Site/excitation factors (equation (1), $P(f)$). (Upper right) Example of three fits to three randomly chosen events. (Middle right) Moment/magnitude relationship (thick solid line; equation (14)) and estimated moments and 95% confidence limits. Dashed line is from universal moment-magnitude scaling (PRIESTLEY and PATTON, 1997). (Bottom right) Observed (open circle) and calculated (dot) amplitude normalized by moment versus distance and theoretical scaling for $\eta = 0.6$. (Lower left) Estimated corner frequency versus m_b (equation (6)) for average stress drop of 0.3 MPa. Upper and lower lines are theoretical curves for 0.1 and 10 Mpa (1 and 100 bar) stress drop. (Middle left) Q versus frequency using power-law model (equation (11)).

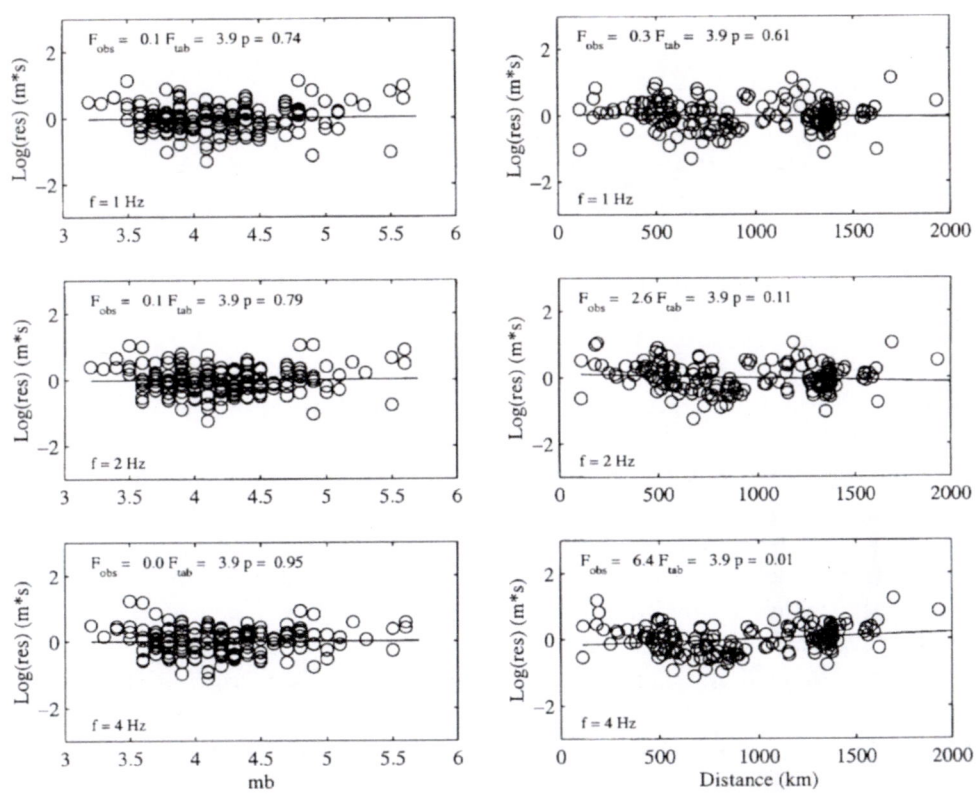

Figure 5

Plots showing results of Lg linear regression versus m_b (left column) and linear distance (right column) for three selected frequency bands. The observed and tabulated F statistics and p values for each of the regressions are shown for $\alpha = 0.05$.

the trends. Using a stress drop of 0.3 MPa and moments determined from the Lg phase, we conducted a search over η, Q_0, and γ. The results for each phase are listed in Table 2. The only phase that was difficult to fit adequately while removing trends with m_b and distance was Pn. As discussed by XIE and PATTON (1999), Pn corner frequencies in central Asia appear to be 4–5 times higher than those for Lg. In our modeling, we assume the same stress drop for all phases, and it is difficult to fit Pn using the same stress drop as Lg. Our best fits (RSS ~ 80 (m s)^2) for Pn appeared to have high Q_0 values (800–2000) and small values of $\gamma(<0.3)$, although none of these models removed m_b and distance trends satisfactorily. However, XIE and PATTON found (1999) interstation Pn Q_0 and γ values of approximately 277 and 0.61, respectively (for $\eta = 1.3$) for the northern part of our study region. Although the misfit for their parameters was not as good as other models, we found it did a reasonable job of removing amplitude trends with m_b (but not distance). Thus, we decided to use values similar to those determined by XIE and PATTON (1999). It is

Table 2

MDAC parameters

Phase	n	ndof	κ	r_0 (km)	σ_b	η	Q_0	γ	F_{mb}	A_{mb}	RSS
Pn	63	421	1/3	1	0.3	1.3	300	0.7	0.5	−10.6	164.0
Pg	61	414	1/3	100	0.3	0.8	500	0.3	0.5	−10.6	39.1
Sn	106	728	1/3	1	0.3	1.1	800	0.3	0.5	−10.4	34.4
Lg	167	980	1/3	100	0.3	0.6	450	0.6	0.5	−10.4	16.9

n – number of events used in computation of MDAC parameters. ndof – number of degrees of freedom. σ_b – (MPa); RSS – (m ∗ s)^2.

difficult to determine why the Pn spectra are so difficult to fit in this region, but the reason may be due to our assumption of frequency-independent geometrical spreading which ignores the effects of complicated upper mantle structure on wave propagation. For example, it is well known that geometric spreading factors for a particular phase can vary with range and frequency depending on the velocity structure (e.g., SERENO and GIVEN, 1990). Because we had difficulty fitting the Pn spectra, we decided to remove the surface for each phase (equation (15)). The correction was minimal for all phases except the Pn versus distance.

We then corrected all amplitudes that passed the signal-to-noise cutoff of 1 using prephase noise including the four Lop Nor nuclear explosions. Figure 6 shows the corrected log amplitudes for each phase as a function of frequency for all four phases. We have also identified four earthquakes occurring within 50 km of the Lop Nor test site.

As expected, the corrected amplitudes for the earthquakes are very close to zero mean, and χ^2 goodness-of-fit tests show most frequency bands are normally distributed. The explosions show a significantly different pattern than the earthquakes. This is not unexpected, since we are using an earthquake model to fit the data. The misfit is particularly severe for Pn as can be seen by the high residuals at intermediate and high frequencies. As will be further illustrated below, the plot in Figure 6 is interesting in that it can be used as a guide in selecting individual discriminants that provide good separation between earthquakes and explosions. For example, it can be seen that a high-frequency Pn to low-frequency Sn or Lg cross-spectral ratio will give excellent separation between the earthquakes and explosions (as noted by HARTSE *et al.*, 1997). Additionally, a Pn, Pg, or possibly even an Sn spectral ratio will provide good separation. These observations reinforce our earlier comments about the potential restrictions on discrimination performance by *a priori* selecting ratios.

The Lop Nor earthquakes show different patterns than the explosions indicating that the corrected amplitudes are showing differences largely due to source or very near-source effects. One of the Lop Nor earthquakes appears to be systematically higher than the others and we suspect this is due to a bias in the m_b values assigned to

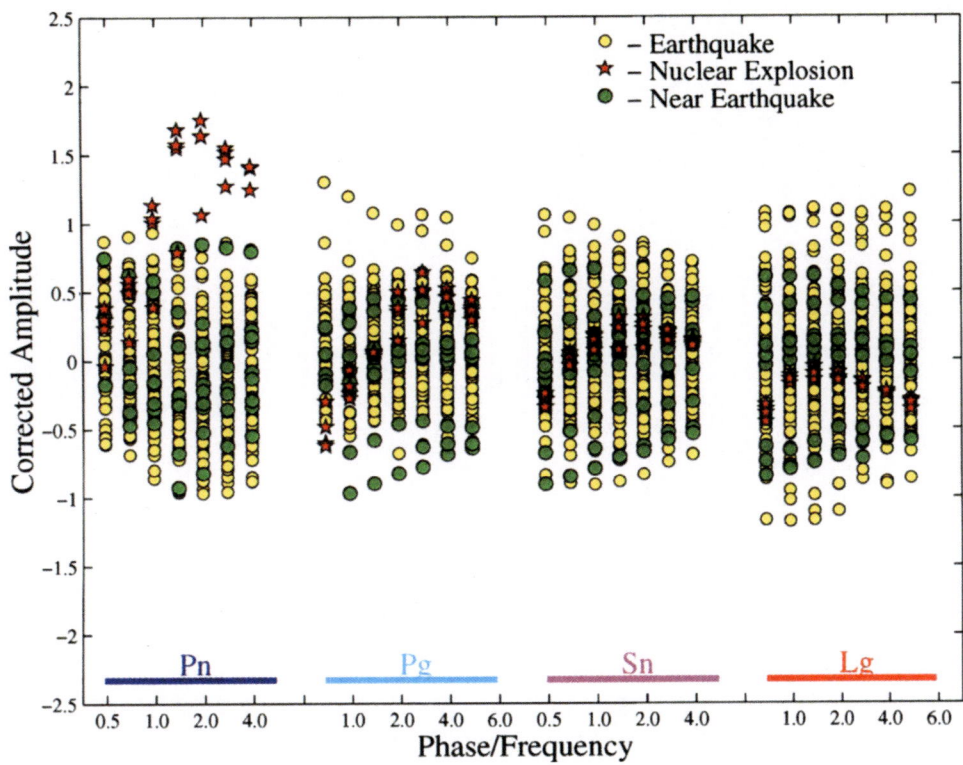

Figure 6

MDAC-corrected log amplitudes for each phase as a function of frequency using the parameters given in Table 2 including the 4 Lop Nor nuclear explosions. Four earthquakes within 50 km of Lop Nor are also identified.

the events. If an m_b value is biased, then the recalculated moments (using equation (14)) will be biased, which in turn biases the corrected amplitudes for each phase (and to some extent the spectral shape).

To illustrate the problem with m_b bias on the corrected amplitudes, we show the corrected amplitudes normalized by the Lg amplitude in the 0.5 to 1 Hz frequency band (the band having the most measurements; Fig. 7). From Figure 7, it can be seen that the overall scatter is reduced considerably and the Lop Nor earthquakes now plot very close to one another.

At this point we can directly make traditional discrimination plots from the corrected amplitudes shown in Figure 6 by simply taking differences (log ratios) using any combination of phase and frequency. This is illustrated in Figure 8 where we show the high-frequency (4–8 Hz) Pn to low-frequency (0.75–1.5 Hz) cross-spectral ratio plotted versus both distance and magnitude. From Figure 8 it can be seen that the trends in the ratio are removed with both distance and magnitude. The separation

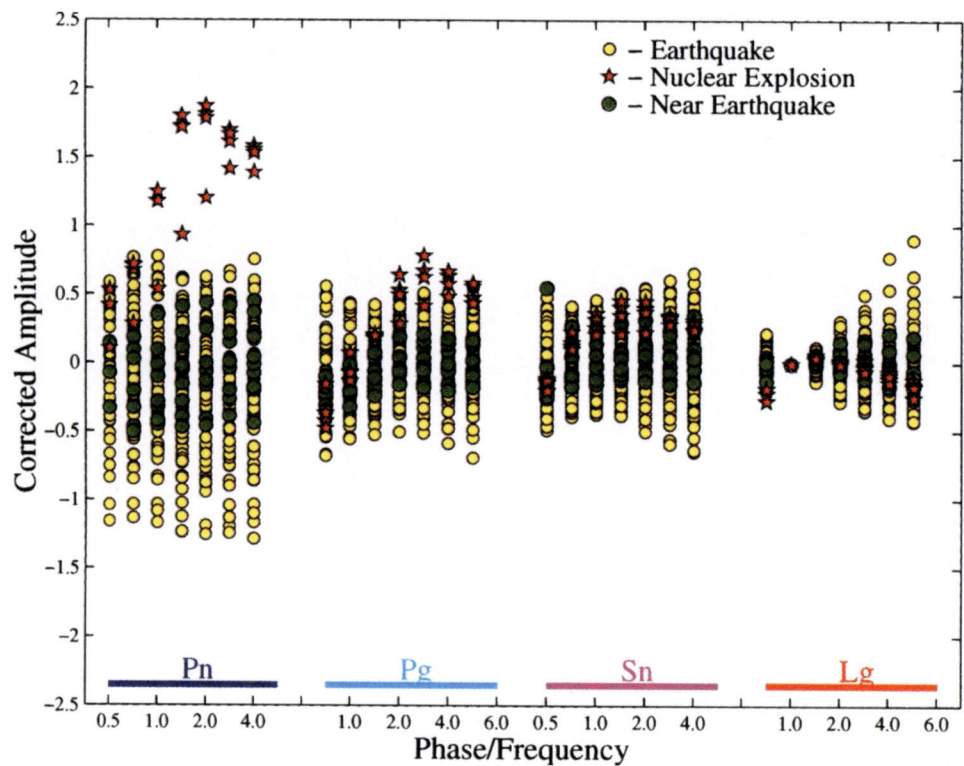

Figure 7
Same as Figure 6, except each amplitude is normalized by the (0.75–1.5 Hz) frequency Lg amplitude.

between the earthquakes and explosions is quite good (including the Lop Nor earthquakes). Further examples of various discrimination ratios computed from the MDAC-corrected amplitudes are shown in Figure 9. Note that in all cases the discrimination performance is excellent and trends are removed with magnitude. The separation between the Lop Nor explosions and earthquakes is very good indicating that the discriminants are based on source or very near-source differences between the two different source types.

The discriminants in Figure 9 illustrate that one-dimensional models for a given station and source-region can lead to good discrimination performance. However, as suggested by TAYLOR and HARTSE (1998) and PHILLIPS *et al.* (1998), the residuals to the one-dimensional model can be spatially interpolated to account for regional variations in attenuation further reducing the scatter in the discrimination plots. PHILLIPS (1999) and RODGERS *et al.* (1999) have suggested kriging (e.g., SCHULTZ *et al.*, 1998) as a viable method to spatially interpolate amplitudes. In Figure 10 we show correction surfaces for the high-frequency Pn and low-frequency Sn amplitudes used to construct ratios in Figure 8. The "surfaces" represent the cross-validated

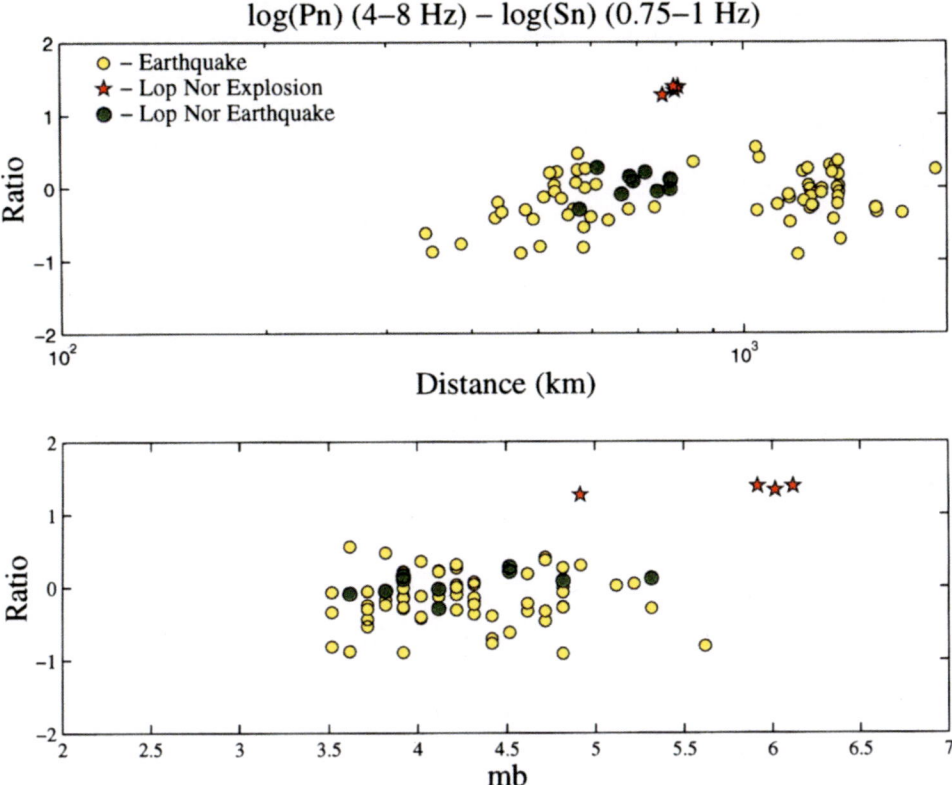

Figure 8
High-frequency Pn (4–8 Hz) to low-frequency (0.75–1.5 Hz) Sn cross-spectral ratio plotted versus log distance (top) and magnitude (bottom).

(leave-one-out) residual values obtained from kriging. Complicated patterns are observed for each phase. It should be noted that some of the magnitude bias problems discussed above and contrasted in Figures 6 and 7 are still manifest in these surfaces. However, we also tried kriging the same amplitude ratio and found that the variance of the ratio was slightly worse than the case where we kriged the individual amplitudes and then formed the ratio. Thus, it appears that the methodology is somewhat robust to the m_b bias problem. This is not totally unexpected by examining the influence m_b has on the MDAC procedure. Using equation (14) the magnitude is used to estimate the moment. The moment is then used to estimate the corner frequency using equation (6). Because the moment is raised to the power of κ (where $\kappa \leq 1/3$) the effects of errors in m_b are reduced somewhat.

Using leave-one-out, we corrected the amplitudes and recomputed the Pn to Sn cross-spectral ratio shown in Figure 8. The results are shown in Figure 11 along with the original MDAC-corrected spectral ratio with just a simple distance correction

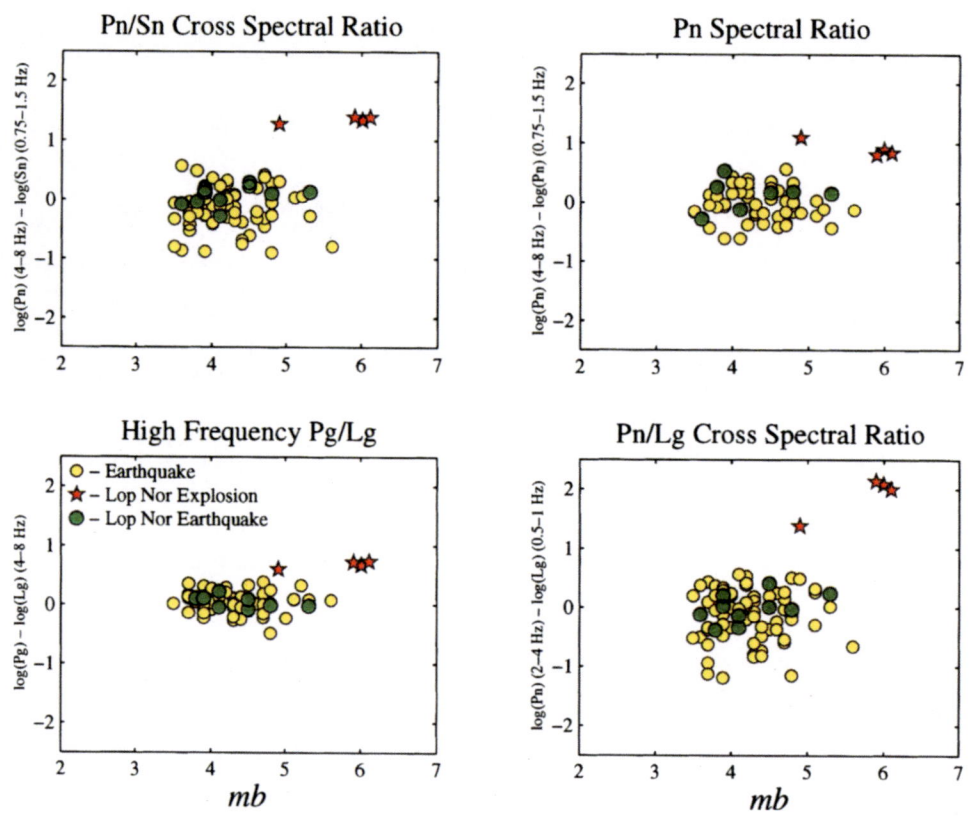

Figure 9
Examples of four discriminants computed from the MDAC-corrected amplitudes shown in Figure 6. Upper left is the Pn/Sn cross-spectral ratio (shown in Fig. 8). Upper right is a Pn spectral ratio. Lower right is a Pn/Lg cross-spectral which from Figure 6 was judged to have the most separation between earthquakes and explosions. Lower left, high-frequency Pg/Lg ratio.

(HARTSE *et al.*, 1997). In each plot, we show the estimate of the variance of the earthquake population and the plug-in or optimal error rate assuming equal prior probabilities of occurence given by (RENCHER, 1998)

$$P^{(0)} = \Phi\left(-\tfrac{1}{2}\Delta\right) \ , \tag{16}$$

where $\Phi[\bullet]$ is the standard normal distribution function and Δ^2 is the Mahalanobis distance metric (RENCHER, 1998). For the lower two plots we show the variance reduction relative to the preceding plot. For the DCR shown in the top portion of Figure 11, we can still see the source scaling effect on the cross-spectral ratio.

The middle plot on Figure 11 shows the MDAC-corrected ratio (same as Fig. 8) showing a variance reduction over the DCR of 35.5%. The error rate has also decreased significantly because of the removal of the ratio versus magnitude trend resulting in better discrimination performance. The bottom portion of Figure 11

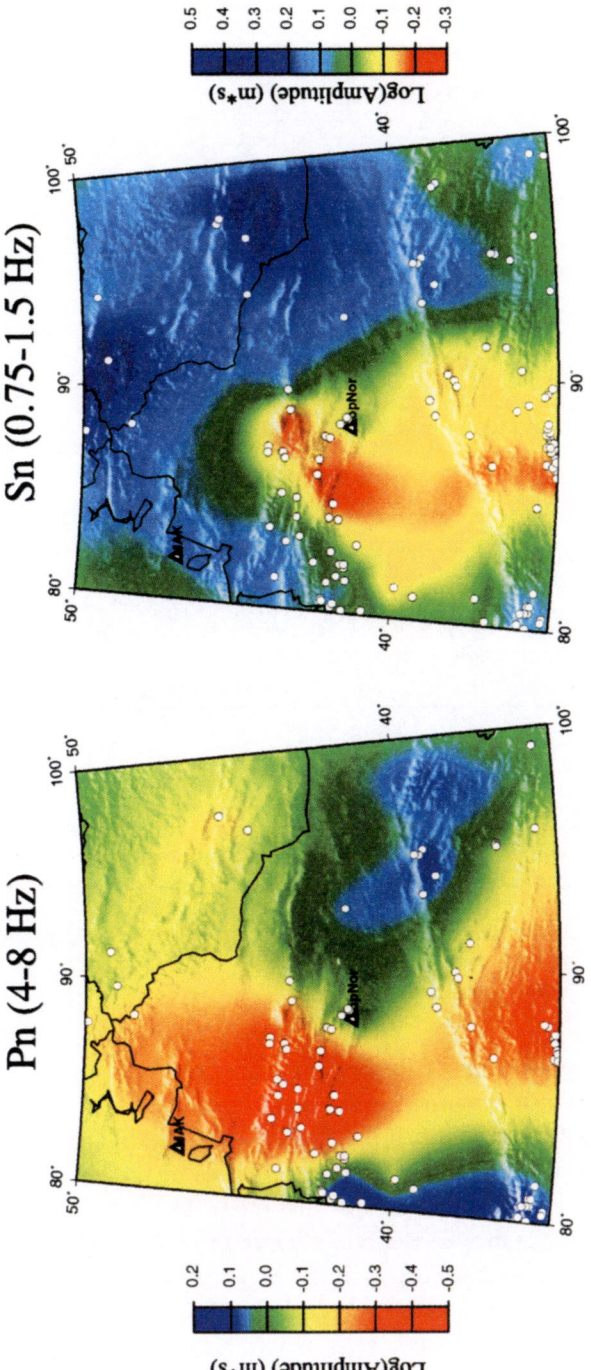

Figure 10

Kriged surface corrections for high-frequency Pn and low-frequency Sn amplitudes. The "surfaces" represent the cross-validated (leave-one-out) residual values obtained from kriging.

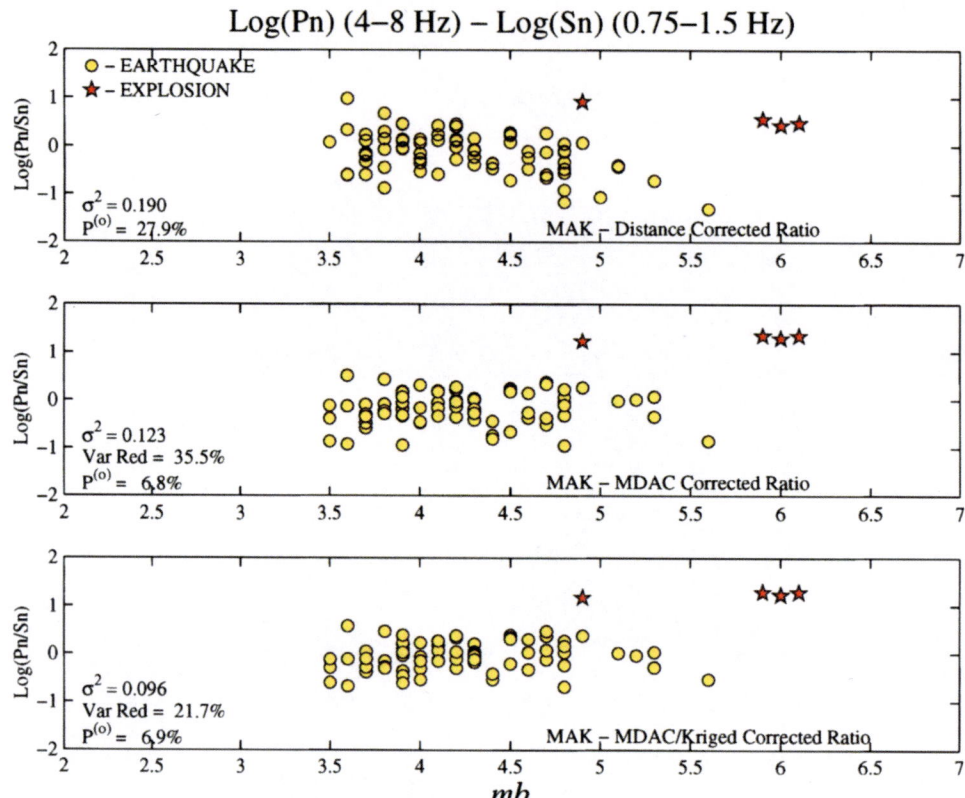

Figure 11

Examples of high-frequency Pn to low-frequency Sn cross-spectral ratio versus magnitude with estimated variance for earthquakes and the plug-in-error rate, $P^{(0)}$. (Top) Distance corrected ratio where it is noted that the source scaling is evident. (Middle) MDAC-corrected ratio (same as Fig. 8) showing a variance reduction over the DCR of 35.5%. The $P^{(0)}$ value has also decreased significantly resulting in better discrimination performance. (Bottom) Results from application of kriged surface corrections (Fig. 10) to individual MDAC-corrected amplitudes and recomputation of ratio. The variance has been reduced an additional 21.7% over the MDAC ratio (49.5% over the DCR).

shows results from application of kriged surface corrections (Fig. 10) to individual MDAC-corrected amplitudes and subsequent recomputation of the ratio. The variance has been reduced an additional 21.7% over the MDAC ratio (49.5% over the DCR). Although the variance of the earthquakes has been reduced significantly, the estimated error rate (Mahalanobis distance) has not changed significantly because the correction has moved the two populations slightly closer together due to the nature of the corrections in the vicinity of Lop Nor. Thus, although the one-dimensional models can be used to correct discriminants and obtain good performance, the spatial surface corrections can be used to improve discriminants even further by reducing the scatter in the earthquake population.

The MDAC methodology presented in this paper can be easily incorporated into an operational pipeline. The concept is illustrated in Figure 12. The MDAC parameters and kriged surfaces can be developed offline and entered into an online database for pipeline processing. Once a new event is located and a magnitude or moment is calculated, the regional phases can be windowed and Fourier transformed. The MDAC-corrected amplitudes for each phase and frequency can be rapidly computed using only the distance and m_b or M_w. The spatial corrections from the kriged surfaces can then be applied to the corrected amplitudes based only on the event location. At this point, a set of multivariate-normal corrected amplitudes is available for event identification.

Discussion and Conclusions

We have presented the MDAC methodology for correcting regional seismic amplitudes for source scaling and propagation using simple physical source and

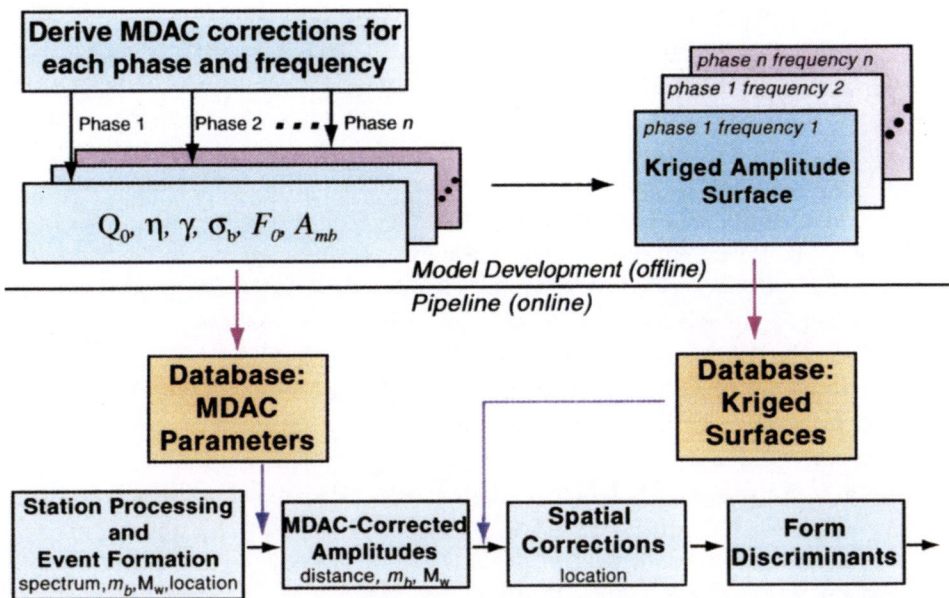

Figure 12

Example flowchart illustrating how the MDAC method could be combined with spatial interpolation (e.g., kriging) to develop database parameters to provide amplitude corrections for regional event identification in near-real time. These parameters can easily be incorporated into an operational pipeline to correct amplitudes from a new event for identification purposes.

propagation models. The resulting corrected amplitudes are independent of m_b and distance and amenable to multivariate discrimination analysis. The amplitudes are corrected using a simple one-dimensional propagation model and the BRUNE (1970) dislocation source model modified to handle non-constant stress drop. The discrimination power in the corrected amplitudes lies in the assumption that the earthquake model will provide a poor fit to the signals from an explosion. Observations of explosions from many different regions have shown significant variability in performance of different discriminants presumably due to the strong effect of linear and nonlinear material-dependent phenomenology acting in the near-source region (e.g., TAYLOR and DENNY, 1991). However, it is expected that there will always be some, possibly unpredictable, differences between earthquake and explosion sources. This fact is the basis for our preference for deriving a set of corrections for seismic amplitudes rather than preselecting traditional amplitude ratios. In an uncalibrated region, it is very difficult to predict *a priori* which combination of discriminants will provide the best performance. Correcting amplitudes allows for much flexibility in the subsequent event identification procedures (Fig. 12) because all the information is retained. Dealing with amplitudes involves a new set of problems (e.g., the m_b bias problem) not encountered as strongly with ratios. However, we feel the advantages outweigh the disadvantages in terms of the flexibility amplitudes give to the source identification problem.

Future work will involve modifying the methodology to reduce some of the problems caused by the simplicity of the models, in particular the one-dimensional propagation models. One potential starting place is the inclusion of attenuation tomography into the Q model. Attenuation tomography is a well-established method that has been used for many years (e.g., YOUNG and WARD, 1980; SINGH and HERRMANN, 1983; MITCHELL *et al.*, 1997). PHILLIPS *et al.* (2001) have recently investigated the utility of attenuation tomography for correcting seismic discriminants using 1 Hz Pg/Lg amplitude ratios and the results appear to be promising. Tomographic methods applied to regional seismic amplitudes can produce attenuation correction surfaces possibly as a function of frequency or frequency parameters (e.g., γ). These background models could be used to place further constraints on lateral variations in the geometric spreading coefficient and stress-drop. Variations in stress drop have been observed in different tectonic regions. For example, stress drops in eastern North America are commonly reported to be about a factor of 2 greater than those in western North America (e.g., HANKS and JOHNSTON, 1992). With constraints on lateral variations in Q_0 (and γ), we may be able to map out regional variations in stress drop. In this way, correction surfaces could be supplied for each of the MDAC parameters used in equation (1).

Work is also needed on obtaining stable estimates of m_b from regional measurements that are not affected by station sampling bias or upper mantle bias. As discussed above, in order to rapidly apply the MDAC parameters a magnitude (or moment) is needed. A bias in magnitude, will cause a bias in the estimated source parameters used to correct the amplitudes. Some potential candidates for regional magnitude scales are $m_b(Lg)$ (e.g., NUTTLI, 1973; PRIESTLEY and PATTON, 1997) and coda magnitude scales (MAYEDA and WALTER, 1996). Tomographic techniques could be used to account for regional variations in propagation inherent to both of the methods.

Appendix A

Estimation of Moment and Moment/magnitude Scaling Parameters

An important part of the MDAC procedure is estimating the parameters F_{mb} and A_{mb} that control the scaling between magnitude and log M_0. This is an important step in the MDAC processing because F_{mb} and A_{mb} have a strong control on the MDAC corrections and their ability to remove magnitude and distance effects. If seismic moments are available on a routine basis, the F_{mb} and A_{mb} are unnecessary. Physically, F_{mb} can be thought of as the frequency at which the m_b scale was computed. The parameter F_{mb} can either be set (typically in the range of 0.5 to 2 Hz) or it can be inverted for. We define the theoretical m_b which is a function of log M_0 to be

$$m_b^{th} = \log M_0 - \log\left[1 + \left(\frac{F_{mb}(KM_0)^\kappa}{c}\right)^2\right] + A_{mb} \quad , \tag{A1}$$

where F_{mb} is defined to be the frequency at which m_b was measured. For a given log $M_0^{(i)}$ we solve for the values of F_{mb} and A_{mb} that minimize the difference between the observed and calculated m_b. To do this, we perform a linearized inversion for F_{mb} and A_{mb} by minimizing the residuals defined to be

$$R = m_b - m_b^{th} = \frac{\partial m_b^{th}}{\partial A_{mb}} \delta A_{mb} + \frac{\partial m_b^{th}}{\partial F_{mb}} \delta F_{mb} \quad , \tag{A2}$$

where the partial derivatives are given by

$$\frac{\partial m_b^{th}}{\partial F_{mb}} = -\frac{2F_{mb}(S_0^{1/3}/c)^2 \log e}{1 + (F_{mb}S_0^{1/3}/c)^2} \tag{A3}$$

and

$$\frac{\partial m_b^{th}}{\partial A_{mb}} = 1 \quad . \tag{A4}$$

Practical experience has shown, that the values of F_{mb} and A_{mb} that minimize the residuals, do not always adequately remove trends of corrected amplitudes with m_b and range. It is often good to examine the trends of corrected amplitudes with m_b and range using a few values of F_{mb} (e.g., $F_{mb} = 0.5$, 1, 2) after final values of η, Q_0, γ and σ_b have been selected. An example of a fit is shown in Figure A1.

A simple linearized inversion is used to find the seismic moments if desired where we minimize the difference between observed and calculated amplitudes (equation (12)) through

$$d_i(f) = \delta_i(f) d \log M_0^{(i)} \tag{A5}$$

where

$$d_i(f) = \log A_i^{(o)}(f) - \log A_i^{(p)}(f) \tag{A6}$$

and

$$\delta_i(f) = \frac{\partial \log A_i(f)}{\partial \log M_0^{(i)}} = 1 - \frac{2\kappa \left(\frac{fK^\kappa}{c}\right)^2 \left(M_0^{(i)}\right)^{2\kappa}}{1 + \left(\frac{f\left(KM_0^{(i)}\right)^\kappa}{c}\right)^2}. \tag{A7}$$

We solve equation (A5) using

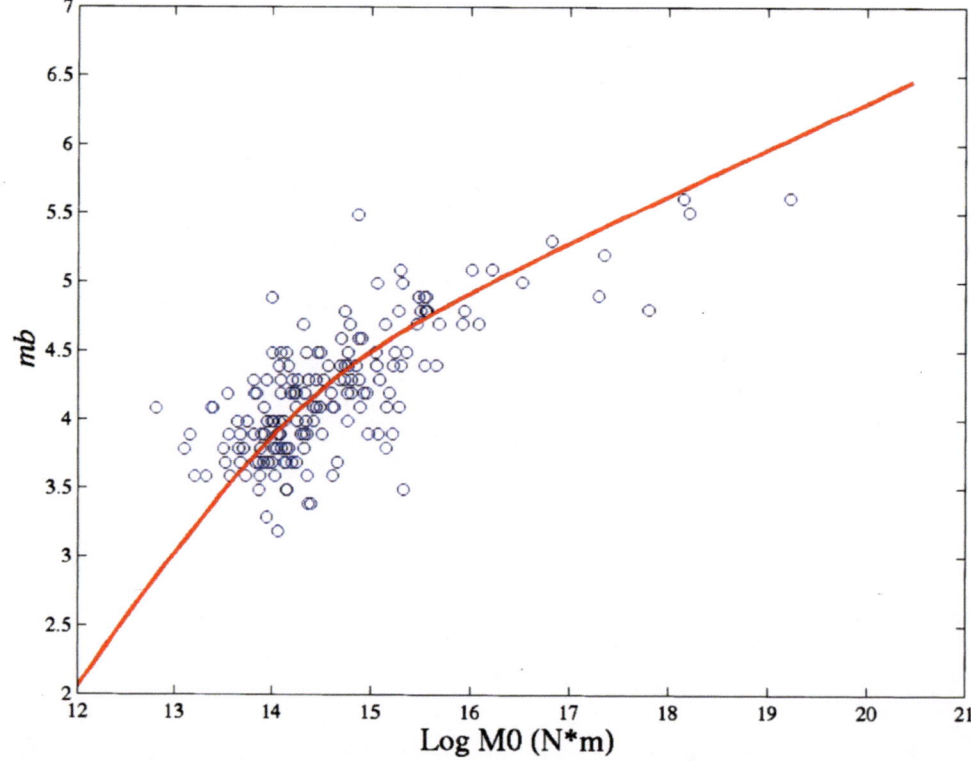

Figure A1

Example plot of m_b versus $\log M_0$ values and fit using equation (A1).

$$dM_0^{(i)} = \frac{\delta_i^T d_i}{\delta_i^T \delta_i} \tag{A8}$$

and compare the variance reduction for each step with the F-statistic for the given number of degrees of freedom to determine when to stop iterating.

Appendix B

Conversion of RMS Amplitudes into Pseudo-Spectral Values

Because MDAC is a spectral method, it is necessary to convert velocity RMS measurements into pseudo-displacement spectrum using Parseval's theorem (e.g., MAISEL, 1971; OPPENHEIM and SCHAFER, 1975; HANKS, 1979). Parseval's theorem states that if $X(\omega)$ is the Fourier transform of $x(t)$, then

$$\int_{-\infty}^{\infty} x^2(t)dt = \frac{1}{2\pi} \int_{-\infty}^{\infty} |X(\omega)|^2 \, d\omega \ . \tag{B1}$$

In computing the RMS measurement, we window a given phase using a window of time length T, and bandpass filter between a low-cut and high-cut frequency $(\omega_l = 2\pi f_l)$ and $(\omega_h = 2\pi f_h)$, respectively (e.g., HARTSE et al., 1997). In this case, equation (B1) becomes

$$\int_0^T x^2(t)dt = \frac{1}{\pi} \int_{\omega_l}^{\omega_h} |X(\omega)|^2 \, d\omega \ , \tag{B2}$$

where the factor of 2 in the right-hand side of equation (B2) originates from the symmetry of the Fourier transform. The mean-square spectrum is given by

$$a_{ms} = \langle x^2(t) \rangle = \frac{1}{T} \int_0^T x^2(t)dt \ . \tag{B3}$$

Using equations (B2) and (B3)

$$T a_{ms} = \frac{1}{\pi} \int_{\omega_l}^{\omega_h} |X(\omega)|^2 \, d\omega \ . \tag{B4}$$

Letting $A^2 = |X(\omega)|^2$ be the average power spectrum in the frequency band between frequencies ω_l and ω_h and integrating equation (B4) gives

$$A^2 = \frac{T}{2(f_h - f_l)} a_{ms} \ . \tag{B5}$$

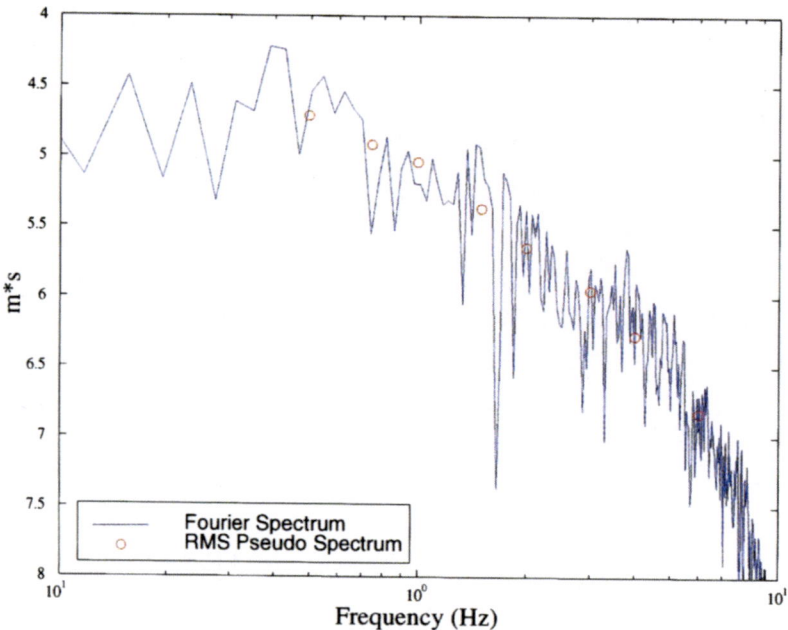

Figure B1
Lg displacement spectrum and RMS pseudo-spectrum computed using equation (B6).

Using $a_{rms} = \sqrt{a_{ms}}$, the base-10 logarithm of the average displacement spectra in the frequency band between frequencies f_l and f_h is (pseudo-displacement spectra)

$$\log A = \log a_{rms} + \tfrac{1}{2}\log\frac{T}{2(f_h - f_l)} \quad . \tag{B6}$$

As discussed by RODGERS *et al.* (1997), the RMS amplitudes can be biased high relative to log averaged frequency-domain amplitudes. To mitigate this effect, we assign the low frequency cutoff, f_l, to be the specified frequency for each band.

We can convert peak-to-peak measurements to pseudo-displacement spectra using a frequency-dependent amplitude scaling parameter. Experiments show that the peak-to-peak measurements for a band-pass filtered signal are typically 5 to 6 times larger than corresponding RMS measurements (which are values expected for a Gaussian random process). Figure B1, shows an example of a Lg displacement spectra and RMS pseudo-spectrum computed using equation (B6). The agreement between the two spectra is very good, illustrating that the method works well.

In cases where we use pre-phase noise, we often use a noise window that is shorter than the signal window in order to minimize contamination from earlier-arriving phases. Using equation (B6) we can see that in estimation of signal-to-noise ratios for unequal window lengths, we need to add the factor of $1/2 \log T_s/T_n$

to the logarithm of the noise, where T_s and T_n are the lengths of the signal and noise, respectively.

Acknowledgments

Helpful discussions with Howard Patton and Dale Anderson are greatly appreciated. This work is performed under the auspices of the U.S. Department of Energy by Los Alamos National Laboratory under contract W-7405-ENG-36.

REFERENCES

AKI, K. and RICHARDS, P. G., *Quantitative Seismology. Theory and Methods* (W.H. Freeman and Co., New York, NY, 1980).

BOORE, D. M. and BOATWRIGHT, J. (1984), *Average Body-wave Radiation Coefficients*, Bull. Seismol. Soc. Am. *74*, 1615–1621.

BRUNE, J. N. (1970), *Tectonic Stress and Spectra of Seismic Shear Waves from Earthquakes*, J. Geophys. Res. *75*, 4997–5009.

CONG, L., XIE, J., and MITCHELL, B. J. (1996), *Excitation and Propagation of L_g from Earthquakes in Central Asia with Implications for Explosion/Earthquake Discrimination*, J. Geophys. Res. *101*, 27,779–27,789.

DEWEY, J. W. (1999), *On the PIDC/USGS m_b Discrepancy*, Seis. Res. Lett. *70*, 211.

HANKS, T. (1979), *b Values and $\omega^{-\gamma}$ Seismic Source Models: Implications for Tectonic Stress Variations Along Active Crustal Fault Zones and the Estimation of High-frequency Strong Ground Motion*, J. Geophys. Res. *84*, 2235–2245.

HANKS, T. and JOHNSTON, A. C. (1992), *Common Features of the Excitation and Propagation of Strong Ground Motion for North American Earthquakes*, Bull. Seismol. Soc. Am. *82*, 1–23.

HARTSE, H. E., TAYLOR, S. R., PHILLIPS, W. S. and RANDALL, G. E. (1997), *Regional Event Discrimination in Central Asia with Emphasis on Western China*, Bull. Seismol. Soc. Am. *87*, 551–568.

HOUGH, S. E. (1996), *Observational Constraints on Earthquake Source Scaling: Understanding the Limits in Resolution*, Tectonophys. *261*, 83–95.

MAISEL, L., *Probability, Statistics and Random Processes* (Simon and Schuster, NY, 1971).

MAYEDA, K. and WALTER, W. R. (1996), *Moment, Energy, Stress Drop, and Source Spectra of Western United States Earthquakes from Regional Coda Envelopes*, J. Geophys. Res. *101*, 11,195–11,208.

MITCHELL, B. J., PAN, Y., XIE, J., and CONG, L. (1997), *Lg Coda Q Variation Across Eurasia and its Relation to Crustal Evolution*, J. Geophys. Res. *102*, 22,767–22,779.

NUTTLI, O. W. (1973), *Seismic Wave Attenuation and Magnitude Relations for Eastern North America*, J. Geophys. Res. *78*, 876–885.

NUTTLI, O. W. (1983), *Average Seismic Source-parameter Relations for Mid-plate Earthquakes*, Bull. Seismol. Soc. Am. *73*, 519–535.

OPPENHEIM, A. V. and SCHAFER, R. W., *Digital Signal Processing* (Prentice-Hall Inc., Englewood Cliffs, NJ, 1975).

PHILLIPS, W. S., RANDALL, G. E., and TAYLOR, S. R. (1998), *Regional Phase Path Effects in Central China*, Geophys. Res. Lett. *25*, 2729–2732.

PHILLIPS, W. S. (1999), *Empirical Path Corrections for Regional-phase Amplitudes*, Bull. Seismol. Soc. Am. *89*, 384–393.

PHILLIPS, W. S., HARTSE, H. E., TAYLOR, S. R., VELASCO, A. A., and RANDALL G. E. (2001), *Regional Phase Amplitude Ratio Tomography for Seismic Verification*, Pure Appl. geophys. *158*, 1189–1206.

PRIESTLEY, K. F. and PATTON, H. J. (1997), *Calibration of $m_b(Pn)$, $m_b(Lg)$ Scales and Transportability of the $M_0:m_b$ Discriminant to New Tectonic Regions*, Bull. Seismol. Soc. Am. *87*, 1083–1099.

RENCHER, A. C., *Multivariate Statistical Inference and Applications* (Wiley, NY, 1998).

RODGERS, A. J., LAY, T., WALTER, W. R., and MAYEDA, K. M. (1997), *Comparison of Regional Phase Amplitude Ratio Measurement Techniques*, Bull. Seismol. Soc. Am. *87*, 1613–1621.

RODGERS, A. J., WALTER, W. R., SCHULTZ, C. A., MYERS, S. C. and LAY, T. (1999), *A Comparison of Methodologies for Representing Path Effects on Regional P/S Discriminants*, Bull. Seismol. Soc. Am. *89*, 394–408.

SERENO, T. J., BRATT, S. R., and BACHE, T. C. (1988), *Simultaneous Inversion of Regional Wave Spectra for Attenuation and Seismic Moment in Scandinavia*, J. Geophys. Res. *93*, 2019–2035.

SERENO, T. J. and GIVEN, J. W. (1990), *Pn Attenuation for a Spherically Symmetric Earth Model*, Geophys. Res. Lett. *17*, 1141–1144.

SINGH, S. and HERRMANN, R.B. (1983), *Regionalization of Crustal Coda Q in the Continental United States*, J. Geophys. Res. *88*, 527–538.

STREET, R., HERRMANN, R., and NUTTLI, O. (1975), *Spectral Characteristics of the Lg Wave Generated by Central United States Earthquakes*, Geophys. J. R. Astron. Soc. *41*, 51–63.

TAYLOR, S. R., DENNY, M. D., VERGINO, E. S., and GLASER, R. E. (1989), *Regional Discrimination Between NTS Explosions and Western U.S. Earthquakes*, Bull. Seismol. Soc. Am. *79*, 1142–1176.

TAYLOR, S. R. and DENNY, H. D. (1991), *An Analysis of Spectral Differences Between NTS and Shagan River Nuclear Explosions*, J. Geophys. Res. *96*, 6237–6245.

TAYLOR, S. R. and HARTSE, H. E. (1997), *An Evaluation of Generalized Likelihood Ratio Outlier Detection to Identification of Seismic Events in Western China*, Bull. Seismol. Soc. Am. *87*, 824–831.

TAYLOR, S. R. and HARTSE, H. E. (1998), *A Procedure for Estimation of source and Propagation Amplitude Corrections for Regional Seismic Discriminants*, J. Geophys, Res. *103*, 2781–2789.

TAYLOR, S. R. and VELASCO, A. A. (1998), *User's Manual for SPAC 1.0: A MATLAB Program for Computing Source and Path Amplitude Corrections*, Los Alamos National Laboratory, Los Alamos, NM, LAUR-98-4363.

YOUNG, C. and WARD, R. W. (1980), *Three-dimensional Q^{-1} Model of the Coso Hot Springs Known Geothermal Resource Area*, J. Geophys. Res. *85*, 2459–2470.

WEISBERG, S., *Applied linear Regression* (Wiley, New York, 1980).

XIE, J. (1998), *Spectral Inversion of Lg from Earthquakes: A Modified Method with Applications to the 1995, Western Texas Earthquake Sequence*, Bull. Seismol. Soc. Am. *88*, 1525–1537.

XIE, J. and PATTON, H. J. (1999), *Regional Phase Excitation and Propagation in the Lop Nor Region of Central Asia and Implications for P/Lg Discriminants*, J. Geophys. Res. *104*, 941–954.

XIE, X. and LAY, T. (1995), *The log $(rmsLg) - m_b$ Scaling Law Slope*, Bull. Seismol. Soc. Am. *85*, 834–844.

(Received July 23, 1999, revised November 3, 2000, accepted November 15, 2000)

To access this journal online:
http://www.birkhauser.ch

Pure appl. geophys. 159 (2002) 651–678
0033–4553/02/040651–28 $ 1.50 + 0.20/0

© Birkhäuser Verlag, Basel, 2002

▌Pure and Applied Geophysics

Path Corrections for Source Discriminants: A Case Study at Two International Seismic Monitoring Stations

GUANG-WEI FAN,[1] THORNE LAY[1] and STEVEN BOTTONE[2]

Abstract — Improving the performance of short-period regional seismic discriminants by applying propagation corrections is explored using observations from two seismic monitoring stations in Asia. Frequency-dependent regional phase amplitude ratio measurements at stations NIL and ZAL for earthquakes and underground nuclear explosions were obtained from the prototype-International Data Center (pIDC) that has been established for developing monitoring capabilities of the Comprehensive Nuclear-Test-Ban Treaty (CTBT). The pIDC discriminant measurements have large scatter, much of which is attributed to wave propagation effects in the heterogeneous crustal waveguide. Linear regressions indicate that the phase ratios are correlated with topographic characteristics along the individual paths, providing an empirical means for correcting for path effects beyond conventional distance corrections. Kriging, a spatial multiple regression algorithm, also reveals coherent spatial patterns in the data indicative of regional path effects. Using available high-resolution topography data, correction of regional P/S ratios for the best models obtained from multivariate regressions systematically reduces the data variance relative to distance corrections alone, as has been observed for other data sets. The reduced scatter in the measurements increases the separation between earthquake and explosion populations in most cases, enhancing the regional discriminant performance. The path-corrected discriminants isolate explosions better for NIL than for ZAL, even though some of the explosion sources are located in a common source area. Kriging achieves comparable or superior variance reduction for the discriminant measures, without requiring knowledge of the path structure, although this may not result in improved discriminant performance. While always desirable, corrections for heterogeneous path effects may prove inadequate in some cases, notably when phase blockage occurs or when strong attenuation eliminates the diagnostic high-frequency energy.

Key words: Seismic discrimination, regional seismic waves, path effects, kriging.

Introduction

A major task in monitoring the Comprehensive Nuclear-Test-Ban Treaty (CTBT) is identification of the sources of seismic signals detected by the International Seismic Monitoring System (ISMS). This requires isolation of

[1] Institute of Geophysics and Planetary Physics, Department of Earth Sciences, University of California, Santa Cruz, CA 95064, U.S.A. E-mail: thorne@emerald.ucsc.edu
[2] Mission Research Corporation, Santa Barbara, CA 93102, U.S.A.

seismic wavefield characteristics that are diagnostic of the source type, which may be an earthquake, quarry blast, mine burst, underground nuclear explosion, or other phenomenon. It is often necessary to use high-frequency seismic signals recorded at regional distances (from a few hundred to 2000 km from the source) for reliable identification of low-yield explosions because more distant signals are either too weak or lack discriminating attributes. Regional seismic waves have complex waveforms and involve multiple reflections and phase conversions within the laterally varying crustal and lithospheric waveguide, and these propagation effects tend to obscure intrinsic source diagnostics such as the relative amount of S-wave versus P-wave energy radiated by the source. Efforts to correct for propagation effects on regional signals are largely empirical, given that earth models and waveform modeling procedures are inadequate for detailed theoretical corrections.

Amplitude ratios for regional phases that involve predominantly S or P energy have been shown to provide promising discrimination of earthquake and explosion sources, as long as the observations have good signal-to-noise ratio and the path effects have been suppressed in some way. Source discrimination using regional P-and S-phase amplitude ratios is found to be most effective at frequencies above 3 Hz (e.g., WALTER *et al.*, 1995; TAYLOR, 1996; HARTSE *et al.*, 1997); however, regional seismic signals are commonly observed to have higher signal-to-noise ratios at frequencies below 3 Hz where intrinsic source differences appear to be more subdued. The lower frequency band also appears to have stronger dependence on path effects (e.g., ZHANG *et al.*, 1994; FAN and LAY, 1998a; PHILLIPS *et al.*, 1998), thus there is potential for enhancing discrimination across the passband by improved path corrections. In general, applying any reliable corrections for propagation effects that can be made is desirable, as discrimination is only trustworthy if the measurements isolate source effects. Some researchers have questioned the value of path corrections when their application reduced the apparent discrimination effectiveness; this is a flawed line of reasoning, for one does not want to base discrimination on what may viably be propagation effects.

Following the historical procedure for seismic magnitude estimation, the most common propagation correction applied to regional phases involves empirical amplitude versus distance relations, determined either for individual phases or for phase ratios. This is essentially a geometric spreading and attenuation correction for a local one-dimensional earth model, which does not take into account variations from path to path. The latter are clearly important, if not dominant, for regional phases, but the specific path effects must be empirically characterized. For example, Lg blockage (which can cause earthquake signals to appear explosion-like) is observed for paths traversing thin oceanic crust (e.g., PRESS and EWING, 1952; ZHANG and LAY, 1995) and for continental paths with abnormally thick crust (e.g., NI and BARAZANGI, 1983; BOSTOCK and KENNETT, 1990; MCNAMARA *et al.*, 1996) or sedimentary basins (e.g., BAUMGARDT, 1990; ZHANG *et al.*, 1994). While the precise physical mechanism of blockage is not always evident, general features of the waveguide structure can often be associated with the likelihood or extent of blockage. This has motivated efforts to use information about individual path topography, crustal thickness, and sedimentary basin structure in developing empirical propagation corrections for regional phases (ZHANG et al., 1994, 1996; FAN and LAY, 1998a,b,c; HARTSE et al., 1998). Multivariate regressions using parametric path properties have been pursued in these studies, seeking to capture the physics of the complex wave propagation effects. The main potential for such methods is in extrapolation to areas where some path properties are known, but seismic data for calibration are sparse.

An alternate strategy has been to forego any model basis for the propagation correction procedure and to simply interpolate the observed patterns using methods such as cap-averaging or kriging (e.g., SCHULTZ et al., 1998; PHILLIPS et al., 1998; PHILLIPS, 1999; RODGERS et al., 1999). As one might expect, spatial interpolation methods such as kriging usually outperform parametric regressions based on model characteristics in terms of driving down the variance in discriminant measurements (e.g., RODGERS et al., 1999; PHILLIPS, 1999), although less physical insight into the path phenomena is obtained, and there is little predictive capability for sparsely sampled regions. Operationally, whatever procedure provides the optimal variance reduction is perhaps the most useful for well-sampled regions, however procedures that extract a physical basis for the discriminant variance are certainly of value for regions that lack sufficient data for independent calibration as well as for guiding modeling studies and for building confidence in the application of path corrections to begin with.

Here, multivariate regression analysis is applied to correct for path effects on regional P/S amplitude ratios observed at two seismic monitoring stations in Asia, NIL (Nilore, Pakistan) and ZAL (Zalesovo, Russia), which are part of the ISMS. The regional phase amplitude ratios recorded at these stations are correlated with regionally-averaged distance effects as well as path-specific effects parameterized by measures of surface topography. A high-resolution topographic database is used because the most reliable waveguide parameters available for any path are those involving surface topography. We consider path-integrated properties (e.g., mean altitude, average surface roughness, etc.) because regional wave propagation effects are likely to involve significant lateral averaging. For comparison, ordinary kriging, which involves multiple regression on spatial parameters, is utilized to estimate path corrections and to visualize the systematic regional path effects. We test whether the reduced scatter in the regional seismic discriminant measures for NIL and ZAL using the topographic models actually yields better separation between earthquake and explosion populations. We also explore whether regionally distinctive characteristics of waveguide structure and geologic setting influence source discrimination performance.

Data

Central Asia is a region of significant interest for monitoring the CTBT. Diffuse seismicity, sparse seismic instrumentation, acute crustal heterogeneity, and recent nuclear testing prioritize the area as a target for calibration of regional discriminants. Observed broadband waveforms for paths in Eurasia display great complexity (e.g., ZHANG *et al.*, 1994), presumably caused by a combination of lateral heterogeneity in waveguide structure and source radiation effects. Our data set involves broadband vertical component recordings at ISMS stations NIL and ZAL for Eurasian earthquakes with magnitudes of $3.5 \leq m_b \leq 6.0$ that occurred between 1995 and 1998. Hypocenter parameters are those determined by the CTBT prototype-International Data Center (pIDC). The sparse ISMS station coverage and the complexity of the crustal and upper mantle structure in the region result in large uncertainties for hypocenter parameters, especially for focal depth, thus many smaller events assigned zero depth in the pIDC location are not well-constrained. We only consider shallow events, with pIDC focal depth estimates less than 50 km. The locations are otherwise deemed sufficient for the path property measurements used in this study, and the location uncertainties are typical of those for discriminant calibration efforts in all regions.

We use the amplitude measurements routinely made by the pIDC. The pIDC algorithm processes regional phases using the Detection and Feature Extraction (DFX) code, computing many amplitude measurements for a variety of filtered traces for specified time windows. Time windows are based on either a predicted travel time for a specific phase, a fixed group velocity window, or some combination of the two. The *Pn* amplitude is calculated within a window starting 8 s before the theoretical arrival time and ending at a group velocity of 6.4 km/s; the *Sn* window starts 5 s before the theoretical arrival time and has a duration of 20 s. The *Pg* and *Lg* windows are between group velocities of 6.3 to 5.8 km/s and 3.7 to 3.0 km/s, respectively. The precise pIDC windows differ from the corresponding windows used in many prior studies of regional discriminants (e.g., WALTER *et al.*, 1995; HARTSE *et al.*, 1997; FAN and LAY, 1998a,b), thus caution in comparison of results is warranted. The default windows can result in spurious amplitude measurements in regions of strong crustal heterogeneity as well as varying proportion of coda contribution with distance, however one of our objectives was to exercise the database generated routinely by the pIDC, as this will ultimately underlie operational applications. In particular, at small epicentral distances there is great difficulty in regional phase isolation, thus only signals at distances greater than 280 km were analyzed. Measurements out to 2100 km were used, recognizing that upper mantle *P*- and *S*-wave triplications overprint true regional guided wave energy at distances exceeding 1400 km. We find that the high-frequency signals vary gradually with distance, therefore the effects of this are subdued, and the number of data is still too limited to further restrict the distance range.

For each phase we consider measurements in four frequency bands: 2 to 4 Hz, 4 to 6 Hz, 6 to 8 Hz, and 8 to 10 Hz. The instrument response was corrected at the center frequency of each passband for all of the phase amplitude measurements. Previous studies have established that rms measures of regional phases like *Lg* have remarkable stability due to the averaging of multiple arrivals (GUPTA *et al.*, 1992; RODGERS *et al.*, 1997a), accordingly we use the pIDC rms amplitude measurements. Noise levels are computed at the pIDC using predicted time and/or group velocity windows around pre-phase noise. In this study only pre-*Pn* noise is considered as a criterion for selecting the data; events with *Pn* signal-to-noise ratios greater than two for the frequency bands of interest are retained. This is somewhat conservative, given that *Pn* is usually the weakest regional phase in the data, except when strong *Lg* blockage occurs (which we do not want to exclude). The number of the selected observations varies with the type of regional phase amplitude measurement observed at each station, ranging between 97 to 109 for NIL, and 73 to 77 for ZAL. The database is growing steadily, and we view our results here as preliminary testing of possible operational implementations with larger data sets. In an operational environment, careful analytical review of automated windows will likely reduce noise in the measurements as well.

The paths traversed by our data sample a large area of the Asian continent. Figure 1 shows epicentral locations for the events, station locations, and surface topography in the region. The topographic relief data derive from a global digital elevation model developed at the U.S. Geological Survey's EROS Data Center. This database provides a spatial resolution of 30 arc-second (one-km) and an overall vertical accuracy of 70 m (GESCH *et al.*, 1999). Many of the paths traverse mountain ranges such as the Tien Shan, Pamirs and Hindu Kush, high plateaus such as the Iranian and Tibetan Plateaus, or sedimentary basins such as the Tarim Basin. Areas of *Lg* blockage, inefficient *Sn* propagation, and low *Pn* velocities have been reported within our region (e.g., NI and BARAZANGI, 1983; RAPINE *et al.*, 1997; RODGERS *et al.*, 1997b). Despite the likelihood that internal crustal structure plays an important role in shaping regional seismic phases (FAN and LAY, 1998a,b,c), the available crustal thickness and sediment thickness models for this region were not utilized because they are very low resolution and involve gross extrapolations. We rely on the high resolution topography data to characterize individual path properties. Although variable isostatic compensation and smoothing effects of sedimentation prevent topography from reflecting details of internal crustal structure, to first order, topography does indicate gross tectonic features of the crust, with high topography in areas of tectonically thickened crust, and laterally variable topography in areas of rapidly changing crust and mantle structure.

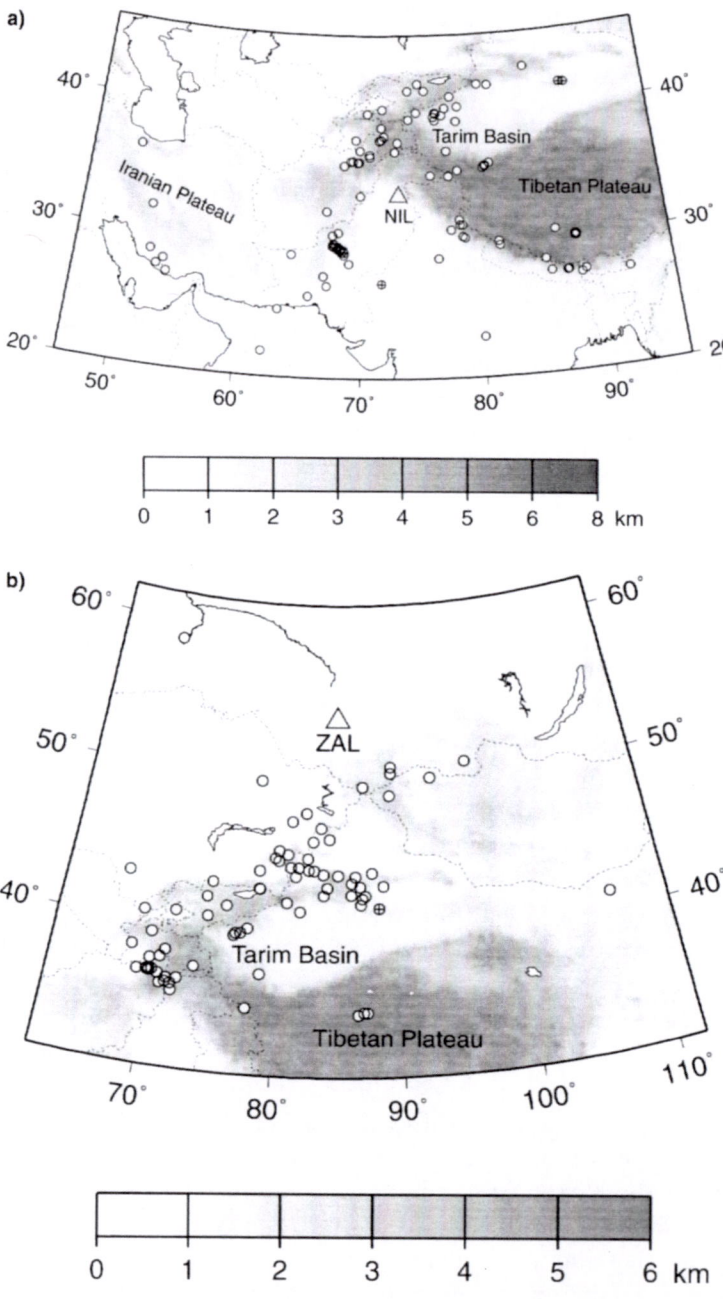

Figure 1
Surface topography in the vicinity of ISMS seismic stations (a) NIL and (b) ZAL. Open circles represent epicenters of earthquakes used in this study; circles with a cross are sites of explosions.

Path Correction

Regional phase amplitude measurements contain a mixture of source and path effects. For CTBT monitoring applications, which often involve very sparse data sets, little can be done regarding source contributions, however empirical procedures can isolate propagation effects to a certain extent. The most standard type of amplitude correction is for propagation path length (e.g., KENNETT, 1993), with amplitude-distance curves being fit to data in a given region. Irregular waveguide properties, as found in tectonically active areas, do produce path-specific effects, and we endeavor to isolate these using multivariate regression analysis and ordinary kriging.

Parametric Multivariate Regression

The strategy of empirical correction for parametric path effects is straightforward. Observed frequency-dependent amplitude ratio measurements for each event recorded by a single station are correlated with various path measurements. High correlations with individual parameters suggest causal influences, but strong covariance amongst parameters mandates a multivariate regression approach to developing a final corrective model. Linear regressions, summarized in Figure 2, illustrate individual path parameter correlations for Pn/Lg and Pn/Sn ratios measured in different frequency bands at NIL and ZAL. Both amplitude ratios show significant distance dependence, with correlation coefficients ranging between 0.1 and 0.8 at frequencies below 8 Hz. The correlation coefficients and slopes of the linear regressions decrease as frequency increases. The average path parameters considered here are mean elevation (mean H), topographic variance (roughness), mean topographic slope (slope), and the second moment of topography (skewness) (ZHANG et al., 1994). NIL exhibits correlations with these path parameters that can be as strong or stronger than correlations with distance. Mean elevation, average roughness, and, for the higher frequencies, mean slope have linear correlation coefficients of 0.3 to 0.6. ZAL has lower correlations in corresponding cases. NIL has strong negative correlations with skewness of topography, similar to observations at WMQ (FAN and LAY, 1998a,b), while weak positive correlations are found at station ZAL. The correlations in Figure 2 motivate a multivariate regression analysis to account for any collinearity of the path properties.

Assuming an empirical model that minimizes the variance in the data, a best set of predictor variables can be determined from a set of parametric path properties. In practice, the initial list of variables includes the path length and the four path-averaged parameters of topography considered above. Elimination of variables is based on two common criteria: (1) the mean square error (MSE) criterion and (2) the C_p-statistic criterion (FAN and LAY, 1998a). In addition, statistical F-tests help to ensure that the final models have satisfactory estimates of all model coefficients at a

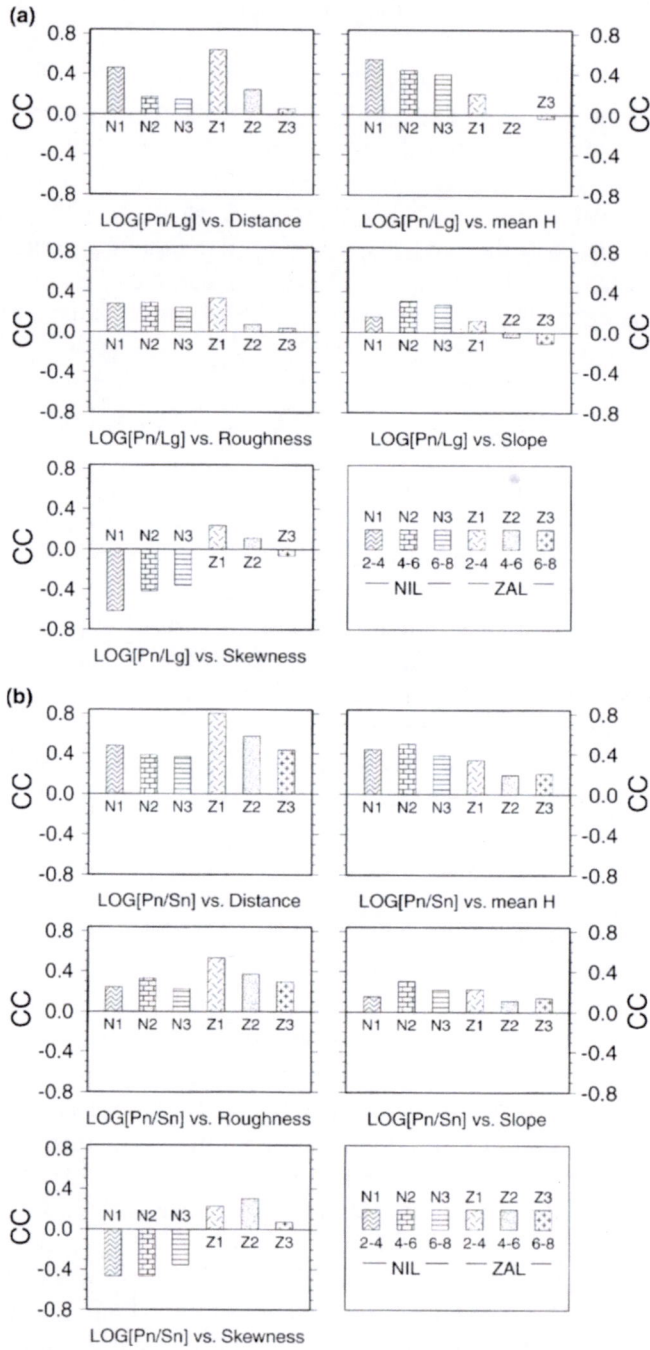

high confidence level. Details of this multivariate regression analysis can be found in FAN and LAY (1998a,b,c).

The regression analysis yields models that involve from one to four path parameters, as shown in Table 1. Path length and mean path elevation appear in most of the models, as was found for observations at WMQ (FAN and LAY, 1998a,b), with surface roughness and rms slope giving some contribution to variance reduction. Figure 3 shows variance reductions achieved by path corrections relative to the raw measurements for NIL and ZAL for four regional discriminant ratios. Known nuclear test signals were omitted. The data for NIL were considered as an entire set (Fig. 3a), and in an azimuthal sector spanning azimuths of 0° to 140° (Fig. 3b), which straddles the azimuth to the Lop Nor test site. Variance reduction for models involving just distance correction are compared to models with the optimal set of path parameters amongst those considered (Table 1). As expected from Figure 2, it is possible to reduce scatter beyond what is achieved by just distance corrections. For the NIL data, increases in variance reduction by 10–20%

Table 1

The best models for various discriminants at station NIL and ZAL

Station	Discriminants	Frequency band (Hz)	Distance	Mean elevation	RMS roughness	RMS slope	RMS skewness
NIL	Log(Pn/Lg)	2.0–4.0	X				X
		4.0–6.0		X			
		6.0–8.0		X			
	Log(Pn/Sn)	2.0–4.0	X	X			
		4.0–6.0	X	X	X	X	
		6.0–8.0	X	X	X	X	
	Log(Pg/Lg)	2.0–4.0		X	X		
		4.0–6.0		X	X	X	
		6.0–8.0		X	X	X	
ZAL	Log(Pn/Lg)	2.0–4.0	X	X			
		4.0–6.0	X	X			
		6.0–8.0	X			X	X
	Log(Pn/Sn)	2.0–4.0	X	X	X		
		4.0–6.0	X	X	X		
		6.0–8.0	X				
	Log(Pg/Lg)	2.0–4.0	X	X			
		4.0–6.0	X		X		
		6.0–8.0			X		

◄

Figure 2

Histograms showing the linear correlation coefficients between various path properties and the regional phase amplitude ratios observed in three frequency passbands at seismic monitoring stations NIL and ZAL. (a) Results for Log(Pn/Lg); (b) results for Log(Pn/Sn). CC represent linear correlation coefficients; the unit of frequency is in Hz.

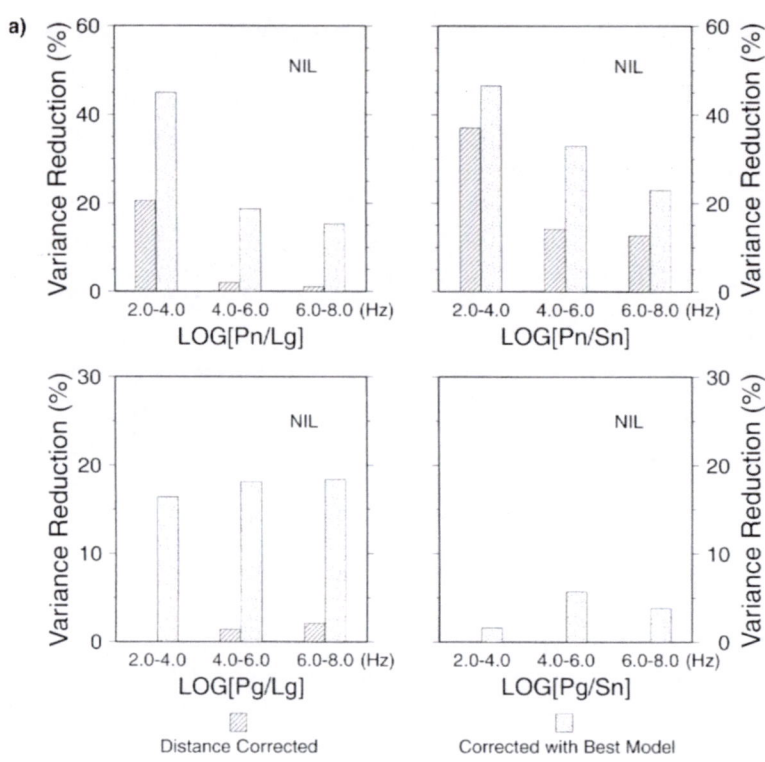

Figure 3

Variance reduction for the *Pn/Lg*, *Pg/Lg*, *Pn/Sn*, and *Pg/Sn* amplitude ratios achieved (a) at NIL; (b) at NIL for an azimuthally restricted data set with azimuths from 0 to 140°; (c) at ZAL. Variance reductions are calculated using corrections for path length only or for the optimal set of parameters in three frequency passbands.

beyond distance correction alone are achieved for Log(*Pn/Lg*), Log(*Pn/Sn*), and Log(*Pg/Lg*) amplitude ratios, and there is a few percent improvement for the Log(*Pg/Sn*) ratios. The azimuthal subset for NIL shows somewhat stronger distance correction effects than for the whole data set, nonetheless there are still significant improvements with the multivariate model. ZAL data experience less variance reduction relative to distance correction alone, involving an extra few percent to 15% (see Fig. 3c). The higher frequency windows generally exhibit less dependence on path properties, but path corrections may prove useful across the passband. The low signal-to-noise ratio 8–10 Hz band was found to have very little correlation with topographic parameters, and results for that frequency band are not shown.

Ordinary Kriging

Kriging involves spatial multiple regression using a covariance function model (ISAAKS and SRIVASTAVA, 1989), and this approach can characterize propagation

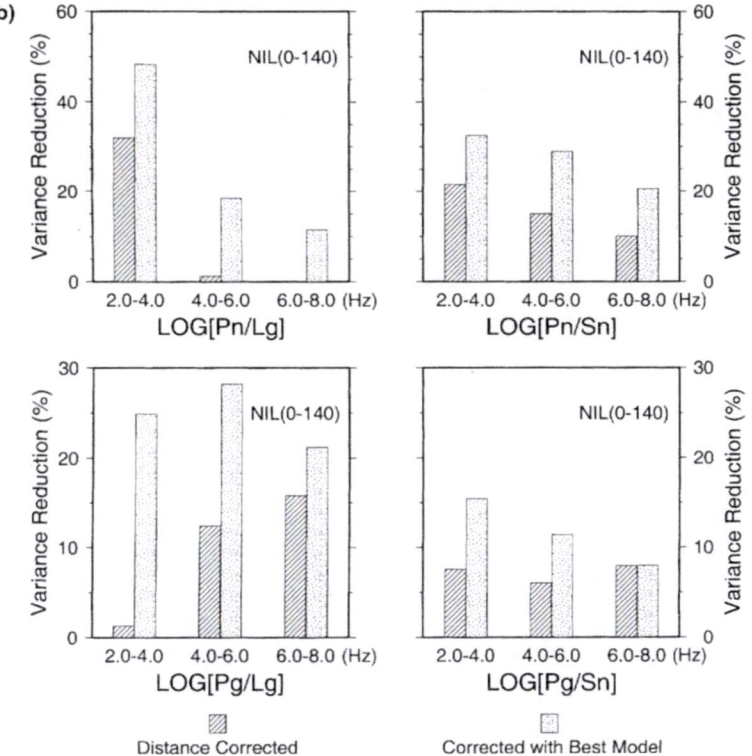

Figure 3b

effects without the need for any information concerning the waveguide. A spatially damped Bayesian kriging (SCHULTZ *et al.*, 1998) has been used successfully as a predictor of arrival-time corrections, with the advantage that it controls estimates in extrapolation zones and allows great flexibility in damping. This algorithm has been applied to regional phase amplitudes, finding that it outperforms parametric regression approaches in terms of variance reduction (RODGERS *et al.*, 1999; PHILLIPS, 1999). A widely available ordinary kriging algorithm is applied to our data, with the constant mean value being replaced by the location-dependent estimate. An advantage of ordinary kriging is that the sum of the weights is constrained to equal 1 so ordinary kriging does not require *a priori* knowledge of the stationary mean for building an estimator, yet it remains unbiased for unsampled values. When applied within a moving data neighborhood, ordinary kriging is a nonstationary algorithm, and it corresponds to a nonstationary random function model with varying mean but stationary covariance. The ability to locally rescale the random function to a different mean value makes ordinary kriging robust. However, extrapolation to poorly sampled areas produces plateaus of varying levels determined

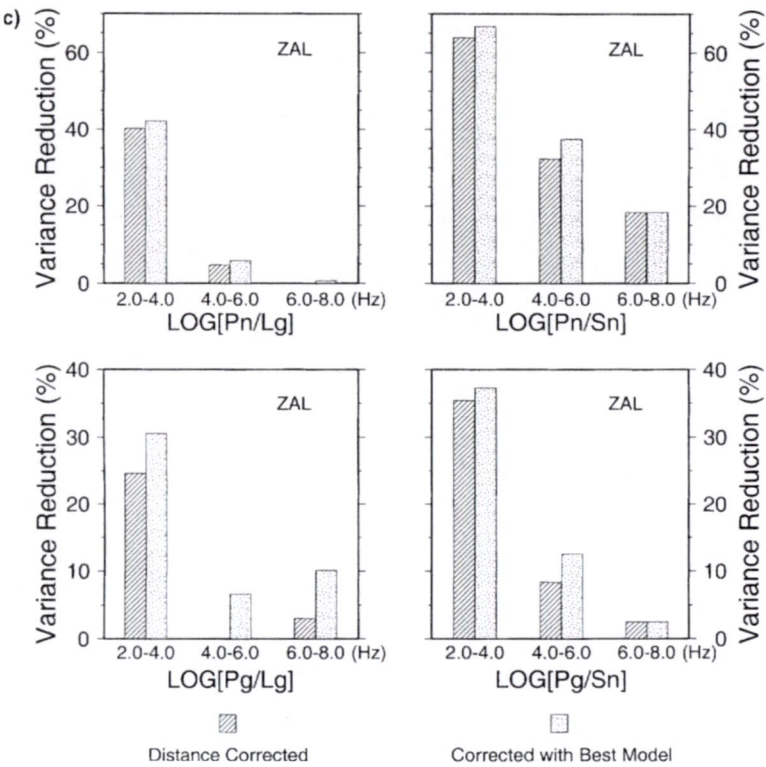

Figure 3c

by proximate data. Since extrapolation is problematic for all kriging algorithms, this is not seen as a unique failing of our algorithm.

We demonstrate the application of ordinary kriging using the Log[*Pn/Lg*] amplitude ratios in the 2–4 Hz frequency band observed at NIL and ZAL. Equal-weighted histograms and cumulative probability plots were created for the *Pn/Lg* amplitude ratios at each station. In Figure 4 the asymmetry of the data seen in the histograms suggests that an assumption of normality would be rejected, while the cumulative probability plots indicate that the amplitude ratios are lognormally distributed (the cumulative frequencies plot as a straight line). Kinks in the cumulative probability plots represent changes in the characteristics of the data over different intervals and may be caused by path effects, as suggested by previous studies in which regional discriminants show better normality in distribution after making multivariate path corrections (FAN and LAY, 1998a,b; RODGERS *et al.*, 1999).

The key to successful kriging estimation is choosing a variogram or a covariance that captures the structural pattern of spatial continuity in the data. We first constructed three types of variograms: semivariogram, covariance and correlogram.

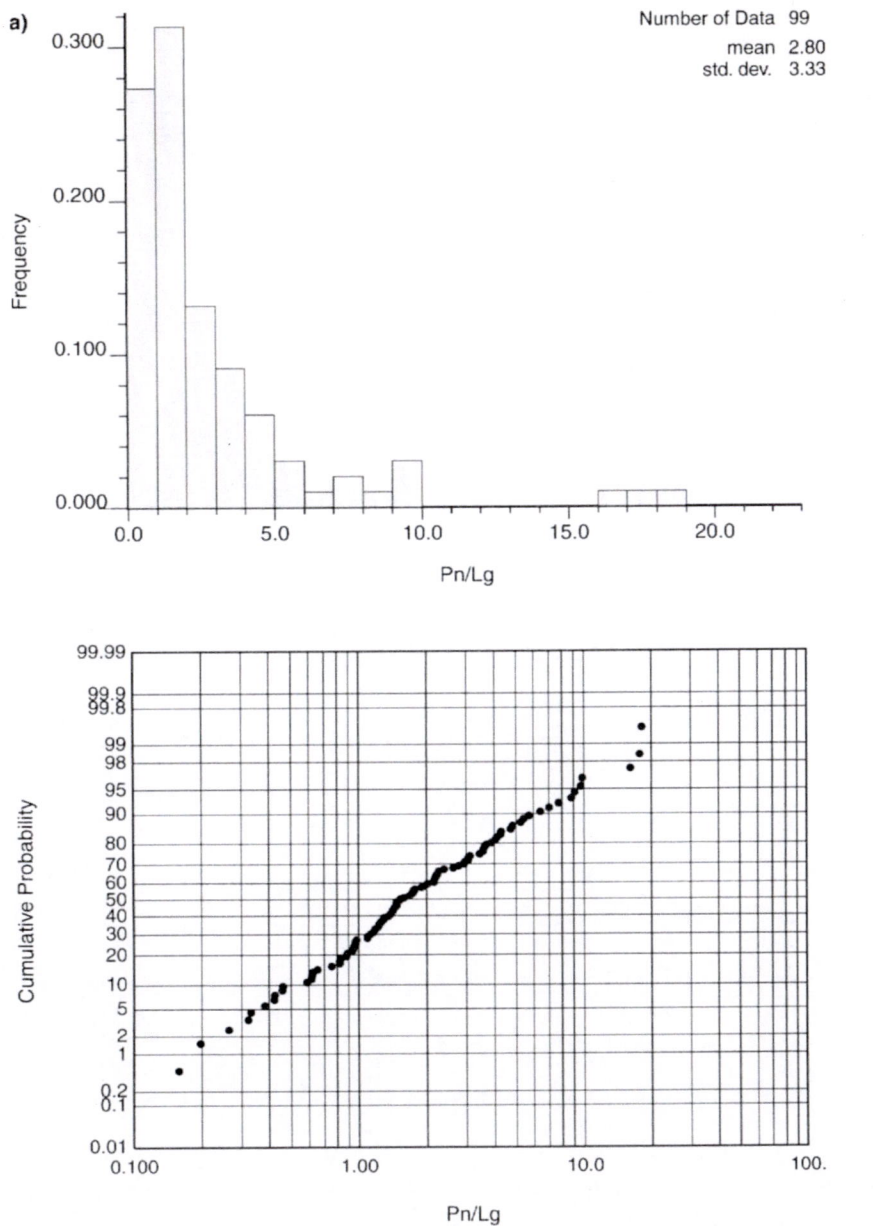

Figure 4
An equal-weighted histogram (top) and lognormal probability plot (bottom) for the *Pn*/*Lg* ratios (2–4 Hz) at (a) NIL and (b) ZAL.

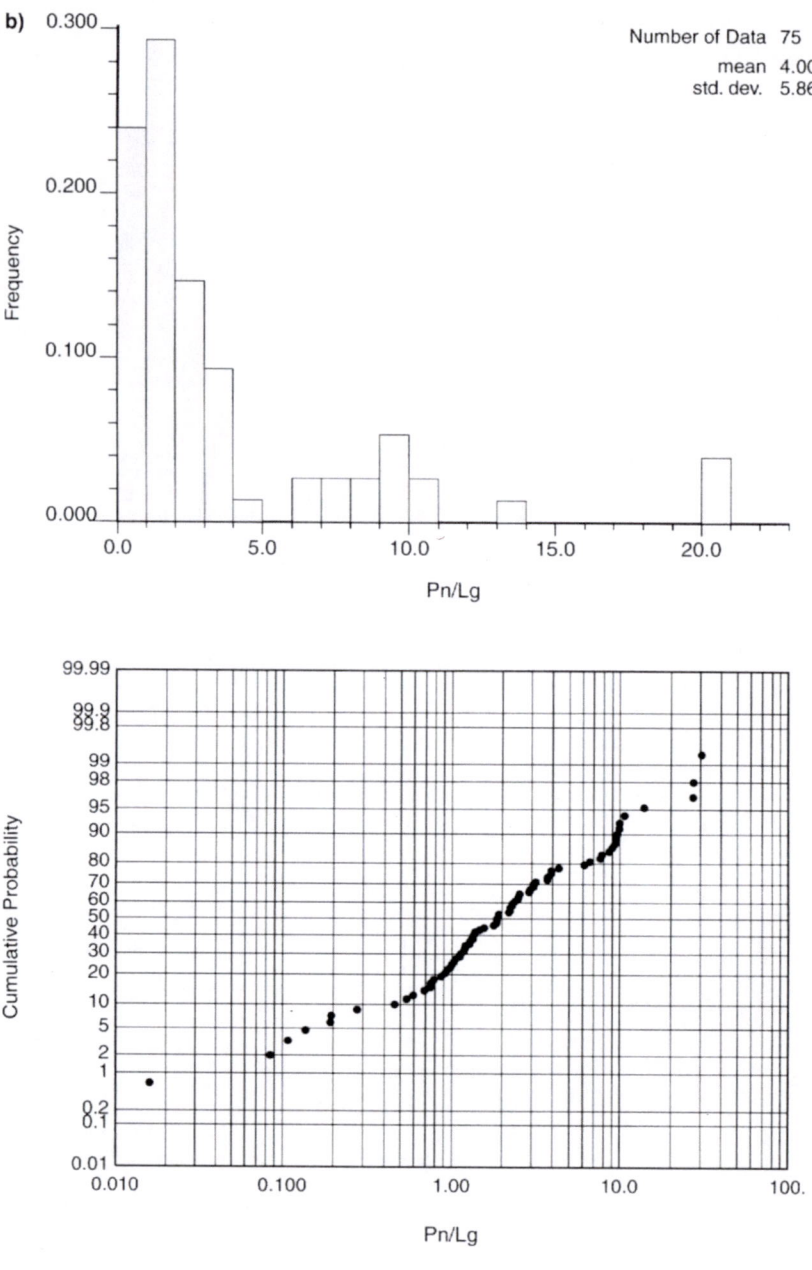

Figure 4b

These variograms typically have been used to describe how spatial variability changes as a function of distance and direction. To explore any geometric anisotropy of the variogram structure we calculate variograms for relative distance along three

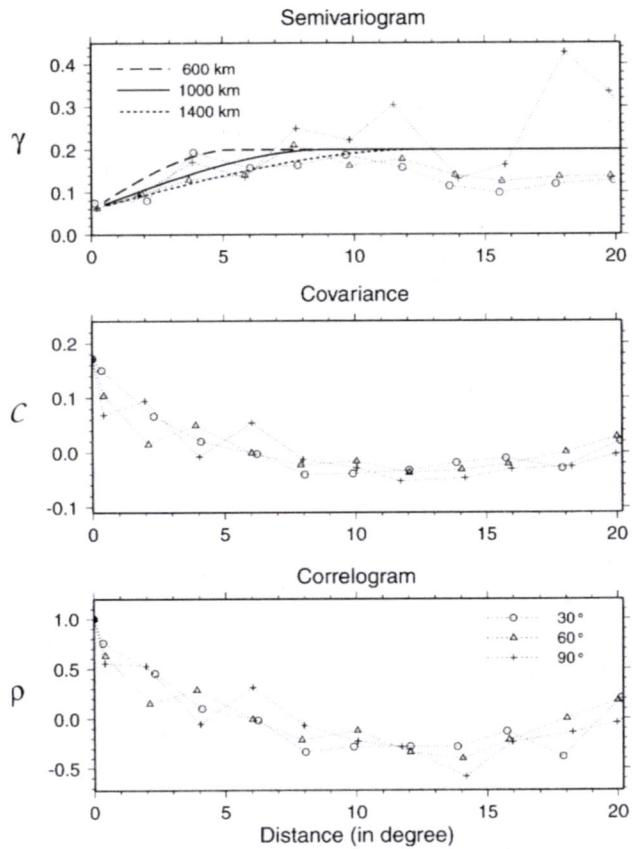

Figure 5

Omnidirectional experimental semivariogram (top), covariance (middle) and correlogram (bottom) for the NIL data set. In the semivariogram, the thick continuous and dashed lines represent an isotropic spherical model with the indicated ranges. The thin lines with different symbols are calculated for directions at angles of 30° (circles), 60° (triangles) and 90° (crosses) from north.

reference directions at angles of 30°, 60° and 90° clockwise from north (Fig. 5). The covariance and correlogram display spatial continuity in the data and have few fluctuations because they are resistant to data sparsity, clustering and outlier values. Based on the behavior of these variograms near the origin, we select a variogram model for the omnidirectional experimental variogram. The spherical and exponential correlation models are commonly used. We adopt a spherical model to match the experimental variogram for its linear behavior at short separation distances near the origin and its flattening out at longer distances. In Figure 5, isotropic models ranging between 800 and 1200 km resemble the observed semivariograms. Because the range has a relatively minor effect on ordinary kriging weights (ISAAKS and SRIVASTAVA, 1989), a range of 1000 km is a good choice for our variogram model. The covariance and correlogram also reach a plateau at this distance, confirming that it is appropriate for our data set.

Based on the selected isotropic model, ordinary kriging was performed by searching a regional grid net. We applied the kriging algorithm to earthquake Log[Pn/Lg] ratios for 2–4 Hz without corrections. As a result of the smoothing effect of kriging, significant variance reduction is achieved. The variance was reduced by 47.1% for the NIL data, and 46.1% for the ZAL data, representing an improvement of 2% (for NIL) and 4% (for ZAL) compared to the parametric multiple regression. The data sets were slightly different for this comparison. For NIL we retained two offshore events in the kriging data set while they were omitted in developing the parametric model. We also removed two isolated events from the kriging analysis for ZAL to avoid unrealistic extrapolation over a large unsampled area. Our purpose in this study is not to compare kriging with parametric models, but to use kriging to help characterize the data sets.

Figure 6 illustrates the distinct spatial patterns displayed by the estimated values from ordinary kriging for these Log[Pn/Lg] ratios. The data clearly display regional coherence, as expected for regionally varying path effects. For NIL (Fig. 6a), low Pn/Lg amplitude ratios are found in Kashmir, while high Pn/Lg amplitude ratios are located in the eastern part of the Tibetan Plateau, where strong Lg attenuation or blockage is observed (RAPINE *et al.*, 1997). A small area between Pakistan and Iran also shows higher Log[Pn/Lg] amplitude ratios. Low Log[Pn/Lg] ratios close to the seismic station perhaps are not reliable because our data do not sample that small region. Correcting for distance effects prior to kriging can reduce this trend. In contrast, at ZAL high Pn/Lg amplitude ratios are found for events in Kashmir, the Pamirs, Hindu Kush, and western Tibet. Low ratios are especially evident for observations from Mongolia and the Baikal rift; these are also regions with low Pg/Lg ratios at LZH (PHILLIPS, 1999). A comparison of these figures indicates that the amplitude ratios are controlled by path effects rather than by localized source region effects. This emphasizes the intrinsic differences between path properties for kriging approaches and parametric regressions. Kriging interpolation can predict the value at any point within the grid net, based on knowledge of the covariance and values at neighboring points. Caution is needed because kriging variance is independent of the data values and generally is not a good measure of local estimation accuracy (JOURNEL, 1986).

For monitoring the CTBT, the mean value of a regional discriminant measure is not of great interest because the sample statistics are often very different from the

▶

Figure 6

Color map showing the observed 2–4 Hz Log[Pn/Lg] measurements and the contoured kriging surface for (a) NIL, and (b) ZAL. The crosses and triangles represent positive and negative log amplitude ratios, respectively, with the symbol size proportional to the absolute values of the amplitude ratios. The dashed lines are 4000-m elevation contours.

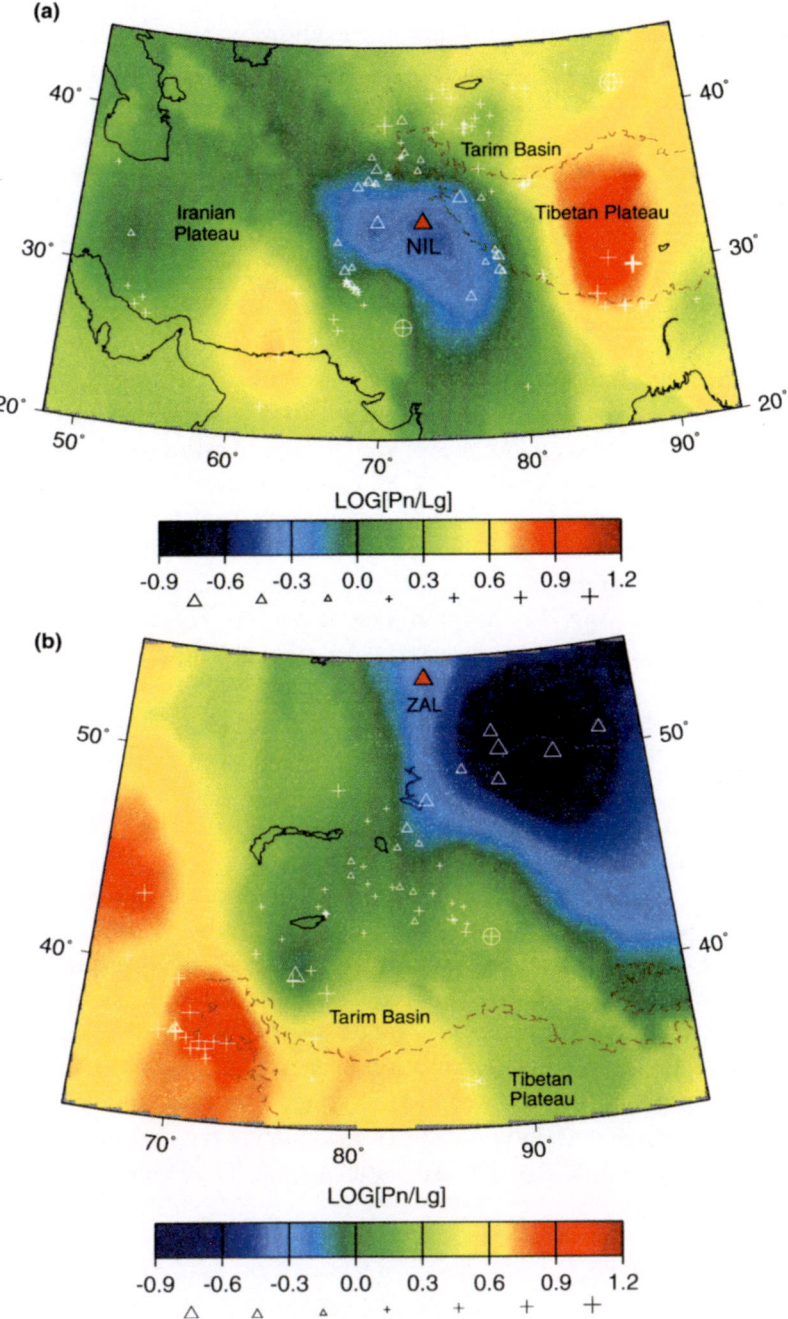

population parameters (Isaaks and Srivastava, 1989); rather we want to know what the sample data set can tell us about the entire population from which they are sampled. Further, the importance of the observed amplitude ratios is for obtaining knowledge of the spatial distribution of the discriminant measures over some large area containing suspicious events to be identified. In this regard, kriging is often known as the best, linear, unbiased estimator, and it produces a set of estimates for which the variance of the errors is minimized.

Source Discrimination

To visualize the effects of ordinary kriging on seismic discrimination, we used the Jackknife for cross-validation to estimate the values of the 2–4 Hz Log(Pn/Lg) ratios at the sites of earthquakes and explosions. The kriging estimated values and observed regional amplitude measurements are shown in Figure 7 for NIL and ZAL. In general, the separation of the explosion population has improved after corrections using ordinary kriging (Fig. 7a). The errors associated with the kriging estimates are relatively small. Apparently, the explosion population is reasonably well separated from the earthquakes at NIL, although there is overlap between the two populations at ZAL. From histograms for observed Log(Pn/Lg) amplitude ratios and kriging residual errors for Log[Pn/Lg] discriminants and estimated values, we found that the kriging residuals of Log[Pn/Lg] for each data set are closer to normally distributed populations than the original values (Fig. 7b). Other kriging algorithms may achieve even better separations, but optimizing this approach is outside the scope of our current paper. The data distribution requires substantial extrapolation for NIL and the coverage is very one-sided at ZAL (Fig. 6).

We also explore whether the separation between earthquake and explosion populations changes after correction for path length and/or other waveguide parameters, i.e., whether the performance of the regional discriminants improves. For our data sets the most effective source discrimination was found for Pn/Lg ratios at NIL. Figure 8a shows Log(Pn/Lg) values as a function of event magnitude for three frequency passbands. Two explosion events at the Lop Nor test site in China are well isolated in the 2–4 Hz and 6–8 Hz frequency bands, while an explosion event at the Pokharan test site in India is close to the earthquake population. In the 4–6 Hz band an earthquake overlaps the explosion group. This event is an outlier of the

▶

Figure 7
(a) Log(Pn/Lg) measurements as a function of event magnitude in uncorrected form (top row) compared with those corrected by ordinary kriging (OK) (bottom row) at NIL (left) and ZAL (right). (b) Histograms of Log(Pn/Lg) measurements in uncorrected form (top row) compared with those of ordinary kriging (OK) residual errors (bottom row) at NIL (left) and ZAL (right). SD, the standard deviation of each population.

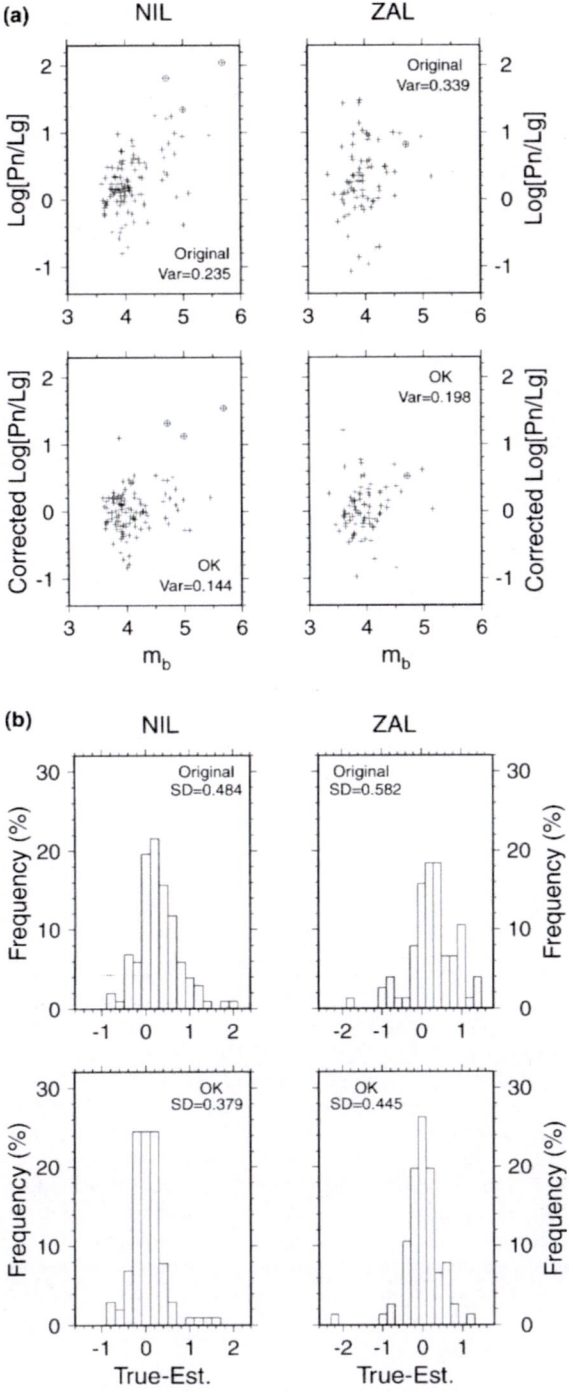

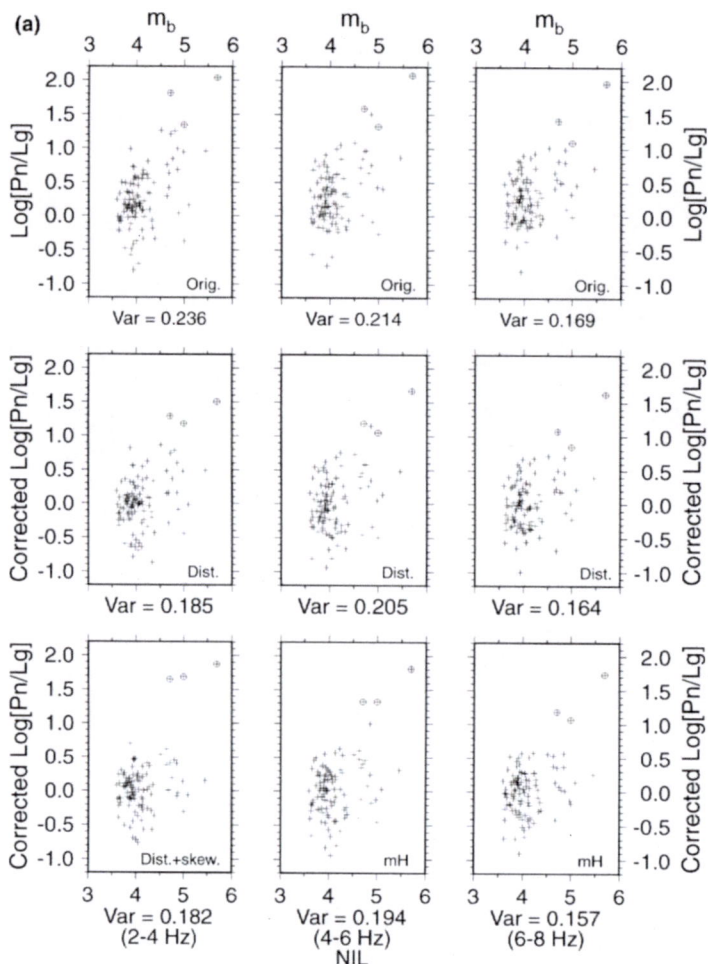

Figure 8

(a) Log(P_n/L_g), (b) Log(Pg/Lg), and (c) Log(Pn/Sn) measurements observed at NIL as a function of event magnitude in uncorrected form (top row) compared with those measurements corrected for path length (middle row) or for the optimal set of topographic parameters (bottom row) in three frequency passbands. Orig., observed data; Dist., corrected for path length; mH, corrected for mean altitude; Dist. + skew., corrected for a two-parameter model involving distance and skewness of topography; mH + rough., corrected for a two-parameter model involving mean altitude and average topography roughness; mH + rough. + slope, corrected for a three-parameter model involving mean altitude, average topography roughness and average slope along the path; Dist. + mH, corrected for a two-parameter model involving distance and mean path elevation; Dist. + mH + rough. + slope, corrected for a four-parameter model involving distance, mean path elevation, average topography roughness and average slope along the path.

Open circles with a cross represent corresponding measurements observed at NIL for explosions.

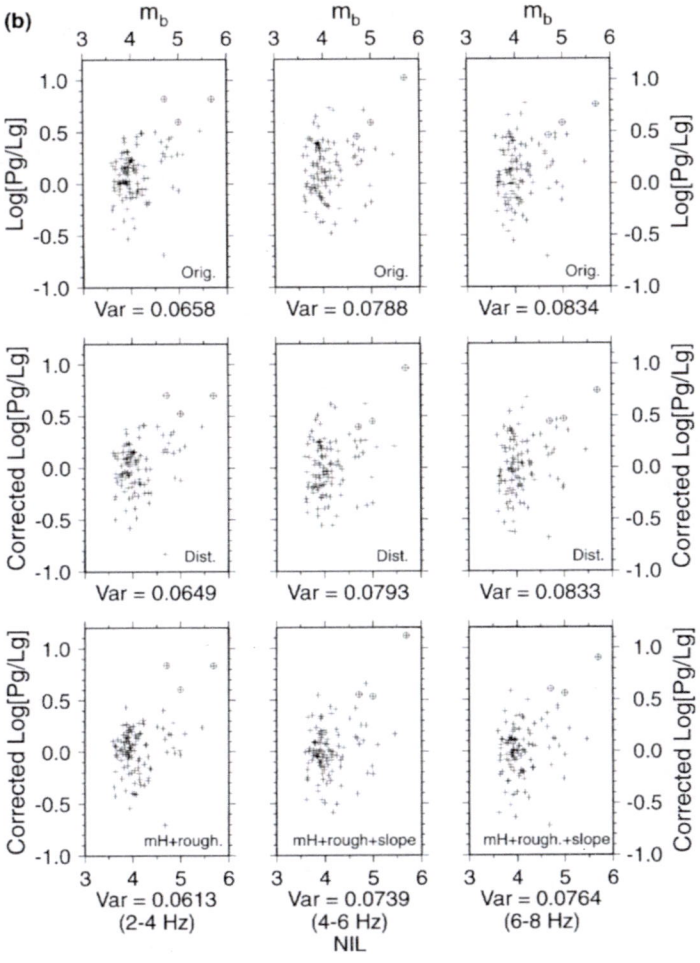

Figure 8b

earthquake population, and is located south of the Tibetan Plateau with a path traveling along the Plateau margin. Distance corrections result in reduced variance as shown in the middle row of Figure 8, and the influence of distance corrections on the separation between the earthquake and explosions populations is more evident in the 2–4 Hz frequency band (given the small number of explosion observations, we do not attempt to apply statistical tests, but simply rely on visual separation). Path correction using the best models from multivariate regression analysis (lower row of Fig. 8) provides further improvement in variance reduction and better source discrimination in all three passbands in this case. This is the ideal goal for path corrections; to tighten up the scatter in the reference earthquake population while enhancing the separation of the explosion population. For the 2–4 Hz observations,

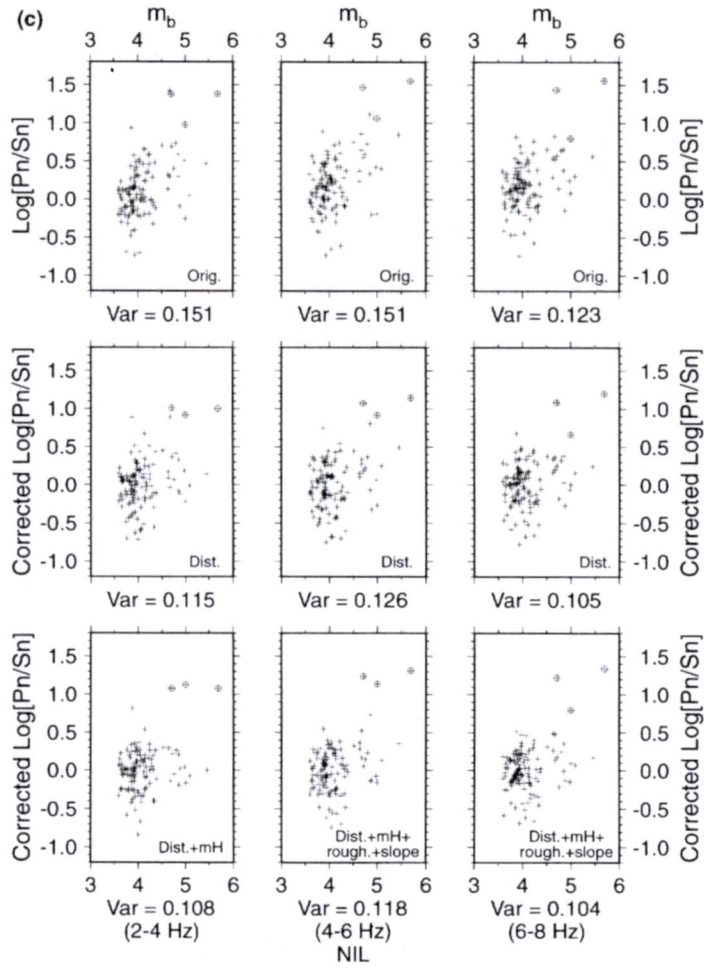

Figure 8c

the parametric approach achieves superior population separation to that achieved with our kriging algorithm (Fig. 7), but possibly other kriging approaches can do as well or better.

Similar behavior is observed for the Log(*Pg/Lg*) measurements at NIL (Fig. 8b). Complete sample separation between earthquake and explosion populations is only found for the frequency band of 2–4 Hz, however the path corrections do provide significant improvement in population separation in the two higher bands. The regional discriminants that involve *Lg* at NIL are more effective for source discrimination at lower frequency. Path corrections for the Log(*Pn/Sn*) ratios (Fig. 8c) also provide significant variance reduction, however, better separation of earthquake and explosion populations is found at higher frequencies.

While not shown here, the discrimination performance for NIL is enhanced by considering the azimuthally restricted subset. Despite the fact that the overall variance reduction of path corrections is no stronger, there is clearer separation of earthquake and explosion populations than for the complete data set. We believe that there are two factors responsible for this. One is that several earthquakes with particularly strong path effects involving Lg blockage were excluded in the azimuthally limited data. Another factor is that by restricting the azimuthal coverage, the data experience more similar travel paths, resulting in more effective path corrections.

For outlier detection algorithms, it is important that the discriminant measures have well defined distributions. Figure 9 displays histograms of the sample distributions for the original $Log(Pn/Lg)$, $Log(Pg/Lg)$, and $Log(Pn/Sn)$ measure-

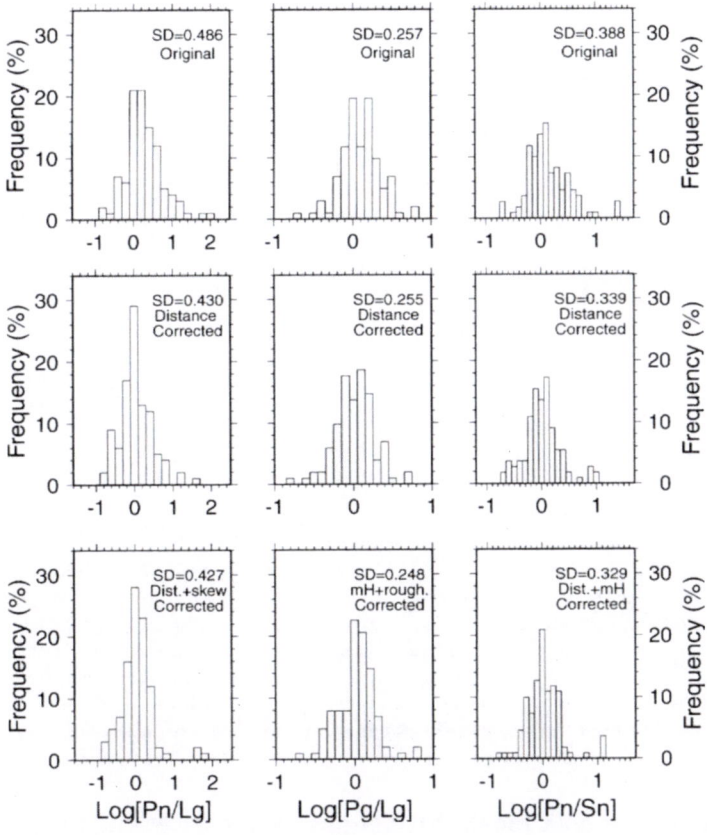

Figure 9

Histograms showing sample distributions for the $Log(Pn/Lg)$, $Log(Pg/Lg)$, and $Log(Pn/Sn)$ measurements in the 2–4 Hz passband at NIL, uncorrected (top row), with path length corrections (middle row) and with best multivariate model corrections (bottom row). SD represents the standard deviation of each population.

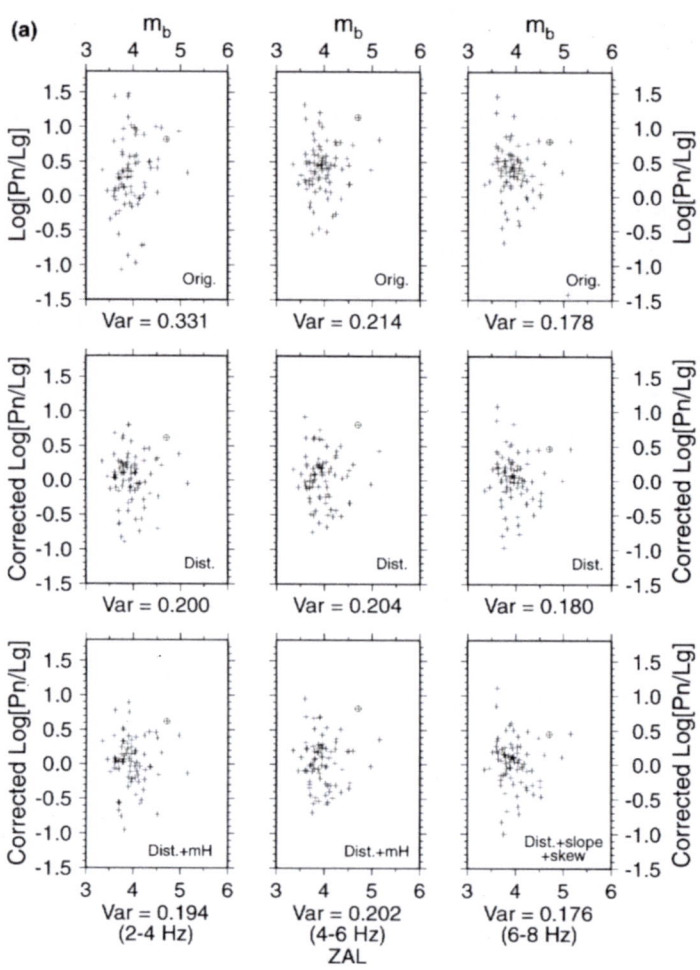

Figure 10

(a) Log(*Pn*/*Lg*) and (b) Log(*Pn*/*Sn*) measurements observed at station ZAL, plotted as functions of event magnitude, for uncorrected data (top row), distance corrected data (middle row), and multivariate regression corrected data (bottom row). Dist. + mH, corrected for a two-parameter model involving distance and mean path elevation; Dist. + slope + skew., corrected for a three-parameter model involving distance, average slope and skewness of topography; Dist. + mH + rough., corrected for a three-parameter model involving distance, mean path elevation and average topography roughness along the path. Open circles with a cross represent corresponding measurements observed for explosions.

ments in comparison with those distributions after corrections for path length and for the best models of waveguide properties in the 2–4 Hz frequency band at NIL. The earthquake population does have a more Gaussian shape and smaller standard deviation after correction, allowing the explosion signals (large positive ratios) to stand out, above the likely tails on the earthquake distributions. Statistical

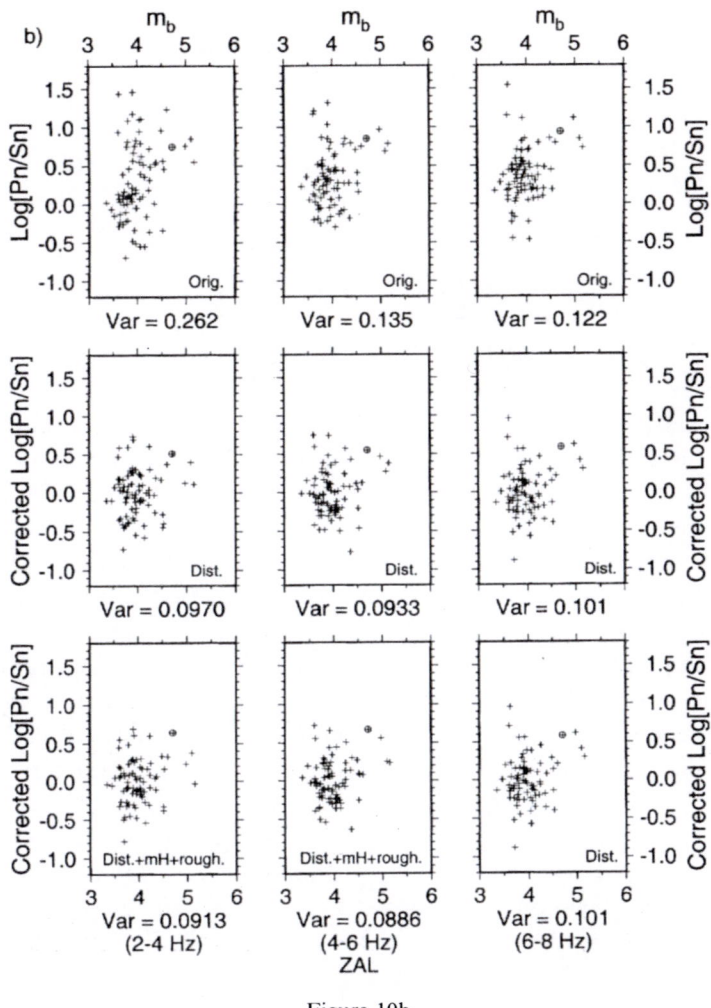

Figure 10b

validation of the identification reliability requires more data, but it appears very promising.

Discriminant measurements at ZAL do not perform as well as those at NIL, even though the available explosion sources are located in the same source region. Figure 10 shows the observations for Log(Pn/Lg) and Log(Pn/Sn). The substantial overlap between earthquake and explosion populations is presumably a result of strong path effects. Correction for path length reduces the overlap, but does not achieve separation between earthquake and explosion populations, as was the case with the kriging application to 2–4 Hz Pn/Lg observations. Using the best parametric regression provides slightly further improvement in source discrimination for Log(Pn/Sn), but has little effect on Log(Pn/Lg). Other stations, such as AAK also do not provide clean discrimination (HARTSE *et al.*, 1998). Combining multiple discriminants and/or multiple station observations may address this failing.

Summary

In this study we made path corrections for regional seismic discriminants observed at two CTBT monitoring stations in Asia. Correction of the path effects is based on empirical models derived from multivariate regression analysis or ordinary kriging. The parametric multiple regression uses a high-resolution topographic database to develop corrections involving path length and/or other topographic parameters. The corrections reduce the variance in the regional discriminants, particularly at lower frequencies. Kriging outperforms the multivariate regression approach in terms of reducing the earthquake population variance, but whether enhanced discrimination results is not yet established. For station NIL, the path corrections enhance discrimination effectiveness for a small set of explosions, while for station ZAL there is improvement, but still substantial overlap of the populations. This variation in discriminant performance presumably reflects acute waveguide heterogeneity that is not captured in the parametric regression analysis. Accumulating larger data sets, with improved sampling of the paths traversed by explosion signals, would likely lead to superior performance for both stations. Improved coverage would also reduce the degree of extrapolation currently required for the kriging analysis to be applied to the specific explosion source region in western China. Overall, there is a promising indication that application of propagation corrections is beneficial for these regional discriminants, nonetheless further work is needed to establish a preferred operational approach.

Acknowledgements

We thank Mission Research Corporation for accessing the data used in this study. Comments from W. Walter, S. Phillips, and A. Rodgers were helpful. We extend our gratitude to Ru-Shan Wu and Xiao-Bi Xie for helpful discussion. GMT software (WESSEL and SMITH, 1991) was used for figures. This research was supported by the Defense Threat Reduction Agency through contract DSWA 01-98-C-0161 (UCSC), and MRC Subcontract SC-4500155967-97-0003 to contract SAIC No. 4500155967. This is the contribution 398, the Center for the Study of Imaging and Dynamics of the Earth, University of California, Santa Cruz.

REFERENCES

BAUMGARDT, D. R. (1990), *Investigation of Teleseismic Lg Blockage and Scattering Using Regional Arrays*, Bull. Seismol. Soc. Am. *80*, 2261–2281.

BOSTOCK, M. G., and KENNETT, B. L. N. (1990), *The Effects of 3-D Structure on L_g Propagation Pattern*, Geophys. J. Int. *101*, 355–365.

FAN, G.-W., and LAY, T. (1998a), *Statistical Analysis of Irregular Waveguide Influences on Regional Seismic Discriminants in China*, Bull. Seismol. Soc. Am. *88*, 74–88.

FAN, G.-W., and LAY, T. (1998b), *Regionalized versus Single-station Waveguide Effects on Seismic Discriminants in Western China*, Bull. Seismol. Soc. Am. *88*, 1260–1274.

FAN, G.-W., and LAY, T. (1998c), *Statistical Analysis of Irregular Waveguide Influences on Regional Seismic Discriminants in China: Additional Results for Pn/Sn, Pn/Lg and Pg/Sn*, Bull. Seismol. Soc. Am. *88*, 1504–1510.

GESCH, D. B., VERDIN, K. L., and GREENLEE, S. K. (1999), *New Land Surface Digital Elevation Model Covers the Earth*, EOS Trans. AGU *80*, 69–70.

GUPTA, I. N., CHAN, W. W., and WAGNER, R. A. (1992), *A Comparison of Regional Phase from Underground Nuclear Explosions at East Kazakh and Nevada Test Sites*, Bull. Seismol. Soc. Am. *82*, 352–382.

HARTSE, H. E., TAYLOR, S. R., PHILLIPS, W. S., and RANDALL, G. E. (1997), *A Preliminary Study of Regional Seismic Discrimination in Central Asia with Emphasis on Western China*, Bull. Seismol. Soc. Am. *87*, 551–568.

HARTSE, H. E., FLORES, R. A., and JOHNSON, P. A. (1998), *Correcting Regional Seismic Discriminants for Path Effects in Western China*, Bull. Seismol. Soc. Am. *88*, 596–608.

ISAAKS, E. H., and SRIVASTAVA, R. M., *An Introduction to Applied Geostatistics* (Oxford University Press, New York 1989).

JOURNEL, A. G. (1986), *Geostatistics: Models and Tools for the Earth Sciences*, Math. Geol. *18*, 119–140.

KENNETT, B. L. N. (1993), *The Distance Dependence of Regional Phase Discriminants*, Bull. Seismol. Soc. Am. *83*, 1155–1166.

MCNAMARA, D. E., OWENS, T. J., and WALTER, W. R. (1996), *Propagation Characteristics of L_g Across the Tibetan Plateau*, Bull. Seismol. Soc. Am. *86*, 457–469.

NI, J., and BARAZANGI, M. (1983), *High-frequency Seismic Wave Propagation Beneath the Indian Shield, Himalayan Arc, Tibetan Plateau and Surrounding Regions: High Uppermost Mantle Velocities and Efficient Sn Propagation Beneath Tibet*, Geophys. J. R. astr. Soc. *72*, 665–689.

PHILLIPS, W. S. (1999), *Empirical Path Corrections for Regional Phase Amplitudes*, Bull. Seismol. Soc. Am. *89*, 384–393.

PHILLIPS, W. S., RANDALL, G. E., and TAYLOR, S. R. (1998), *Path Correction Using Interpolated Amplitude Residuals: An Example from Central China*, Geophys. Res. Lett. *25*, 2729–2732.

PRESS, F., and EWING, M. (1952), *Two Slow Surface Waves Across North America*, Bull. Seismol. Soc. Am. *42*, 219–228.

RAPINE, R. R., NI, J. F., and HEARN, T. M. (1997), *Regional Wave Propagation in China and its Surrounding Regions*, Bull. Seismol. Soc. Am. *87*, 1622–1636.

RODGERS, A. J., LAY, T., WALTER, W. R., and MAYEDA, K. M. (1997a), *A Comparison of Regional Phase Amplitude Ratio Measurement Techniques*, Bull. Seismol. Soc. Am. *87*, 1613–1621.

RODGERS, A. J., NI, J. F., and HEARN, T. M. (1997b), *Propagation Characteristics of Short-period Sn and Lg in the Middle East*, Bull. Seismol. Soc. Am. *87*, 396–413.

RODGERS, A. J., WALTER, W. R., SCHULTZ, C. A., MYERS, S. C., and LAY, T. (1999), *A Comparison of Methodologies for Representing Path Effects on Regional P/S Discriminants*, Bull. Seismol. Soc. Am. *89*, 394–408.

SCHULTZ, C. A., MYERS, S. C., HIPP, J., and YOUNG, C. J. (1998), *Nonstationary Bayesian Kriging: A Predictive Technique to Generate Spatial Corrections for Seismic Detection, Location and Identification*, Bull. Seismol. Soc. Am. *88*, 1275–1288.

TAYLOR, S. R. (1996), *Analysis of High-frequency Pn/Lg Ratios from NTS Explosions and Western U.S. Earthquakes*, Bull. Seismol. Soc. Am. *86*, 1042–1053.

WALTER, W. R., MAYEDA, K. M., and PATTON, H. (1995), *Phase and Spectral Ratio Discrimination between NTS Earthquakes and Explosions. Part I: Empirical Observations*, Bull. Seismol. Soc. Am. *85*, 1050–1067.

WESSEL, P., and SMITH, W. H. F. (1991), *Free Software Helps Map and Display Data*, EOS Trans. *72*, 441–445.

ZHANG, T.-R., and LAY, T. (1995), *Why the Lg Phase Does not Traverse Oceanic Crust*, Bull. Seismol. Soc. Am. *85*, 1665–1678.

ZHANG, T.-R., SCHWARTZ, S., and LAY, T. (1994), *Multivariate Analysis of Waveguide Effects on Short-period Regional Wave Propagation in Eurasia and its Application in Seismic Discrimination*, J. Geophys. Res. *99*, 21,929–21,945.

ZHANG, T.-R., LAY, T., SCHWARTZ, S., and WALTER, W. R. (1996), *Variation of Regional Seismic Discriminants with Surface Topographic Roughness in the Western United States*, Bull. Seismol. Soc. Am. *86*, 714–725.

(Received July 9, 1999, revised October 6, 2000, accepted October 15, 2000)

 To access this journal online:
http://www.birkhauser.ch

Pure appl. geophys. 159 (2002) 679–700
0033–4553/02/040679–22 $ 1.50 + 0.20/0

▌Pure and Applied Geophysics

Seismic Discrimination of the May 11, 1998 Indian Nuclear Test with Short-period Regional Data from Station NIL (Nilore, Pakistan)

ARTHUR J. RODGERS[1] and WILLIAM R. WALTER[1]

Abstract — Regional seismic discriminants for the May 11, 1998 Indian underground nuclear test(s) and earthquakes recorded at station NIL (Nilore, Pakistan) provide new data to test strategies that can be used to monitor the Comprehensive Nuclear-Test-Ban Treaty (CTBT). Three categories of regional discriminants (ratios of P- and/or S-wave energy) were measured on short-period (0.5–6 Hz) seismograms: P/S amplitude ratios (phase ratios) measured in the same frequency band, P- and S-wave spectral ratios (i.e., low frequency to high frequency for the same phase) and P/S cross-spectral ratios (i.e., low frequency S-wave to high frequency P-wave). The P/S amplitude ratios show good separation of the Indian nuclear test and regional earthquakes for Pn/Lg and Pn/Sn, however Pg/Lg does not discriminate as well. Pn/Lg and Pn/Sn discriminate well at frequencies as low as 0.5–2 Hz, especially after accounting for path effects. This observation differs from previous studies that report poor separation of earthquakes and explosions at lower frequencies. The P/S amplitude ratios do not show any magnitude dependence, suggesting that forming the ratios in a fixed frequency band cancels the effects of source size-corner frequency scaling. Spatial variability of the observed discriminants arises from variations in crustal waveguide and/or attenuation structure (path propagation effects). Grouping amplitude ratios for earthquakes with paths similar to the Indian test greatly improves discrimination. Removing distance trends does not generally improve discrimination. Accounting for path effects with Bayesian kriging significantly improves discrimination. Spectral ratios (e.g., Pn [0.5–1 Hz]/Pn [4–6 Hz]) and cross-spectral ratios (e.g., Lg [1–2 Hz]/Pn [4–6 Hz]) show distance and magnitude dependence. We developed a technique for simultaneously removing the effects of distance and source size-corner frequency scaling on amplitude and spectral ratios. The technique uses a gridsearch to find several parameters that characterize the observed distance and magnitude dependence. Discrimination of the Indian test improved dramatically after the distance and magnitude trends were removed from the spectral and cross-spectral ratio data.

Key words: Regional discrimination, path effects.

Introduction

The Government of India announced that it detonated three underground nuclear tests on May 11, 1998 at its Pokharan test site in the northwestern Rajasthan desert (Fig. 1). The events were routinely located by the United States Geological

---
[1] Geophysics and Global Security Division, Lawrence Livermore National Laboratory, Livermore, CA 94551, U.S.A.
    Correspondence: Arthur Rodgers, LLNL, L-205, P.O. Box 808, Livermore, CA 94551, U.S.A.
E-mail: rodgers7@llnl.gov

Survey–Preliminary Determination of Epicenters (USGS–PDE) and by the proto-type International Data Center (pIDC). The similarity of teleseismic *P* waves of the May 11, 1998 Indian tests with the previous 1974 Indian nuclear test suggests that the May 11 test appears seismically as a single large detonation (WALLACE, 1998; BARKER *et al.*, 1998; WALTER *et al.*, 1998). This suggests either near-simultaneous, closely-spaced detonations or one of the tests was much larger than the other two and dominates the response. In either case we can treat the May 11 events as a single source. Any complexity that may have arisen from multiple sources is not critical for the purposes of the short-period regional discrimination analysis presented here. This test provides new data for which to investigate regional discrimination strategies in support of Comprehensive Nuclear-Test-Ban Treaty (CTBT) monitoring. In this article we investigate a variety of short-period discriminants using data recorded at station NIL (Nilore, Pakistan).

Many studies have shown that short-period regional discriminants (amplitude or spectral ratios of *P*- and *S*-wave energy) are effective at identifying explosion sources

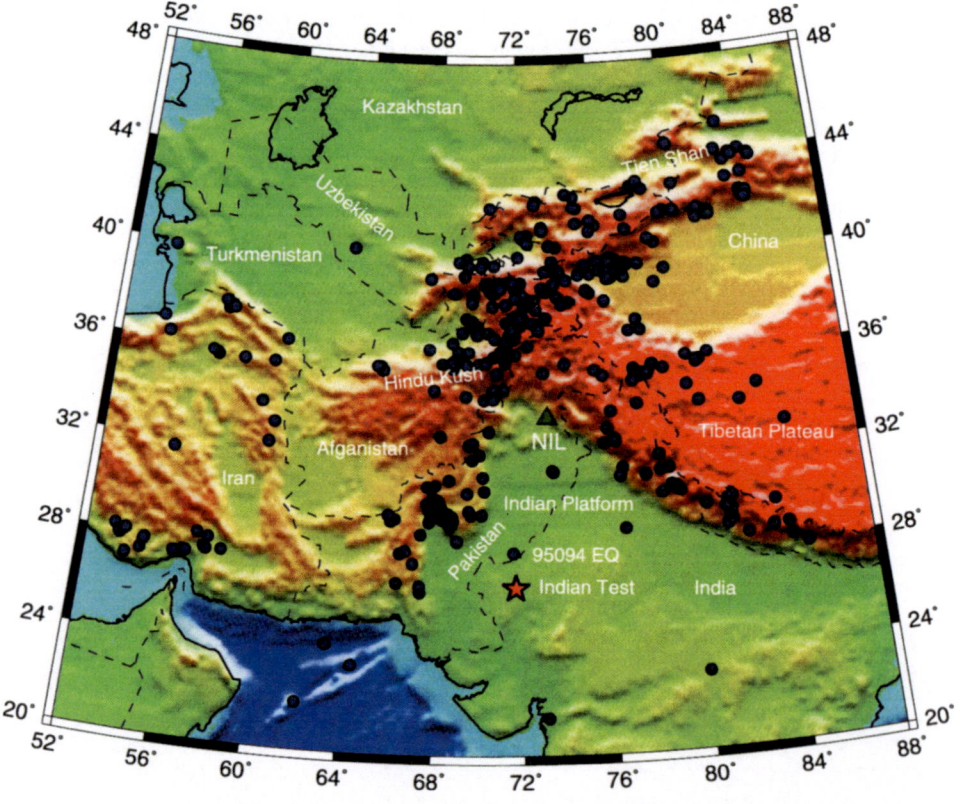

Figure 1

Map of station NIL (green triangle), the Indian test (red star) and regional earthquakes (blue circles) used in this study. The waveform shown in Figure 2 is for the earthquake marked 95094.

(e.g., TAYLOR et al., 1988, 1989; WALTER et al., 1995; HARTSE et al., 1997). However, the specific discriminants and frequency bands which best separate explosions from earthquake populations vary from region to region due to near source and path propagation effects. Thus, one of the challenges of discrimination at regional distances is to determine which discriminants perform best in a given region. This is difficult to do globally because nuclear tests have been conducted in only a few places since the current network of modern broadband seismic stations has been operational. The May 11 event was well recorded at regional distance by station NIL (Nilore, Pakistan; see Fig. 1). Figure 2 shows the NIL recording of the Indian test. This event had a body-wave magnitude, m_b, of 5.2 and a surface wave magnitude, M_S, of 3.5 as reported in the USGS-PDE. Although this event was large enough to be identified as an explosion by the teleseismic $M_s:m_b$ discriminant, nuclear tests smaller than m_b 4.5 in the future may not be discriminated teleseismically due to poor signal-to-noise at distances greater than 2000 km. In this paper, we show that short-period (0.5–6 Hz) regional discriminants (P/S amplitude ratios, phase-spectral ratios and cross-spectral ratios) measured at NIL clearly identify the May 11 Test as an explosion. We also show that procedures which account for distance effects, laterally variable path effects and source size effects can dramatically improve discrimination.

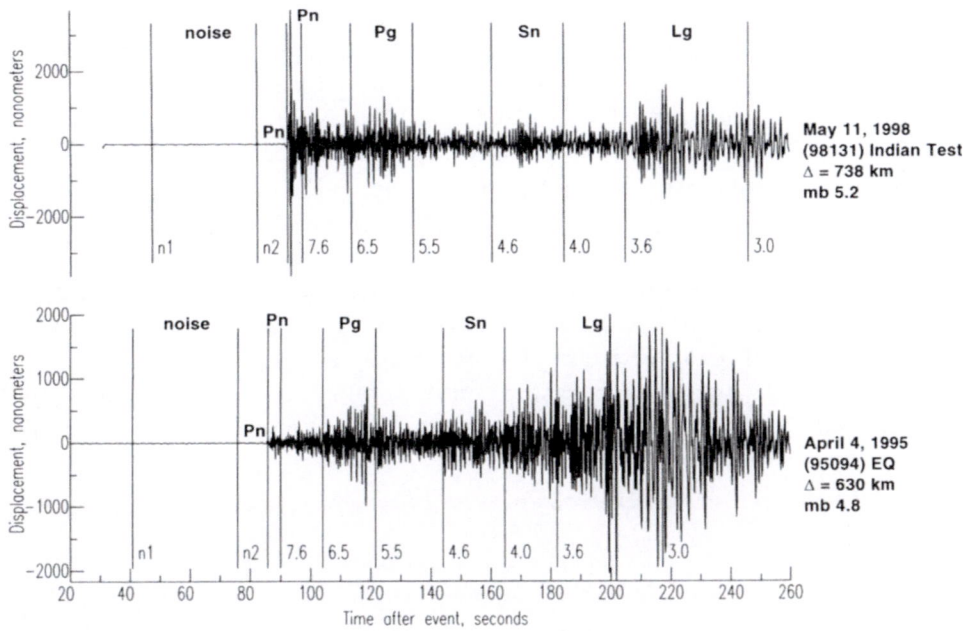

Figure 2
NIL vertical component recordings of the Indian test (top) and a nearby (95094) earthquake (bottom, location shown in Fig. 1). Both traces were bandpass filtered 0.5–6 Hz. Vertical lines indicate phase and noise windows.

Data and Amplitude Measurements

Waveform data for regional earthquakes recorded at station NIL were selected from our research database (RUPPERT et al., 1998). Event locations and origin times were taken from the USGS-PDE. We selected events with reported depths less than 50 km. Distances were limited to 300–1500 km to avoid problems with Pg-Pn interference at close distances and upper mantle triplication body waves at far-regional distances. The earthquake and Indian test locations are shown in Figure 1. All waveforms were previewed and data with glitches and/or poor signal-to-noise were discarded. The first arriving P wave was picked on vertical components. In this study only vertical component waveforms were used. The SHZ and BHZ channels at NIL recorded broadband velocities at 40 samples/second and 20 samples/second, respectively. The instrument response was removed, the amplitude spectra measured from acceleration seismograms and recorded as the Fourier transform of ground displacement. The NIL amplitude ratios were taken from the SHZ channel. An additional 39 events were recorded on the BHZ channel that were not available on the SHZ channel. In order to provide the most complete data set for station NIL, we merged the amplitude measurements for both SHZ and BHZ for this study. For frequencies below the anti-aliasing filter of the BHZ channel (∼6.0 Hz) the spectra of any phase on the SHZ and BHZ components are identical. Thus we limited the maximum frequency of amplitude and spectral ratios to 6.0 Hz.

Regional phases were isolated with the following group velocity windows: Pn 8.25–7.7 km/s; Pg 6.5–5.5 km/s; Sn 4.6–4.0 km/s and Lg 3.6–3.0 km/s. The phase windows were uniformly shifted so that the P wave arrives at 8.25 km/s plus a static delay time of 10.0 s, nominal Pn velocity and crustal leg times for the region, respectively. This time shift accounts for small errors in the location, origin time and station clock accuracy and ensures that the regional phases are windowed uniformly. We discarded traces for which this time shift was greater than ±10 seconds, as these events have more significant timing and/or location problems. Amplitudes were measured in four frequency bands (0.5–1.0, 1.0–2.0, 2.0–4.0, and 4.0–6.0 Hz) from the smoothed amplitude spectrum of each phase, as in previous studies (e.g., RODGERS et al., 1999). Noise measurements were made on 30 second pre-Pn windows. Figure 2 illustrates the windows used to isolate the phases and noise for the Indian test and a nearby earthquake.

Discriminants

Discriminants were formed from the amplitude measurements. Ratios were formed only when the amplitudes of both phases had pre-Pn signal-to-noise ratio greater than 5:1 (within a given frequency band and normalized by window length). Three classes of short-period discriminants were formed from the observed amplitudes: P/S amplitude or phase ratios (e.g., Pn/Sn, Pn/Lg and Pg/Lg in the same frequency band); spectral ratios (low-frequency to high-frequency ratios of the same phase such as Pn [0.5–1 Hz]/Pn [4–6 Hz]); and cross-spectral ratios (low frequency S to high frequency P ratios such as Lg [1–2 Hz]/Pn [4–6 Hz]). We investigated the distance and magnitude dependence of each discriminant along with its spatial variability. Figures 3a and 3b shows the magnitude and distance dependence for one of each class of discriminant, respectively. A linear regression on distance was computed using the earthquake data (shown in Fig. 3b). This trend was removed from both the earthquake and explosion data and the resulting magnitude dependence is shown in Figure 3c. These plots show that two of the three discriminants have a strong distance dependence and that discriminants involving a spectral ratio have a magnitude dependence, especially the Lg [1–2 Hz]/Pn [4–6 Hz]

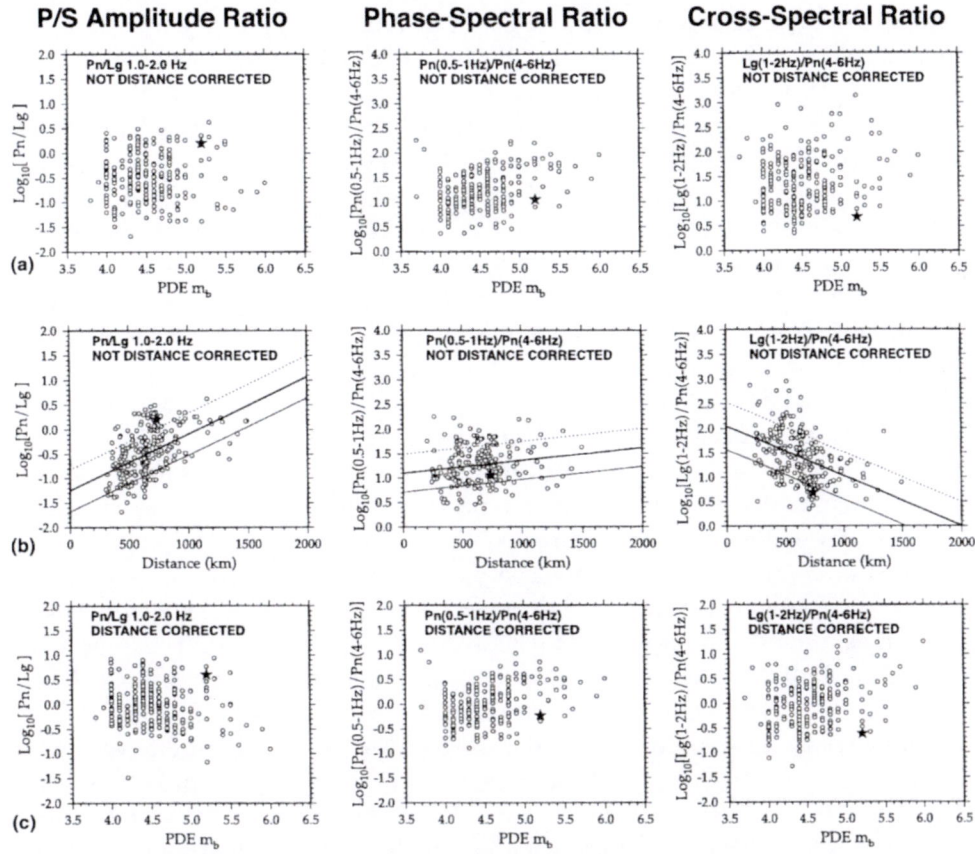

Figure 3

(a) Magnitude dependence, (b) distance dependence and (c) distance-corrected magnitude dependence for three classes of discriminants for the earthquakes (circles) and the Indian test (star) recorded at NIL. The discriminants shown are Pn/Lg [1–2 Hz] (left column); Pn [0.5–1 Hz]/Pn [4–6 Hz] (center column); and Lg [1–2 Hz]/Pn [4–6 Hz] (right column).

discriminant. This magnitude trend results from source size-corner frequency scaling. Correction of this source size-corner frequency scaling effect is discussed later.

P/S Amplitude Ratios

The Pn/Lg [1–2 Hz] amplitude ratios shown in Figure 3 show that the Indian test has on average a higher P/S ratio than the earthquakes recorded at station NIL. This is consistent with the absence of S-wave energy (Lg) from the explosion source, whereas the earthquakes have larger S-wave amplitudes (lower Pn/Lg ratios). The Pn/Lg [1–2 Hz] amplitude ratios plotted in Figures 3a and 3c show no magnitude dependence. We found this to be true for other P/S ratios and frequency bands. This indicates that any source size-corner frequency scaling affects both P and S waves identically, or that any source size effect is smaller than the scatter that arises from propagation path effects and/or source depth or radiation pattern effects. There is a strong distance dependence for the Pn/Lg [1–2 Hz] amplitude ratios seen in Figure 3b (left). Removing this distance trend reduces the scatter of the earthquake population but does not necessarily improve the separation of the Pn/Lg [1–2 Hz] discriminant for the Indian test and the earthquakes. The remaining scatter in the earthquake population is due to path effects and/or source depth and radiation effects. In order to visualize these propagation path effects, we map the observed amplitude ratios projected to their event locations in Figure 4. The ratios in Figure 4 show that events to the north tend to have higher ratios (weaker Lg energy relative to Pn) than those from the south, however it is important to note that at this stage these data were not corrected for their distance trend.

Large variations in topography (Fig. 1), crustal thickness and attenuation structure may weaken Lg amplitudes for the northern paths relative to those from the south (see for example FAN and LAY, 1998). Limiting backazimuths to isolate events having similar paths resulted in better representations of the distance trends (i.e., greater scatter reduction) for earthquake P/S ratios observed at station ABKT (RODGERS et al., 1999). That study compared various methods for reducing the scatter in earthquake P/S ratios that arises from path effects and used variance reduction as a metric of performance. Thus for the current data, it may be wise to compare the Indian nuclear test with earthquakes having similar paths to NIL. In Figure 4 we show azimuthal rays at 100° and 255° which isolate the southerly events with paths restricted more or less to the Indian Platform. In Figure 5 we show the Pn/Lg, Pg/Lg and Pn/Sn amplitude ratios of the Indian test and the southerly earthquakes for several frequency bands. When only events from the southerly backazimuths are considered, the Pn/Lg ratios above 2 Hz and Pn/Sn in all bands clearly separate the Indian test from the earthquakes. The Pg/Lg ratios do not discriminate as well as the ratios involving Pn. Note that the Pg/Lg ratios are less scattered than the Pn/Lg and Pn/Sn ratios. Pg amplitudes by themselves are typically less scattered than Pn amplitudes probably because Pg is composed of multiply

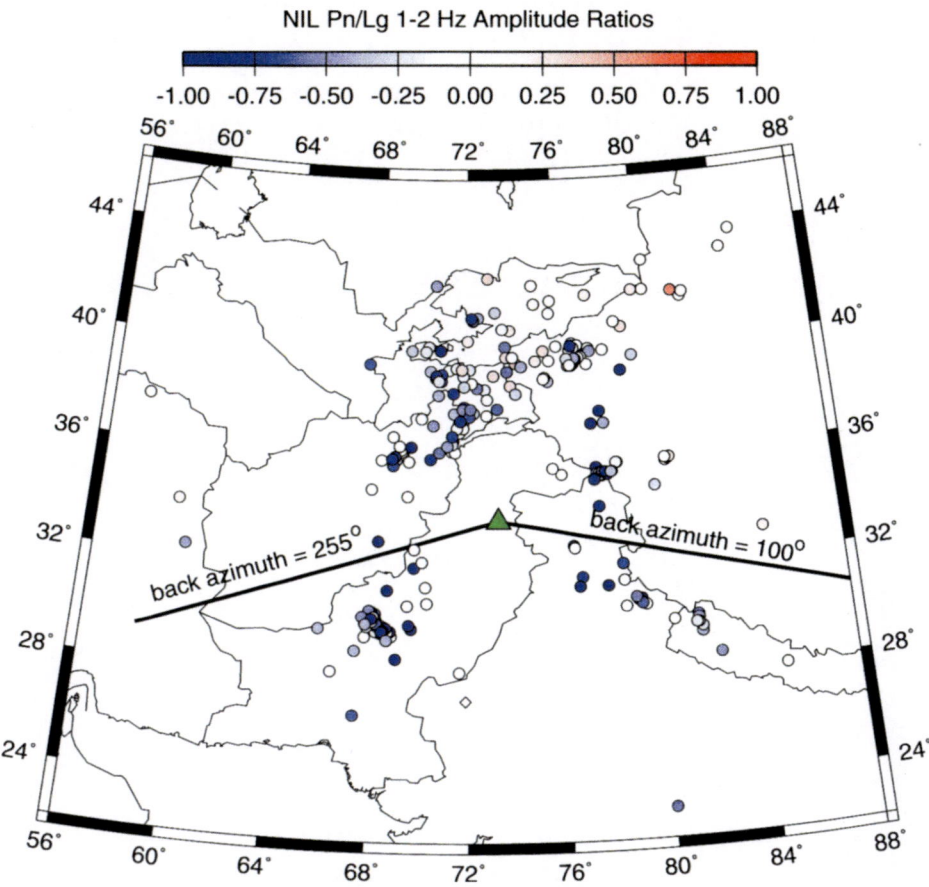

Figure 4

Map of Pn/Lg [1–2 Hz] amplitude ratios for the regional earthquakes (circles) and the May 11 Indian test (diamond). Ratios are plotted at the event location and color-coded according to the scale. Rays drawn from station NIL indicate the back-azimuthal limits used to isolate events from the south (100°–255°).

reflected waves, which sample a large part of the focal sphere. It is remarkable that the lower frequency Pn/Sn ratios (0.5–2 Hz) for southern azimuths discriminate almost as well as higher frequency ratios. It is usually the case that discrimination performance improves as frequency increases, especially for frequencies above 3.0 Hz (e.g., WALTER et al., 1995; HARTSE et al., 1997).

To further illustrate the importance of path effects on discrimination, we show the Mahalanobis distance as a function of frequency band for the P/S ratios for two cases: firstly where earthquakes from all azimuths are included and secondly where earthquakes from just the southerly azimuths are included (Fig. 6). The Mahalanobis distance, D^2, is a measure of the separation of two populations. It is computed using the following equation:

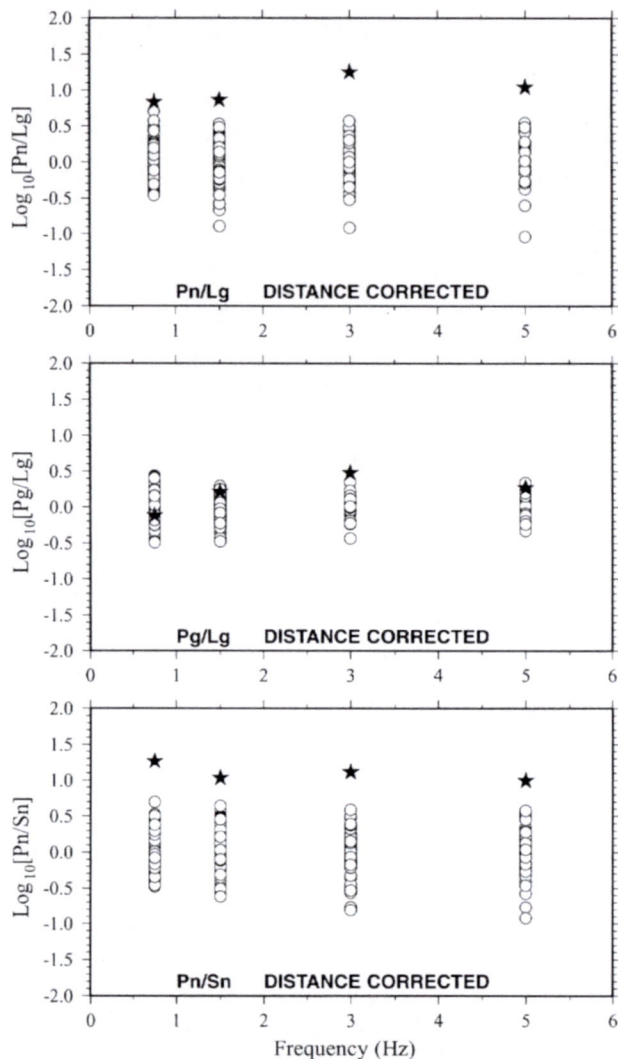

Figure 5
Distance corrected amplitude ratios plotted at the center frequency for Pn/Lg (top); Pg/Lg (middle) and Pn/Sn (bottom). Only events from southerly backazimuths (100°–255°) from NIL are shown. Circles indicate earthquakes and the stars indicate the Indian test values.

$$D^2 = (\mu_{ex} - \mu_{eq})^2 / (\sigma_{ex}^2 + \sigma_{eq}^2) \qquad (1)$$

where μ and σ^2 are the mean and variance, respectively, for the explosion (ex) and earthquake (eq) populations. We assume the (single) Indian test represents the mean of the (unknown) explosion population and that the variance of the explosion population is equal to that of the earthquakes for each ratio. Using the one explosion

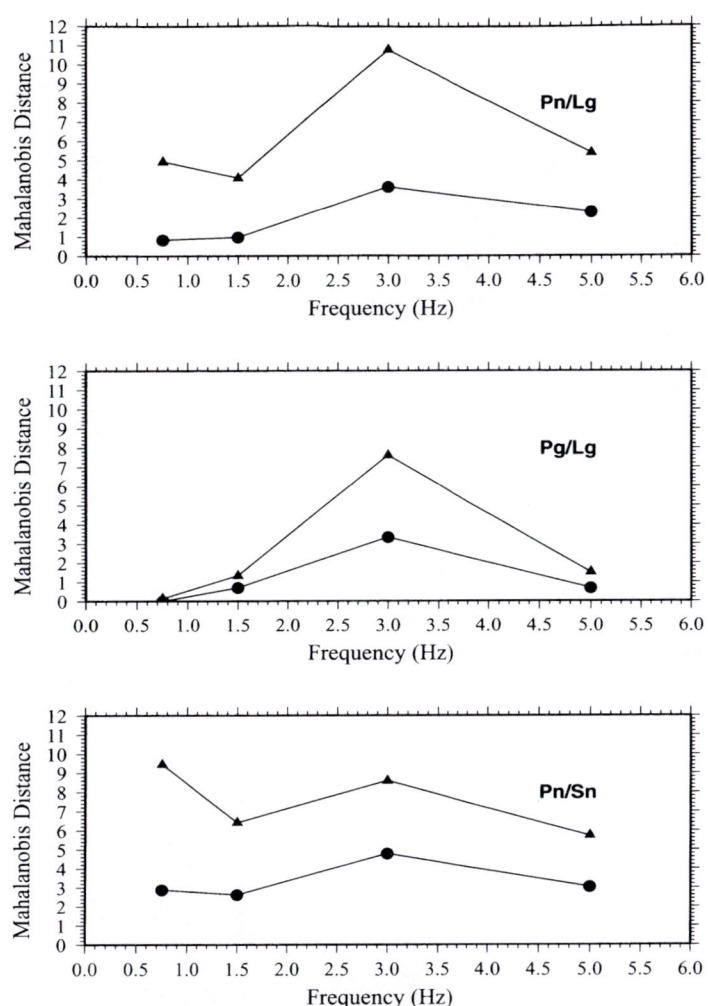

Figure 6

Mahalanobis distance (see text) as a function of frequency band for all azimuths (circles) and the southerly azimuths (triangles).

as the mean is the simplest assumption we can make, while using the earthquake variance is probably conservative. Despite the uncertainty of these assumptions, using equation (1) provides a simple objective measure to compare the separation of each of the various discriminants for the earthquakes and the Indian test. Figure 6 shows that the Mahalanobis distance is much greater (i.e., better separation) when the Indian test is compared with earthquakes from the south of NIL, emphasizing the importance of path effects. We found that when the distance trends were removed there was little change in the Mahalanobis distance regardless of how the data were grouped.

The path variability seen in the map of Pn/Lg ratios (Fig. 4) can be represented by a correction surface generated with Bayesian kriging (SCHULTZ *et al.*, 1998). This technique has been shown to be superior to other strategies for representing regional *P/S* ratios because it provides a smooth, continuous correction surface with uncertainty estimates and results in the largest scatter reduction of the data (RODGERS *et al.*, 1999; PHILLIPS, 1999). Figure 7 shows a map of the distance corrected Pn/Lg [1–2 Hz] ratios plotted on top of the correction surface obtained from the kriging algorithm. Notice that the surface predicts low Pn/Lg ratios (<0.0) near the Indian test site, corresponding to strong Lg energy relative to Pn. Applying the corrections predicted by the kriged surface results in improved discrimination. Figure 8 shows the Pn/Lg [1–2 Hz] for all azimuths as a function of magnitude without any correction,

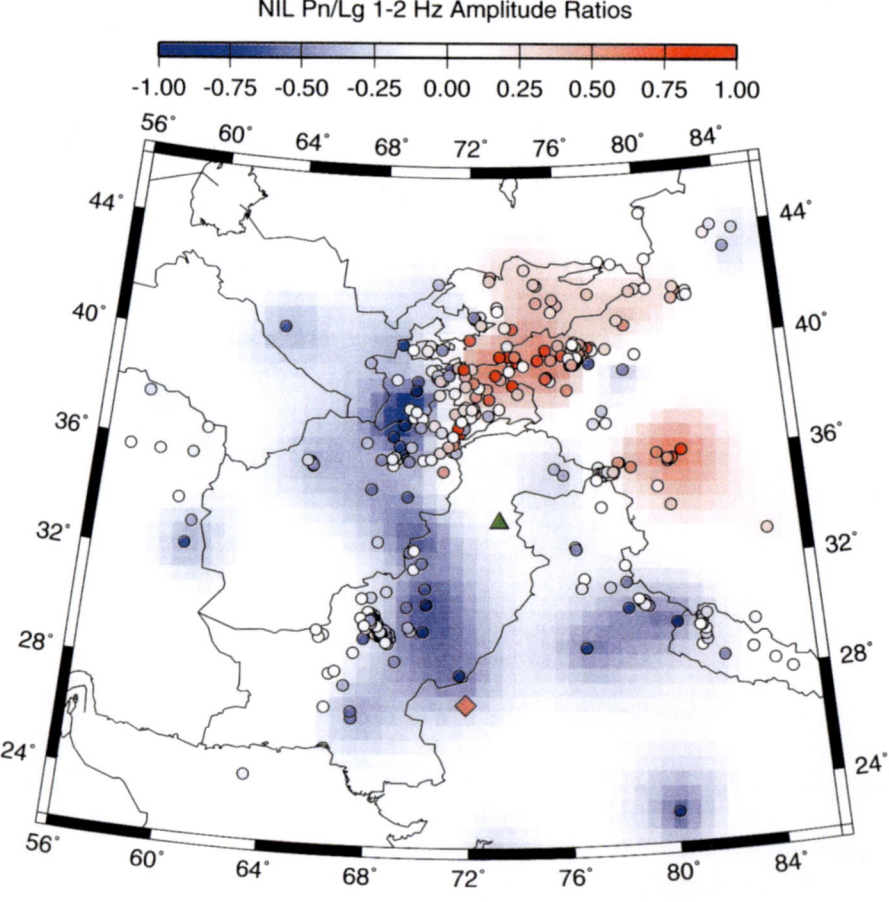

Figure 7

Distance corrected Pn/Lg [1–2 Hz] earthquake ratios (circles) plotted over the correction surface obtained from kriging. Also shown is the distance corrected Pn/Lg ratio of the Indian test (diamond).

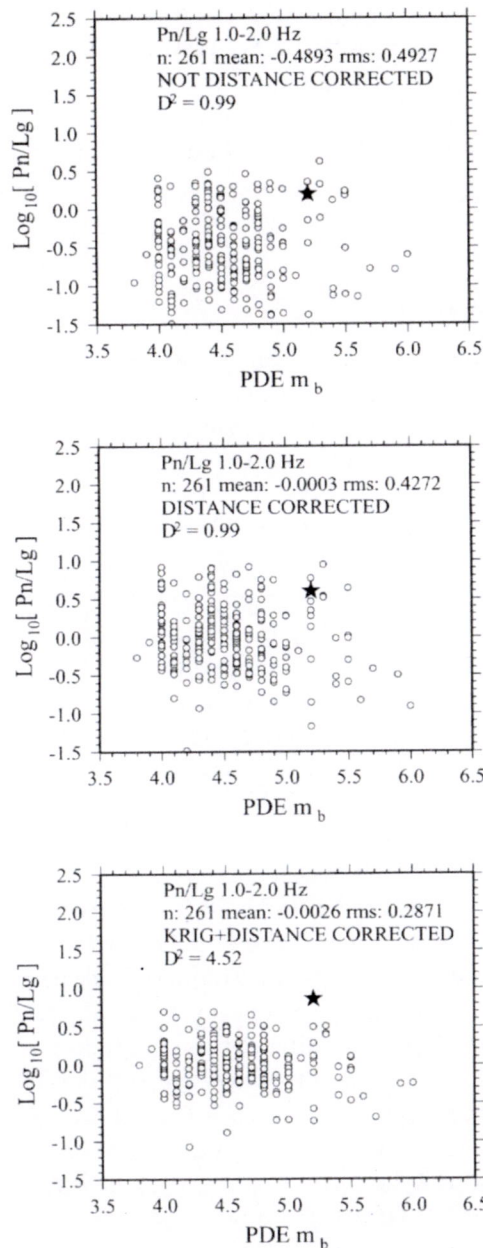

Figure 8

Pn/Lg [1–2 Hz] versus PDE m_b for all azimuths: (top) uncorrected ratios; (center) ratios corrected for the distance trend and (bottom) ratios corrected with the kriged correction surface. Circles indicate earthquakes and the star indicates the Indian test. Also shown are the number, mean and standard deviation (rms) of the earthquake population and the Mahalanobis distance, Δ^2, in each case.

with distance corrections and with the distance and path corrections from kriging. To quantitatively assess the separation, the Mahalanobis Distance is given in each case. Note that the separation of the Indian test from the mean of the earthquakes is greater in Figure 8c than in Figures 8a and 8b and that the scatter of the earthquake population is reduced significantly (rms reduction of 42%) after the path corrections were applied. Reducing the scatter and separating the means of the earthquake and explosion populations is necessary for good discrimination.

Spectral and Cross-spectral Ratios

The distance corrected spectral and cross-spectral ratios shown in Figure 3c discriminate the Indian test from events of about the same magnitude. However, this is not the case when the Indian test is compared to events of smaller magnitude. For models of earthquake or explosion generated seismic phase spectra, forming the ratio of amplitudes in two different frequency bands leads to a magnitude dependence (TAYLOR and DENNY, 1991; WALTER and PRIESTLEY, 1991; TAYLOR and HARTSE, 1998). This arises because the corner frequency of the source amplitude spectrum is inversely proportional to the seismic moment. The magnitude dependence of spectral ratios (e.g., Pn [0.5–1 Hz]/Pn [4–6 Hz]) and cross-spectral ratios (e.g., Lg [1–2 Hz]/Pn [4–6 Hz]) leads to two undesirable consequences. Firstly, there is excess scatter, which may inhibit discrimination performance when events of different sizes are compared. Secondly, and more importantly, the spectral ratio discriminants are not normally distributed and thus fail a basic assumption necessary for multivariate discrimination strategies. The magnitude dependence can be removed if a model of the source size-corner frequency scaling is obtained.

Recently, TAYLOR and HARTSE (1998) described a procedure for modeling absolute amplitude spectra of regional phases (the source path amplitude correction or SPAC method). They based their method on a model of regional phase amplitudes that depends on the seismic moment (low-frequency spectral level), a BRUNE (1970) earthquake source model, a source size-corner frequency scaling relation, geometric spreading and frequency-dependent attenuation. An iterative inversion algorithm fits all observed spectra of a given phase to several model parameters. Because of severe tradeoffs between some model parameters (especially geometric spreading and attenuation), some model parameters are fixed while the remaining parameters are estimated by inversion and checked to see if they are physically reasonable. This method has been generalized to the magnitude distance amplitude correction (MDAC, TAYLOR et al., this volume).

We developed a grid-search method to correct regional spectral and cross-spectral ratios for the combined dependence on distance (through geometric spreading and attenuation) and magnitude (through source size-corner frequency scaling) such as that seen in Figure 3. The theory is described in detail in the Appendix. The method is based on the same theory of regional phase amplitude spectra that is used by SPAC

(TAYLOR and HARTSE, 1998). However, we chose to model the distance and magnitude (log seismic moment) dependence of the discriminant ratios, rather than model the absolute amplitude behavior of the individual phases. In this way we explicitly link the source parameters for each phase in the ratios, rather than fit phases independently as in the SPAC method (TAYLOR and HARTSE, 1998). Our hope was that by dealing with only the relative source and path terms in ratios we could better bound the source-path tradeoffs.

Given the model of regional phase amplitude spectra, the 10-base log of a cross-spectral ratio of an S wave at frequency f_1, $S(f_1)$, and P wave at frequency f_2, $P(f_2)$, can be expressed as:

$$\text{Log10}[S(f_1)/P(f_2)] = C + f(f_1, f_2, F, a, \kappa; M_0) + G\log_{10}\Delta + D\Delta \; ; \qquad (2)$$

where C is a constant, F is the source moment-corner frequency scaling parameter, a is the P-wave to S-wave corner frequency scaling parameter, κ is the source moment-corner frequency scaling exponent, M_0 is the seismic moment, G is the $\log_{10}$ distance parameter (related to geometric spreading), Δ is the distance and D is the linear distance parameter (related to attenuation). The function $f(f_1, f_2, F, a, \kappa; M_0)$ describes the variation of the spectral ratio with source size (moment), and is plotted in the Appendix (Fig. A1b). The PDE body-wave magnitudes for the events considered were converted to seismic moments using the scaling relation from CONG et al. (1996): $\log_{10} M_0 = 10.66 + 1.04 m_b$. We found with several data sets, including both amplitude and spectral ratios, that the $\log_{10} \Delta$ term in equation (2) is not well resolved when both linear and log distance terms are sought. Therefore we only consider the linear distance term in our parameter search ($G = 0$). For spectral ratios, the $\log_{10} \Delta$ term and the P-wave to S-wave corner frequency scaling constant, a, drop out. To constrain the search, we chose to fix the source moment corner frequency scaling exponent, $\kappa = 1/3$, and the P-wave to S-wave corner frequency scaling constant, $a = 1.7$ (for cross-spectral ratios only). We searched over ranges of C, D and F and minimized the error. Error was measured as the data normalized sum of the squared residuals between the observed spectral ratios and the model predictions. Best-fitting parameters are those that provide the minimum error (misfit). Many experiments were run to ensure that the ranges and increments of parameter values for the grid-search were sufficient to resolve the best-fitting values. Grid-searches run with $\kappa = 1/4$ and/or $a = 1.0$ resulted in a similar fit to the spectral ratio data.

Figure 9 shows the contoured error surfaces for the Lg [1–2 Hz]/Pn [4–6 Hz] spectral ratios. The model uncertainties and tradeoff between parameters can be inferred from these plots. Note that the moment scaling constant, F, is poorly resolved and tradeoffs to some extent with the constant term, C, and the distance term, D. The distance term, D, obtained from the grid-search (-0.001 km^{-1}, is similar to the value obtained from linear regression on distance alone (-0.0008 km^{-1}), which suggests that the distance and magnitude dependence are more or less independent. The fact that the moment scaling constant, F, is poorly resolved is not surprising

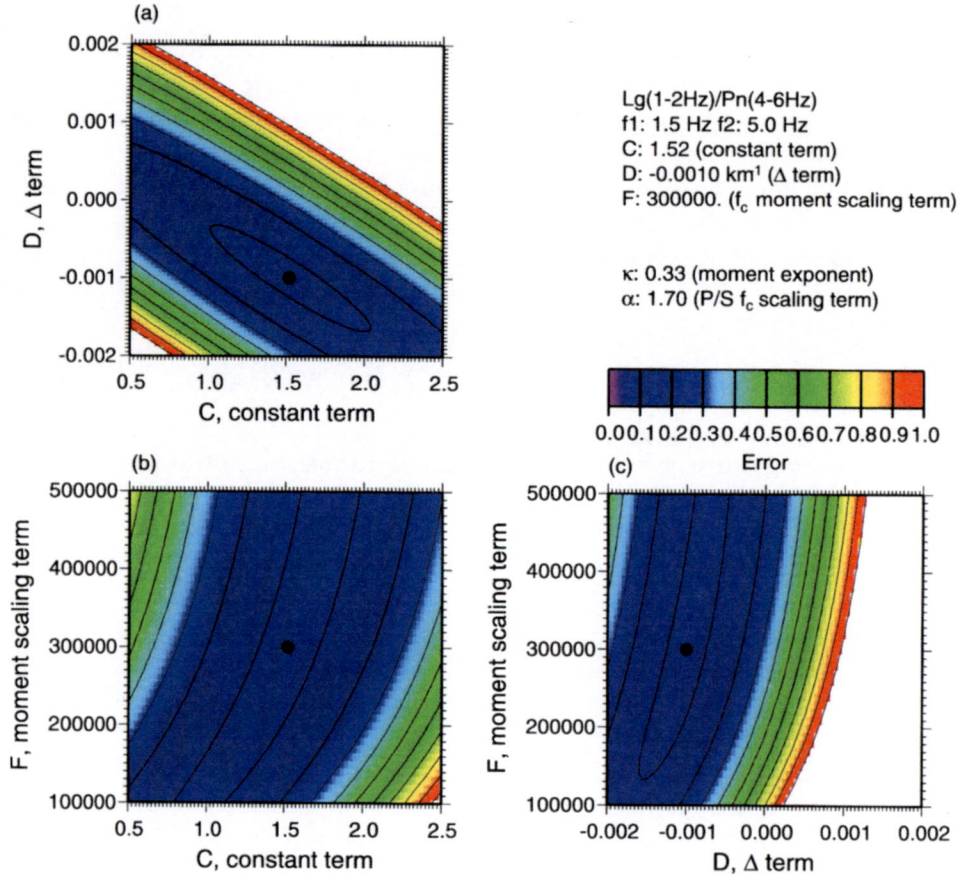

Figure 9
Slices through the model error (misfit) surfaces: (a) *D* vs. *C*; (b) *F* vs. *C*; and (c) *F* vs. *D*. Each surface is plotted with the third parameter held at its best-fitting value. Best-fitting parameters are indicated by black dots at the minimum misfit.

given that the function $f(f_1, f_2, F, a, \kappa; M_0)$ is only weakly dependent on this parameter. Viewing the model predictions plotted over the data demonstrates the performance of this technique. Figure 10 shows the moment corrected Lg [1–2 Hz]/ Pn [4–6 Hz] spectral ratios as a function of distance and the distance corrected spectral ratios as a function of moment, along with the best-fitting model predictions. The variance of the earthquake population is reduced by 33% by the joint distance and moment modeling, compared to a variance reduction of 18% for the (linear) distance dependence alone. The resulting distance and moment dependence after the model predictions were applied, shown in Figure 11, indicates that the distance and moment trends are removed by our method. The spectral ratio for the Indian test was corrected by the same model and is also shown in Figure 11. The discrimination of

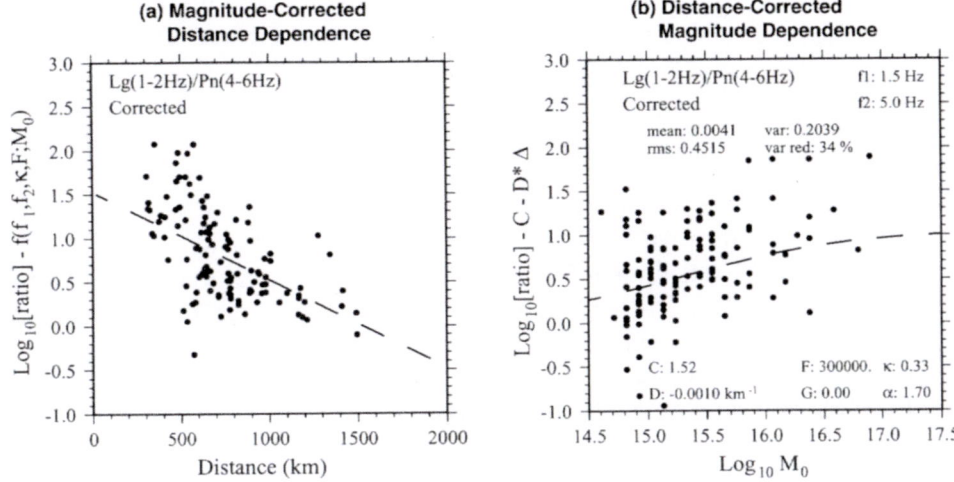

Figure 10

Lg [1–2 Hz]/Pn [4–6 Hz] spectral ratios (a) corrected for the moment scaling and plotted versus distance and (b) corrected for the distance dependence and plotted versus log moment. The model predictions are plotted as the dashed lines.

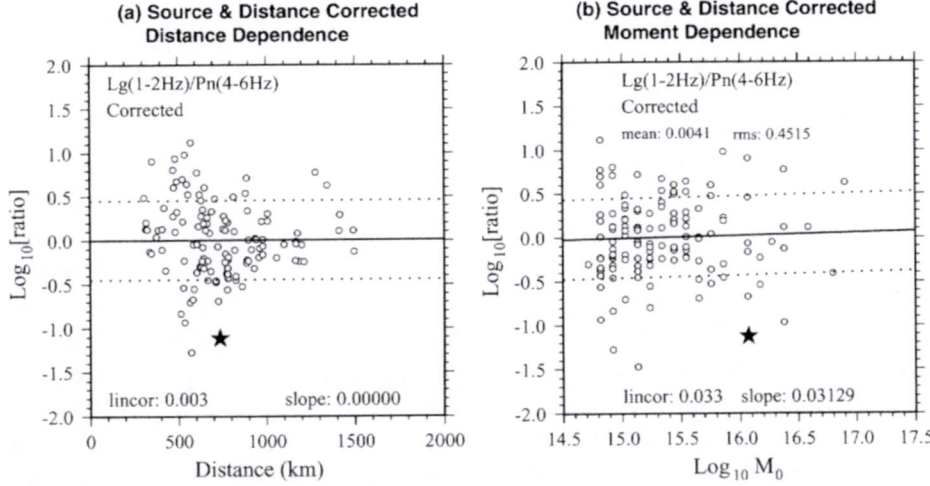

Figure 11

The Lg [1–2 Hz]/Pn [4–6 Hz] spectral ratios corrected for the best-fitting model and plotted versus (a) distance and (b) log moment. The earthquakes are indicated as circles and the star indicates the Indian test. Linear regressions and $1-\sigma$ uncertainties are also plotted, showing that the distance and log moment dependence is removed after correction. The linear correlation (lincorr) and slope of the regression line for each case are also shown.

the test is better for the corrected data than the uncorrected data ($D^2 = 0.96$ for the uncorrected case and $D^2 = 2.99$ for the source and distance corrected case).

Discussion and Conclusions

The May 11, 1998 Indian nuclear test provides valuable new data to calibrate regional discrimination in support of Comprehensive Nuclear-Test-Ban Treaty (CTBT) monitoring. The Indian test is well discriminated from regional earthquakes observed at station NIL (Nilore, Pakistan). The test shows high values for P/S ratios and small values for low-frequency/high-frequency spectral ratios, consistent with an explosion source. Particularly remarkable is that even relatively low frequency (<4 Hz) P/S ratios, especially Pn/Sn and Pn/Lg, show good separation between the test and regional earthquakes. However, P/S ratios in the highest band considered in this study (4–6 Hz) do not as well as the same measures in the 2–4 Hz band. Discrimination studies at the U.S. Nevada test site (e.g., WALTER et al., 1995), the Former Soviet Union Semipalitinsk test site in Kazakhstan and the Lop Nor test site in China (e.g., HARTSE et al., 1997) generally showed that separation between explosion and earthquake P/S ratios increases with frequency. Similar results have been found in studies discriminating mining explosions from earthquakes in Scandinavia (e.g., BAUMGARDT and YOUNG, 1990), the eastern U.S. (e.g., KIM et al., 1993) and the Caucasus (e.g., KIM et al., 1997). The cause of the different discriminant behavior of the Indian Test at NIL warrants more research.

One remarkable observation from this study is the excellent performance of the Pn/Sn discriminant at all frequencies considered. This could be due to an absence of Sn energy for the Indian Test or efficient generation and propagation of Sn energy for regional earthquakes. The propagation of S-wave phases is efficient and attenuation is probably low for the paths through the Indian Platform to the south of NIL (NI and BARAZANGI, 1983). We believe the S waves from the earthquakes south of NIL are clearly observed. The source of explosion S waves from explosions has been the subject of numerous investigations over the years and remains an area of active research. One possible source of shallow explosion S waves in the 0.5–4 Hz frequency band is from Rg-to-S scattering (e.g., GUPTA et al., 1992; PATTON and TAYLOR, 1995; MAYEDA and WALTER, 1996). MYERS et al. (1999) recently examined the 1997 Kazakhstan depth-of-burial experiment and found strong evidence to support this idea. In particular they observe strong Rg decay in the 0.7 to 5 Hz frequency band within 20 km of the explosions. They note the regional S waves and S-wave coda show increases in amplitude in this same band, as the shot depth shallows, consistent with an Rg scattering as the source for the observed explosion S waves. One possibility is that in the relatively flat lying terrain and geology of the Indian Test site such Rg-to-S scattering is much less efficient that at other test sites such as NTS, Lop Nor and Semipalitinsk.

This paper demonstrates that proper calibration of distance and path effects is very important to effective regional discrimination. The trends shown in Figure 3b indicate that distance effects can be significant. Removing the distance trends by regression reduces the scatter in the earthquake population, however this often does not improve the separation of earthquakes and explosions. Path effects on discriminants arise from lateral variations in elastic and anelastic structure associated with tectonic features. Kriging has been shown to be the best method to account for path effects on discriminants (RODGERS et al., 1999; PHILLIPS, 1999). We show that applying the path corrections predicted from kriging improves the separation of the Indian test and regional earthquakes (Fig. 8).

Spectral ratios have a magnitude dependence that results from source size-corner frequency scaling. We show that magnitude and distance trends of spectral ratios can be simultaneously removed. Our method is based on a simple theoretical model of regional phase amplitudes, a Brune earthquake source model and source size-corner frequency scaling relationship (Appendix). The method uses a grid-search to estimate three parameters that control the magnitude and distance trends. Removing the magnitude and distance trends of spectral ratios improves the separation of regional earthquakes and the Indian Test. More importantly, the resulting corrected spectral ratios are more normally distributed, which is essential for multivariate discrimination algorithms such as linear discriminant analysis (LDA). The corrected spectral ratios have no appreciable magnitude or distance dependence (Fig. 11). A drawback of our method is that the earthquake data often span a limited magnitude range (m_b 3.7–6.0, e.g., Fig. 3c). Small events ($m_b < 4.0$) will be of particular interest to CTBT monitoring. The ability of our correction procedure to account for the magnitude dependence of these small events will depend critically on having a training set of small magnitude earthquakes. In regions with very few earthquakes and/or limited magnitude ranges of data the correction procedure will have to rely heavily on expert knowledge to properly extrapolate from the fits of limited data to the magnitudes, distances and locations of interest. In this regard having a good understanding of both the source scaling and the underlying attenuation in the region will be important. Other types of structural or hazard investigations (attenuation tomography, refraction line interpretations, stress drop studies, etc.) provide some information on these topics in regions where seismicity is limited. Because the method outlined here uses simple theoretic formulations for source scaling, geometrical spreading and attenuation, it is straightforward to incorporate and evaluate such results from other studies.

In this paper, we showed that the May 11, 1998 Indian nuclear test could be clearly identified as an explosion using regional discrimination techniques developed in other areas of the world (e.g., western U.S., Kazakhstan, China). This result shows the strength of regional discrimination techniques and the value of regional distance data for CTBT monitoring down to fairly small magnitudes. These results also generate confidence that regional discriminants developed at former test sites can be transported to other areas of the world if the proper path and source calibrations are done. This is important since in many regions of the world regional CTBT monitoring will need to be done without any empirical nuclear explosion calibration data.

Appendix

Magnitude and Distance Corrections for Spectral Ratios

Following TAYLOR and HARTSE (1998), the amplitude spectrum of a short-period regional seismic phase (e.g., Pn, Pg, Sn, Lg), $A(f)$, can be written as the product of three terms:

$$A(f) = S_0(f)G(\Delta)Q(f) \; ; \tag{A1}$$

where f is frequency, Δ is distance, $S_0(f)$ is the source spectrum, $G(\Delta)$ is the (frequency independent) geometric spreading and $Q(f)$ is the attenuation operator. Note that a frequency-dependent site effect can also be included in equation (A1). We assume that the BRUNE (1970) model represents the source spectrum:

$$S_0(f) = cM_0[1 - (f/f_c)^2]^{-1} \; ; \tag{A2}$$

where M_0 is the seismic moment, f_c is the corner frequency and c is a constant that depends on source and receiver material properties and the radiation pattern. The source corner frequency scales with the moment as:

$$f_c = FM_0^{-\kappa} \; ; \tag{A3}$$

where F is the source moment corner frequency scaling parameter and κ is the source moment corner frequency scaling exponent. Examples of Brune model source spectra with the above corner frequency scaling are plotted in Figure A1a. The corner frequency of P and S waves can be different and related by:

$$f_c \; (P \text{ wave}) = af_c \; (S \text{ wave}) \; ; \tag{A4}$$

where $1.0 \leq a \leq v_P/v_S \approx 1.7$. The geometric spreading is represented by the distance to a power:

$$G(\Delta) = \Delta^{-\gamma} \; ; \tag{A5}$$

where γ is the geometric spreading exponent. The attenuation is represented as:

$$Q(f) = \exp[-\omega t/2 \; q(f)] = \exp[-\pi f^{1-\eta}\Delta/UQ_0] \; ; \tag{A6}$$

where $q(f) = Q_0 f^{\eta}$ is the frequency-dependent attenuation and U is the velocity of the phase in question.

Forming the S-wave to P-wave cross-spectral amplitude ratio, $S(f_1)/P(f_2)$, and ignoring the site effect and taking the base-10 log yields:

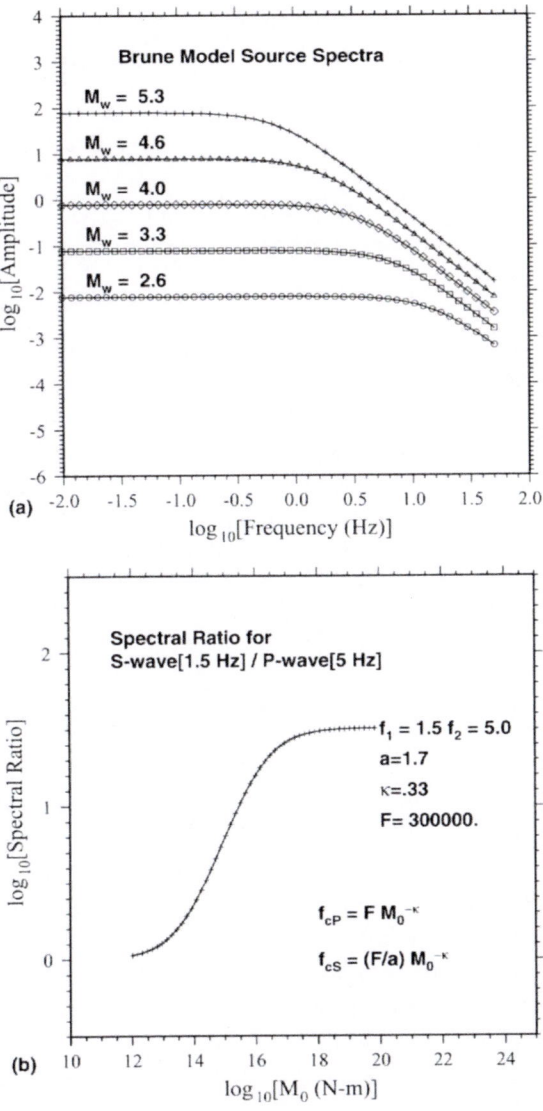

Figure A1
(a) Examples of Brune model source spectra using the source moment-corner frequency scaling parameters: $\kappa = 1/3$, $a = 1.7$ and $F = 300,000$. Spectra are shown for various moments with moment magnitudes, M_W, identifying each curve. (b) Moment dependence of spectral ratios specified by the function $f(f_1, f_2, F, a, \kappa; M_0)$ using the best-fitting parameters from our grid-search modeling for the Lg [1–2 Hz]/Pn [4–6 Hz] ratios.

$$\log_{10}[S(f_1)/P(f_2)] = s - \log_{10}[1 - (f_1 M_0^{-k}/aF)^2] - \gamma^S \log_{10} \Delta - \pi \log_{10} e f^{1-\eta S} \Delta/Q_{0S} U^S$$
$$+ \log_{10}[1 - (f_2 M_0^{-k}/F)^2] + \gamma^P \log_{10} \Delta + \pi \log_{10} e f^{1-\eta P} \Delta/Q_{0P} U^P \ .$$

$$(A7)$$

Collecting terms results in:

$$\log_{10}[S(f_1)/P(f_2)] = \log_{10}[1 - (f_2 M_0^{-k}/F)^2] - \log_{10}[1 - (f_1 M_0^{-k}/aF)^2]$$
$$+ [\gamma^P - \gamma^S] \log_{10} \Delta + \pi \log_{10} e[f_2^{1-\eta P}/Q_{0P} U^P - f_1^{1-\eta S}/Q_{0S} U^S]\Delta \ ;$$

$$(A8)$$

where superscripts P and S identify P- and S-wave values. Note that the $\log_{10} M_0$ term common to each amplitude cancels. We have assumed that radiation pattern effects are common to both phases. Strictly speaking this may not be true, however the effects of focal mechanism on short-period regional phase amplitudes is largely unknown. The site effects for each phase at the given frequencies, f_1 and f_2, lead to a constant term (independent of source moment and distance). Including the site effect constant, the above form (A8) can be represented as:

$$\log_{10}[S(f_1)/P(f_2)] = C + f(f_1, f_2, F, a, \kappa; M_0) + G \log_{10} \Delta + D\Delta \ ; \qquad (A9)$$

where

$$f(f_1, f_2, F, a, \kappa; M_0) = \log_{10}[1 - (f_2 M_0^{-k}/F)^2] - \log_{10}[1 - (f_1 M_0^{-k}/aF)^2] \ . \quad (A10)$$

The function $f(f_1, f_2, F, a, \kappa; M_0)$ is plotted in Figure A1b for the Lg [1–2 Hz]/Pn [4–6 Hz] spectral ratios modeled in the text. The term D can be related to the attenuation parameters Q_0, η and U for P and S waves. The term G is identically equal to $[\gamma^P - \gamma^S]$. We have not found this term to be resolvable from the linear distance term, D, probably because scatter in the data from path propagation effects has greater influence than the effects of the log-distance term.

There are two distinct advantages of this method. Firstly, we only need to specify two parameters (κ and a) and search for three parameters (C, D and F). And secondly, the dependence of the spectral ratios on distance and magnitude (log moment) can be completely removed.

Acknowledgments

Waveform data were obtained from the Incorporated Research Institutions for Seismology – Data Management Center (IRIS–DMC). Waveform data were collected and organized by Stan Ruppert, Terri Hauk and Jennifer O'Boyle. Comments by Hans Hartse and Mark Tinker improved the original manuscript. Research was performed under the auspices of the U.S. Department of Energy by the

Lawrence Livermore National Laboratory under contract W-7405-ENG-48. This is LLNL journal contribution UCRL-JC-134453.

REFERENCES

BARKER, B., CLARK, M., DAVIS, P., FISK, M., HEDLIN, M., ISRAELSSON, H., KHALTURIN, V., KIM, W.-K., MCLAUGHLIN, K., MEADE, C., MURPHY, J., NORTH, R., ORCUTT, J., POWELL, C., RICHARDS, P., STEAD, R., STEVENS, J., VERNON, F., and WALLACE, T. (1998), *Monitoring Nuclear Tests*, Science *281*.

BAUMGARDT, D. R. and YOUNG, G. B. (1990), *Regional Seismic Waveform Discriminants and Case-based Event Identification Using Regional Arrays*, Bull. Seismol. Soc. Am. *80*, 1874–1892.

BRUNE, J. (1970), *Tectonic Stress and the Spectra from Seismic Shear Waves Earthquakes*, J. Geophys. Res. *75*, 4997–5009.

CONG, L., XIE, J. and MITCHELL, B. (1996), *Excitation and Propagation of Lg from Earthquakes in Central Asia with Implications for Explosion/Earthquake Discrimination*, J. Geophys. Res. *101*, 27,779–27,789.

FAN, G. and LAY, T. (1998c), *Regionalized Versus single-station Waveguide Effects on Seismic Discriminants in Western China*, Bull. Seismol. Soc. Am. *88*, 1260–1274.

GUPTA, I. N., CHAN, W. W., and WAGNER, R. A. (1992), *A Comparison of Regional Phases from Underground Nuclear Explosions at East Kazakh and Nevada Test Sites*, Bull Seismol. Soc. Am. *82*, 352–382.

HARTSE, H., TAYLOR, S., PHILLIPS, S., and RANDALL, G. (1997), *A Preliminary Study of Regional Seismic Discrimination in Central Asia with Emphasis on Western China*, Bull. Seismol. Soc. Am. *87*, 551–568.

KIM, W.-Y., SIMPSON, D., and RICHARDS, P. (1993), *Discrimination of Earthquakes and Explosions in the Eastern United States Using Regional High-frequency Data*, Geophys. Res. Lett. *20*, 1507–1510.

KIM, W.-Y., AHARONIAN, V., LERNER-LAM, A. L., and RICHARDS, P. G. (1997), *Discrimination of Earthquakes and Explosions in Southern Russia Using Regional High-frequency Three-component Data from the IRIS/JSP Caucasus Network*, Bull. Seismol. Soc. Am. *87*, 569–588.

MAYEDA, K. and WALTER, W. (1996), *Moment, Energy, Stress Drop and Source Spectra of Western United States Earthquakes from Regional Coda Envelopes*, J. Geophys. Res. *101*, 11,195–11,208.

MYERS, S. C., WALTER, W. R., MAYEDA, K., and GLENN, L. (1999), *Observations in Support of Rg Scattering as a Source of Explosion S Waves: Regional and Local Recordings of the 1997 Kazakhstan Depth of Burial Experiment*, Bull. Seismol. Soc. Am. *89*, 544–549.

NI, J., and BARAZANGI, M. (1983), *High-frequency Seismic Wave Propagation Beneath the Indian Shield, Himalayan Arc, Tibetan Plateau and Surrounding Regions: High Uppermost Mantle Velocities And Efficient Propagation Beneath Tibet*, Geophys. J. R. astr. Soc. *72*, 665–689.

PATTON, H. J. and Taylor, S. R. (1995), *Analysis of Lg Spectral Ratios from NTS Explosions: Implications for the Source Mechanisms of Spall and the Generation of Lg Waves*, Bull. Seismol. Soc. Am. *85*, 220–236.

PHILLIPS, W. S. (1999), *Empirical Path Correction for Regional-phase Amplitudes*, Bull. Seismol. Soc. Am. *89*, 384–393.

RODGERS, A., WALTER, W., SCHULTZ, C., MYERS, S., and LAY, T. (1999), *A Comparison of Methodologies for Representing Path Effects on Regional P/S Discriminants*, Bull. Seismol. Soc. Am. *89*, 394–408.

RUPPERT, S., HAUK, T., LEACH, R., and O'BOYLE, J. (1998), *LLNL Middle East and North Africa Research Database*, Proc. 20th Annual Seismic Res. Symp. on Monitoring a Comprehensive Test-Ban-Treaty, Santa Fe, NM, Department of Defense, Nuclear Treaty Programs Report, 727–735.

TAYLOR, S., SHERMAN, N., and DENNY, M. (1988), *Spectral Discrimination Between NTS Explosions and Western United States Earthquakes at Regional Distances*, Bull. Seismol. Soc. Am. *78*, 1563–1579.

TAYLOR, S., DENNY, M., VERGINO, E., and GLASER, R. (1989), *Regional Discrimination between NTS Explosions and Western U.S. Earthquakes*, Bull. Seismol. Soc. Am. *79*, 1142–1176.

TAYLOR, S., and DENNY, M. (1991), *An Analysis of Spectral Differences between NTS and Shagan River Nuclear Explosions*, J. Geophys. Res. *96*, 6237–6245.

TAYLOR, S. and HARTSE, H. (1998), *A Procedure for Estimation of Source and Propagation Corrections for Regional Seismic Discriminants*, J. Geophys. Res. *103*, 2781–2789.

TAYLOR, S., VELASCO, A., HARTSE, H., PHILLIPS, W. S., WALTER, W., and RODGERS, A. *Amplitude Corrections for Regional Seismic Discriminants*, Pure appl. geophys. *159*, 623–678.

SCHULTZ, C., MYERS, S., HIPP, J., and YOUNG, C. (1998), *Nonstationary Bayesian Kriging: Application of Spatial Corrections to Improve Seismic Detection, Location and Identification*, Bull. Seismol. Soc. Am. *88*, 1275–1288.

WALLACE, T. (1998), *The May 1998 Indian and Pakistan Nuclear Tests*, Seismol. Res. Lett. *69*, 386–393.

WALTER, W. and PRIESTLEY, K., *High-frequency P-wave spectra from explosions and earthquakes*. In *Explosion Source Phenomenology* (eds. S. Taylor, H. Patton, and P. Richards.) (1991).

WALTER, W., MAYEDA, K., and PATTON, H. (1995), *Phase and Spectral Ratio Discrimination Between NTS Earthquakes and Explosions. Part I: Empirical Observations*, Bull. Seismol. Soc. Am. *85*, 1050–1067.

WALTER, W., RODGERS, A., MAYEDA, K., MYERS, S., PASYANOS, M., and DENNY, M. (1998), *Preliminary regional seismic analysis of nuclear explosions and earthquakes in southwest Asia*, Proc. 20th Annual Seismic Res. Symp. on Monitoring a Comprehensive-Test-Ban-Treaty, Santa Fe, NM, 23–25 September, 1998, Department of Defense, Nuclear Treaty Programs Report, 387–396.

(Received August 6, 1999, revised July 12, 2000, accepted July 20, 2000)

 To access this journal online:
http://www.birkhauser.ch

Pure appl. geophys. 159 (2002) 701–719
0033–4553/02/040701–19 $ 1.50 + 0.20/0

❚ Pure and Applied Geophysics

Observed Characteristics of Regional Seismic Phases and Implications for P/S Discrimination in the European Arctic

FRODE RINGDAL,[1] ELENA KREMENETSKAYA[2]
and VLADIMIR ASMING[2]

Abstract — In this paper, we use data from seismic stations operated by NORSAR, the Kola Regional Seismological Centre (KRSC) and IRIS to study the characteristics of regional phases in the European Arctic, with emphasis on the P/S ratio discriminant. While the detection and location capability of the regional station network is outstanding, source classification of small seismic events has proved very difficult. For example, the $m_b = 3.5$ seismic event near Novaya Zemlya on 16 August, 1997 has been the subject of extensive analysis in order to locate it reliably and to classify the source type. We consider the application of the P/S discriminant in the context of this event and other events observed at regional distances in the European Arctic. We show that the P/S ratios of Novaya Zemlya nuclear explosions measured in the 1–3 Hz filter band scale with magnitude, indicating a need for caution and further research when applying P/S discriminants. Using mainly data from the large NORSAR array, we note that observed P/S amplitude ratios in the European Arctic show large variability for the same source type and similar propagation paths, even when considering closely spaced observation points. This effect is most pronounced at far regional distances and relatively low frequencies (typically 1–3 Hz), but it is also significant on closer recordings (around 10 degrees) and at higher frequencies (up to about 8 Hz). Our conclusion from this study is that the P/S ratio at high frequencies (e.g., 6–8 Hz) shows promise as a discriminant between low-magnitude earthquakes and explosions in the European Arctic, but its application will require further research, including extensive regional calibration and detailed station-source corrections. Such research should also focus on combining the P/S ratio with other short-period discriminants, such as complexity and spectral ratios.

Key words: Seismic sources, discrimination, wave propagation regional phases.

Introduction

NORSAR and Kola Regional Seismological Centre (KRSC) of the Russian Academy of Sciences have for many years cooperated in the continuous monitoring of seismic events in northwest Russia and adjacent sea areas. The overall objective is to characterize the seismicity of this region, to investigate the detection and location

[1] NORSAR, Kjeller, Norway. E-mail: frode@norsar.no
[2] Kola Regional Seismological Centre (KRSC), Apatity, Russia.

capability of regional seismic networks and to study various methods for screening and identifying seismic events in order to improve monitoring of the Comprehensive Test-Ban Treaty (CTBT). The research has been based on data from a network of sensitive regional arrays which has been installed in northern Europe during the last decade in preparation for the CTBT monitoring network. This regional network, which comprises stations in Fennoscandia, Spitsbergen and NW Russia (see Fig. 1), provides a detection capability for the Barents/Kara Sea region that is close to $m_b = 2.5$ (RINGDAL, 1997).

The Kara Sea seismic event on 16 August, 1997 that was reported by the prototype International Data Center has caused considerable and renewed interest in the seismicity of the region surrounding the Novaya Zemlya Islands. Historically, registered earthquake activity in this region has been virtually nonexistent, with the exception of one event in the Kara Sea close to the Novaya Zemlya coast on 1 August 1986. This event was classified as an earthquake, based upon the focal solution and depth estimate (19 km) of MARSHALL *et al.* (1989). The event on 16 August 1997 ($m_b = 3.5$) was studied by several investigators (RICHARDS and KIM,

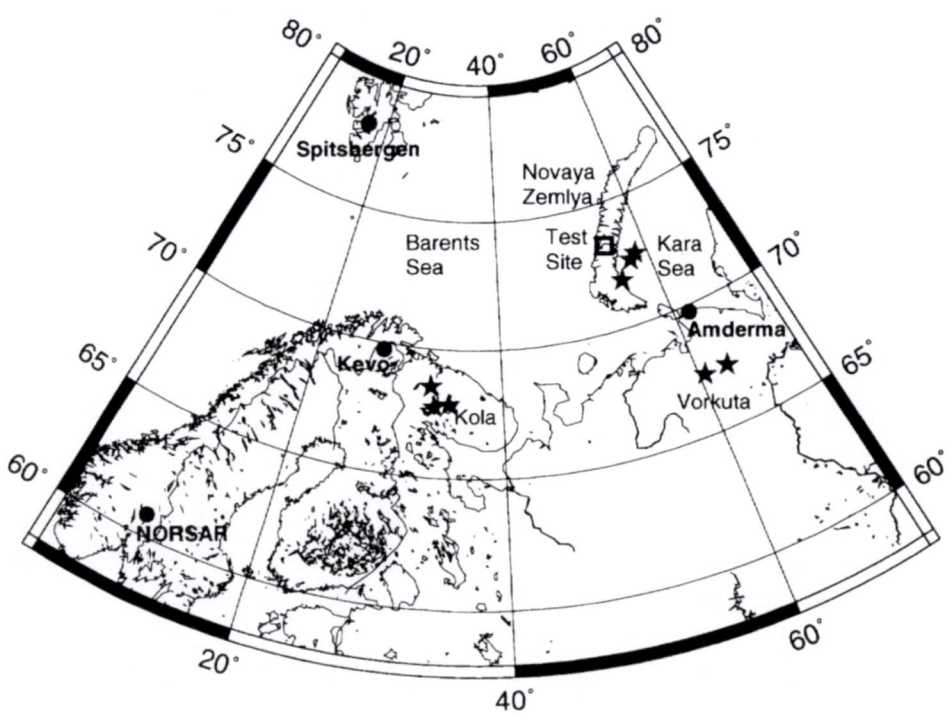

Figure 1

Regional network of seismic stations and seismic events analyzed in this study. The location of the Novaya Zemlya nuclear test site is indicated.

1997; HARTSE, 1998), with particular attention to the P/S ratio observed at high frequencies.

This paper addresses the possibilities and limitations of utilizing the P/S ratio to characterize seismic events at low magnitudes in this region. We note that the P/S and other similar discriminants (e.g., Pn/Lg) have been extensively studied in many areas of the world (e.g., WALTER et al., 1995; TAYLOR, 1996; HARTSE et al., 1997), but at present there is no consensus on the applicability of such discriminants on a global basis.

Data

The seismicity of the Barents/Kara sea region has been discussed by RINGDAL (1997). Nuclear and chemical explosions were conducted at Novaya Zemlya until 1990, however the availability of regional data for these events is quite limited because most of the high-quality regional arrays in Fennoscandia and adjacent areas were established after this time. In addition, the Novaya Zemlya explosions were generally large, except for two smaller nuclear explosion in 1977 and 1984, and two low-magnitude chemical explosions in 1978 and 1987. A small presumed earthquake (MARSHALL et al., 1989) occurred on Novaya Zemlya near the nuclear test site in 1986. To our knowledge, there are no available digital recordings at near-regional distances (less than 12 degrees) for any of the above-mentioned smaller events. Although there have been several low-magnitude seismic events detected near Novaya Zemlya in recent years, they are difficult to use for establishing or testing regional discriminants, since there is no confirmed evidence available as to their source type.

In other parts of the European Arctic, there is a good selection of reference earthquakes and mining explosions. For example, there are well-known mining areas in the Kola Peninsula and Vorkuta south of Novaya Zemlya. The seismic event occurrence is also very high in the Spitsbergen area and offshore Norway (to the north and west). These events are presumably mostly earthquakes.

We have made a selection of known nuclear explosions, known earthquakes and some unknown events as a basis for this study. The events are listed in Table 1 and shown in Figure 1 together with the station network. Some of the smaller events have been located by KREMENETSKAYA et al. (1999).

Regional Phases

Wave propagation characteristics of short-period seismic phases in the European Arctic have been studied by many authors (e.g., SERENO et al., 1988; BAUMGARDT, 1990). This region is characterized by very efficient P- and S-phase propagation at

Table 1

List of seismic events used in this study

Date/time	Location	m_b	Comment
04.09.72/07.00.00	67.75 N, 33.10 E	4.3	Nuclear explosion, Kola Peninsula
20.10.76/08.00.00	73.40 N, 54.85 E	4.9	Nuclear explosion, Novaya Zemlya
09.10.77/11.00.00	73.414 N, 54.935 E	4.5	Nuclear explosion, Novaya Zemlya
11.10.82/07.15.00	73.33 N, 54.60 E	5.5	Nuclear explosion, Novaya Zemlya
10.08.78/08.00.00	73.293 N, 54.885 E	6.0	Nuclear explosion, Novaya Zemlya
27.08.84/06.00.00	67.75 N, 33.00 E	4.3	Nuclear explosion, Kola Peninsula
01.08.86/ 13.56.38	72.945 N, 56.549 E	4.3	Located by MARSHALL *et al.* (1989) (presumed to be an earthquake)
16.04.89/ 06.34.42	67.61 N, 33.81 E	3.5	Earthquake in Khibiny mine (KREMENETSKAYA and ASMING, 1995)
16.06.90/ 12.43.28	68.52 N, 33.09 E	4.0	Earthquake, felt in Murmansk, Kola Peninsula
24.10.90/14.58.00	73.360 N 54.670E	5.6	Nuclear explosion, Novaya Zemlya
23.02.95/ 21.50.00	71.856 N, 55.685 E	2.5	Located by KREMENETSKAYA *et al.* (1999)
26.06.96/21.32.15	67.73 N, 32.92 E	3.0	Earthquake Imandra, near Khibiny mines
31.01.97/ 04.23.53	67.3 N, 60.6 E	2.5	Mining explosion – Vorkuta region
16.08.97/02.11.00	72.510 N, 57.550 E	3.5	Located by RINGDAL *et al.* (1997)
16.08.97/06.19.10	72.5 N, 58 E	2.6	Probably colocated with preceding event
14.02.98/ 00.49.37	67.34 N, 62.9 E	2.4	Mining explosion – Vorkuta region
17.08.99/04.44.36	67.865 N, 34.454 E	4.3	Earthquake/mine collapse, felt in Revda, Kola Peninsula

high frequencies, whereas the Lg phase, which is often the largest observed phase at low frequencies, is usually strongly attenuated at high frequencies (Fig. 2a). Furthermore, the Lg phase is subject to blockage on certain paths (BAUMGARDT, 1990). For example, Lg is not observed in Fennoscandia for nuclear explosions at Novaya Zemlya, for which the travel path crosses the thick sedimentary layers of the central Barents Sea. This has been shown for the ARCESS array (distance 10 degrees) by BAUMGARDT (1990), and is likewise apparent on the more distant NORSAR array recordings as illustrated in Figure 2b. For these reasons, we focus in this paper on the *P* and *S* phases, in particular by considering the *P/S* ratio at high frequencies.

The NORSAR large array (BUNGUM *et al.*, 1971) is a particularly valuable data source for regional discrimination studies in the European Arctic. This array has an extensive database of digital recordings dating from about 1970, including a number of earthquakes and nuclear explosions in this region. The large aperture of NORSAR, combined with the large number of short-period seismometers (initially 132, now 42) makes it possible to study the spatial variability of signal characteristics for the same seismic event over an area extending up to 100 km across.

Since we want to use the NORSAR data at high frequencies, we need to discuss the response and digitizing procedure employed at the NORSAR array. As described by BUNGUM *et al.* (1971), the *SP* channels are sampled at 20 Hz, which in principle

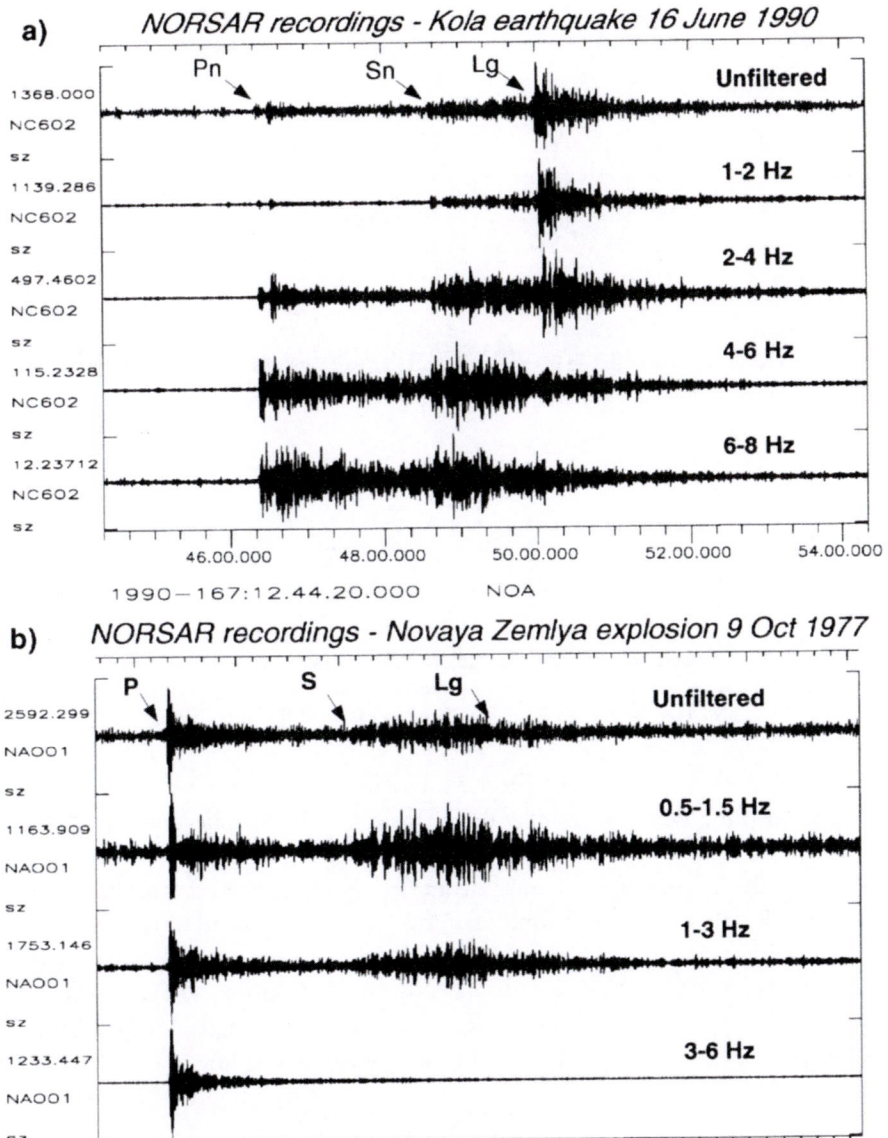

Figure 2

Illustration of regional phase propagation as seen using NORSAR data. Part a) shows recordings by the NORSAR SP seismometer 06C02 unfiltered and filtered in four frequency bands for the Kola earthquake of 16 June 1990 (distance 11 degrees). Note the differences in frequency contents of the various seismic phases and that the Lg phase is not visible above 4 Hz. Part b) shows the NORSAR center seismometer data for a nuclear explosion at Novaya Zemlya (distance 20 degrees). In this case, the Lg phase is essentially absent, and there is no visible S-wave energy above 3 Hz.

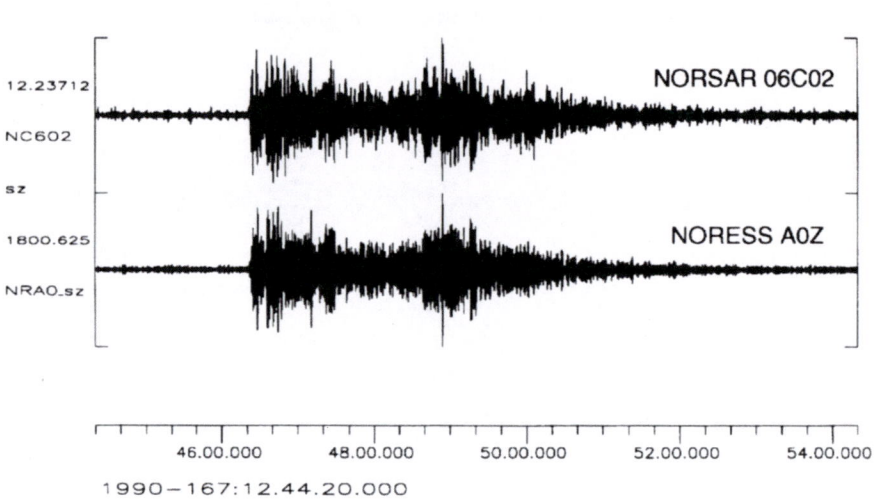

Figure 3
Comparison of narrow-band waveforms (6–8 Hz) for the NORSAR seismometer 06C02 (sampled at 20 Hz, with attenuated response above 5 Hz) and the colocated NORESS A0 seismometer (sampled at 40 Hz). There is a high consistency between the two traces.

should provide useful data up to the Nyquist frequency of 10 Hz. However, above 5 Hz, a presampling (analog) filter with a slope of 24 db/octave is applied, and this reduces the sensitivity and resolution at the highest frequencies. We have investigated the effect of this problem by comparing NORSAR 06C02 sensor data to the colocated NORESS center seismometer (A0) in various filter bands and at various signal-to-noise ratios. Since NORESS data (available since 1985) are sampled at 40 Hz, this array has useful data up to at least 16 Hz frequency. The comparison has shown a high degree of consistency, and thus indicates that the NORSAR data can indeed be used up to about 8 Hz (see Fig. 3 for an example).

Data Analysis

Novaya Zemlya Events

The NORSAR array has numerous recordings from events near Novaya Zemlya, including some nuclear explosions of magnitudes similar to those of the 16 August event and the nearby presumed earthquake of 1 August 1986. We will in the following compare the P/S ratios (based on maximum amplitudes) as recorded by individual sensors in the array. Figures 4a and b show, as examples, recordings at one

a) NORSAR Amplitude Pattern - Novaya Zemlya event 10/09/77
Filter 1-3 Hz

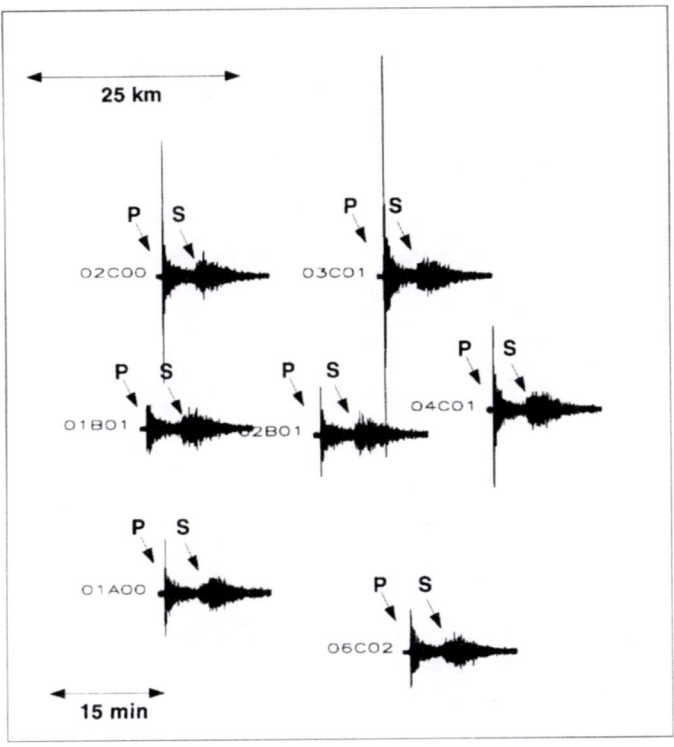

Figure 4

Recordings at the center seismometer of the 7 current NORSAR subarrays a) for the Novaya Zemlya nuclear explosion of 9 October 1977 and b) for the 1 August 1986 event. The magnitudes are 4.5 and 4.3 and the epicentral distance is about 20 degrees in both cases. The data have been filtered in the band 1.0–3.0 Hz. Note the large variation in P/S ratios.

seismometer of each of the 7 NORSAR subarrays for the nuclear explosion of 9 Oct. 1977 and the 1 August 1986 presumed earthquake. The magnitudes are 4.5 and 4.3 respectively, and the epicentral distance is about 20 degrees in both cases. The data have been filtered in the band 1.0–3.0 Hz. The following observations can be made:

- The P/S ratios display considerable variability (about an order of magnitude) across the array.
- This variability is dominated by strong P-wave focusing effects across NORSAR (see also RINGDAL, 1990).
- The amplitude patterns for the two events are rather similar, although not identical.

It may be concluded from the variability shown in this figure that the P/S ratio in the 1–3 Hz frequency band is not a very promising discriminant when using data

b) NORSAR Amplitude Pattern - Novaya Zemlya event 08/01/86
Filter 1-3 Hz

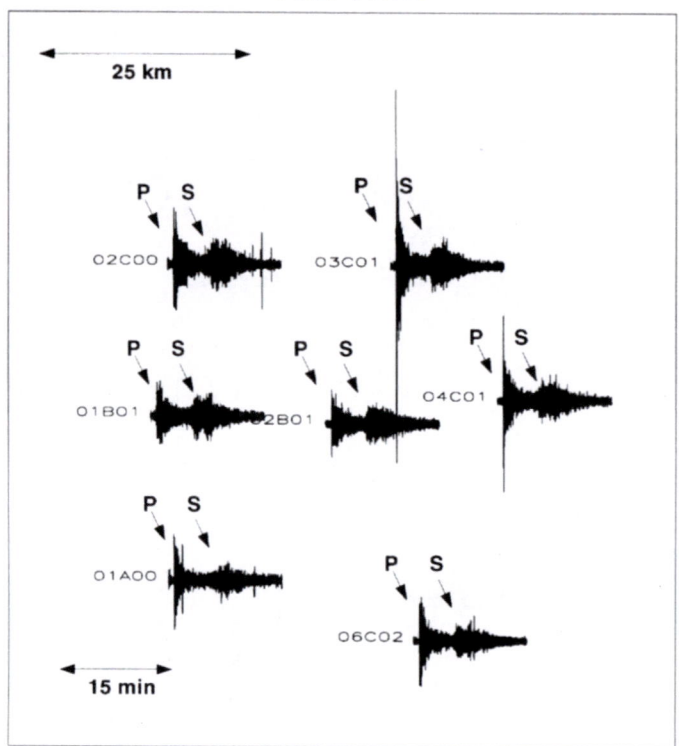

Figure 4b

recorded at a single station. Recent studies for Central Asia (HARTSE *et al.*, 1997), have shown that the P/S discriminant for that region appears effective at frequencies above 4 Hz, but performs poorly for frequencies below 4 Hz. At NORSAR, there is almost no significant S- wave energy above 3 Hz for nuclear explosions at Novaya Zemlya, thus we are confined to consider the lower frequencies for Novaya Zemlya events.

Source Scaling of the P/S Ratio

The NORSAR array data base includes digital recordings of both large and small nuclear explosions from Novaya Zemlya. It is instructive to study the P/S pattern of these explosions as a function of the event size. In order to accomplish this, we have used the one NORSAR sensor (01A01) that has dual gain recording (the usual high-gain channel and a channel that is attenuated by 30 dB). The attenuated channel has been available since 1976, and therefore provides a good database of unclipped short-period recordings of Novaya Zemlya explosions.

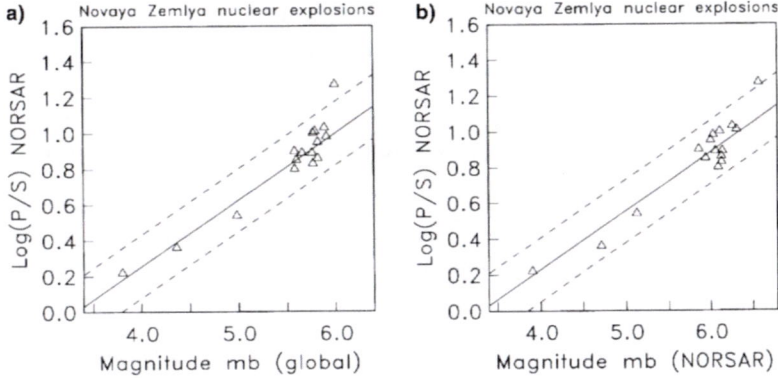

Figure 5

NORSAR P/S ratios (seismometer 01A01) in the 1.5–2.5 Hz filter band for a suite of 16 Novaya Zemlya nuclear explosions. Parts a) and b) show log(P/S) as a function of worldwide m_b and NORSAR m_b, respectively. Note the similar trend for the two cases.

Figures 5a-b show the P/S ratio (ratios of maximum amplitudes) as a function of magnitude m_b (worldwide m_b as well as NORSAR m_b) for 16 Novaya Zemlya nuclear explosions for which attenuated channel data were available. A magnitude-dependent trend can be clearly seen, and can be approximated by

$$\log(P/S) = 1.35\,m_b + c \ .$$

The slope is similar in the two plots, whereas the value of the constant c is slightly different due to a NORSAR m_b bias relative to worldwide m_b. The similarity of the two plots indicates that source-to-station specific P-wave focusing effects at NORSAR are not a dominant cause of the observed trend. There could be other possible explanations, such as systematic differences in depth of burial or source corner frequency effects, nonetheless we do not intend to extensively examine this topic. We note, however, that a somewhat similar observation was made by TAYLOR (1996) for Nevada Test Site explosions. He found that the Pg/Lg ratio increases with magnitude for all frequency bands considered, consequently earthquake explosion separation is better at high magnitudes (see also WALTER et al., 1995). In our case, we have no data to investigate the P/S ratio as a function of magnitude for frequencies above 3 Hz, and we have made no attempt to examine whether a similar magnitude dependency exists for earthquakes. For our purposes, it is sufficient to note that comparing the P/S ratios of large and small events in a discrimination context could easily give misleading conclusions. An illustration, in an expanded scale, for four of the explosions in our data set is shown in Figure 6.

Kola Peninsula Events

In order to illustrate the behavior of the P/S discriminant at higher frequencies, we show in Figure 7 the pattern of recorded signals (P and S phase) across the full

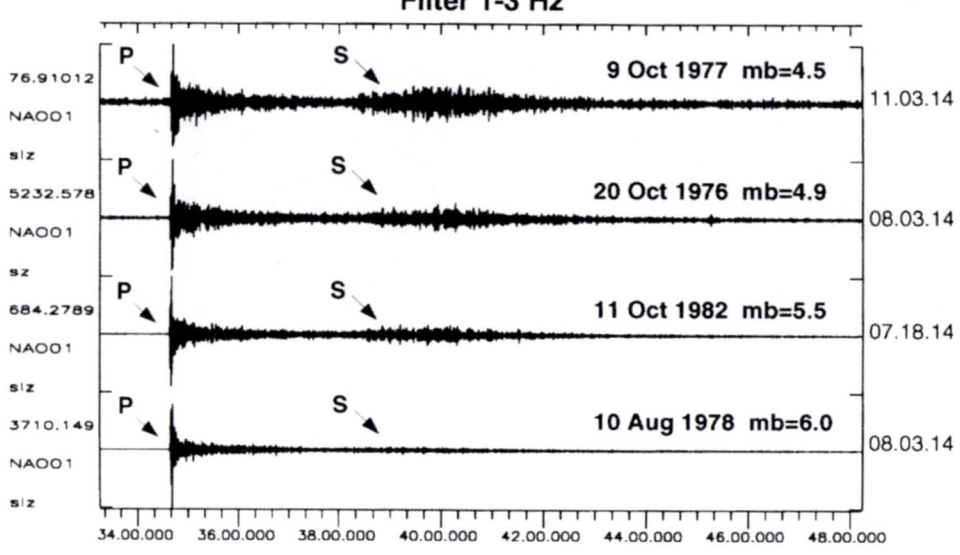

NORSAR data - selected NZ nuclear explosions
Filter 1-3 Hz

Figure 6

NORSAR recordings (seismometer 01A01) of four Novaya Zemlya nuclear explosions, filtered in the 1–3 Hz band. The traces show a selection of magnitudes, with the smallest explosion ($m_b = 4.5$) at the top, and the largest ($m_b = 6.0$) at the bottom. Note the systematic increase in P/S ratios with magnitude.

NORSAR array (22 subarrays, center sensors) for the nuclear explosion in the Kola Peninsula on 4 September 1972. The filter band is 4–6 Hz. This explosion had an epicentral distance of only about 11 degrees, and consequently has a considerable amount of high-frequency energy both for the P and the S phase. The P-wave amplitudes (and consequently the P/S ratio) vary considerably across NORSAR, even in this frequency range, however the variation is considerably less than for the 1–3 Hz recordings of the Novaya Zemlya events shown earlier. This indicates that the P/S ratio may be more stable in higher frequency bands such as the band shown.

In order to assess further the discrimination potential of the P/S ratio, we show in Figure 8 selected NORSAR traces, filtered in the 6–8 Hz band, for three Kola Peninsula events of similar magnitude: a nuclear explosion in 1984, colocated with the 1972 explosion, an earthquake near Murmansk in 1990 and an earthquake in 1999 in Revda. All are at an epicentral distance of between 11 and 12 degrees and similar azimuths. In general, the P/S ratios are higher for the explosion, although there are cases where the difference is slight.

We have also plotted the same three events as recorded by the much closer Kevo station, together with two additional, smaller earthquakes in the Kola Peninsula (Fig. 9). In this case we show three filter band, and we note that the nuclear explosion

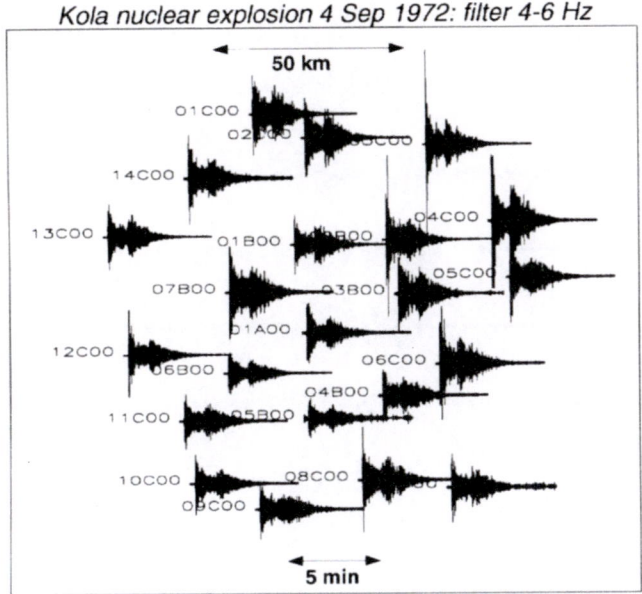

Figure 7

Amplitude pattern across the 100 km aperture original NORSAR array for the *P* and *S* phase of the Kola nuclear explosion on 4 September 1972 (distance 11–12 degrees). The data have been filtered in the 4–6 Hz band. Note that there is a strong variation in *P/S* ratios, although less than what was shown in Figure 4 for the 1–3 Hz band.

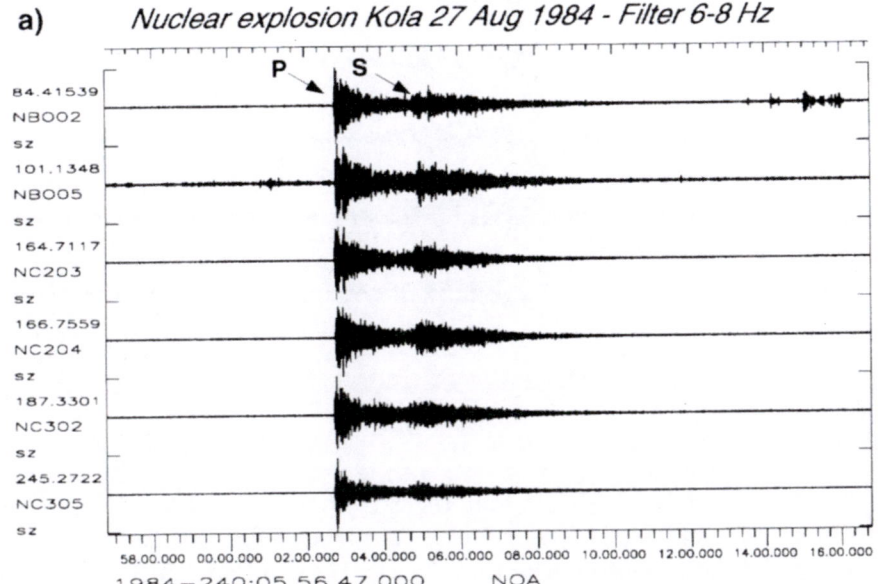

Figure 8

Selected NORSAR traces for a) the Kola nuclear explosion in 1984 (colocated with the 1972 explosion), b) the earthquake in the Kola Peninsula in 1990 (felt in the Murmansk district) and c) the Revda earthquake in 1999. All are at an epicentral distance of between 11 and 12 degrees. The data have been filtered in the 6–8 Hz band.

b)   *Earthquake Kola 16 Jun 1990 - Filter 6-8 Hz*

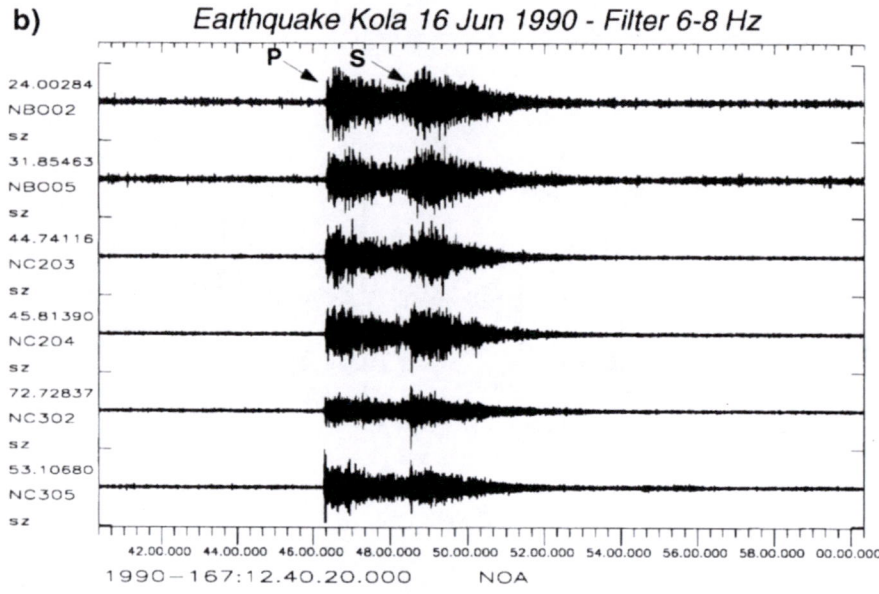

c)   *Earthquake Kola 17 Aug 1999 - Filter 6-8 Hz*

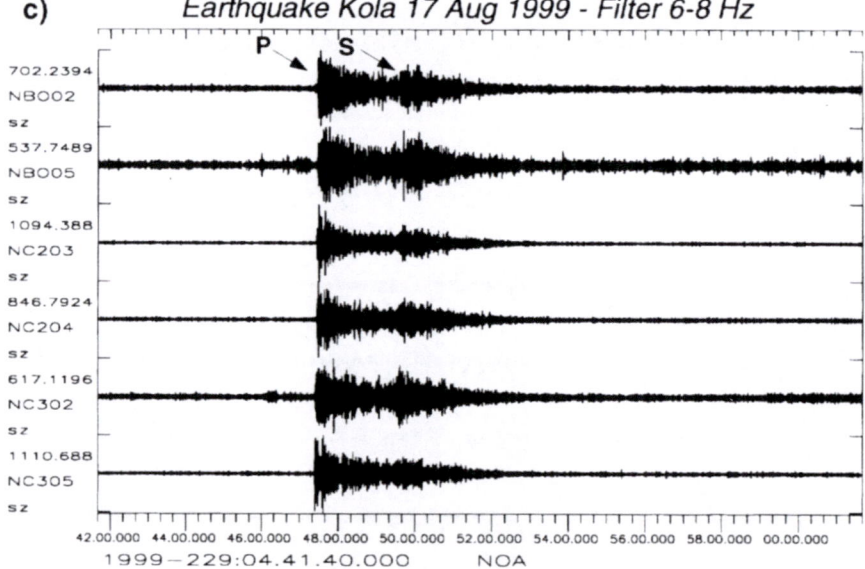

Figure 8b,c

has a considerably higher P/S ratio than the earthquakes in the 5–7 Hz band. On the other hand, there appears to be no clear separation in P/S ratios between the earthquakes and the explosion in neither the 3–5 Hz band nor the 7–9 Hz band. The data indicates that the P/S ratio discriminant has promise, but that considerable

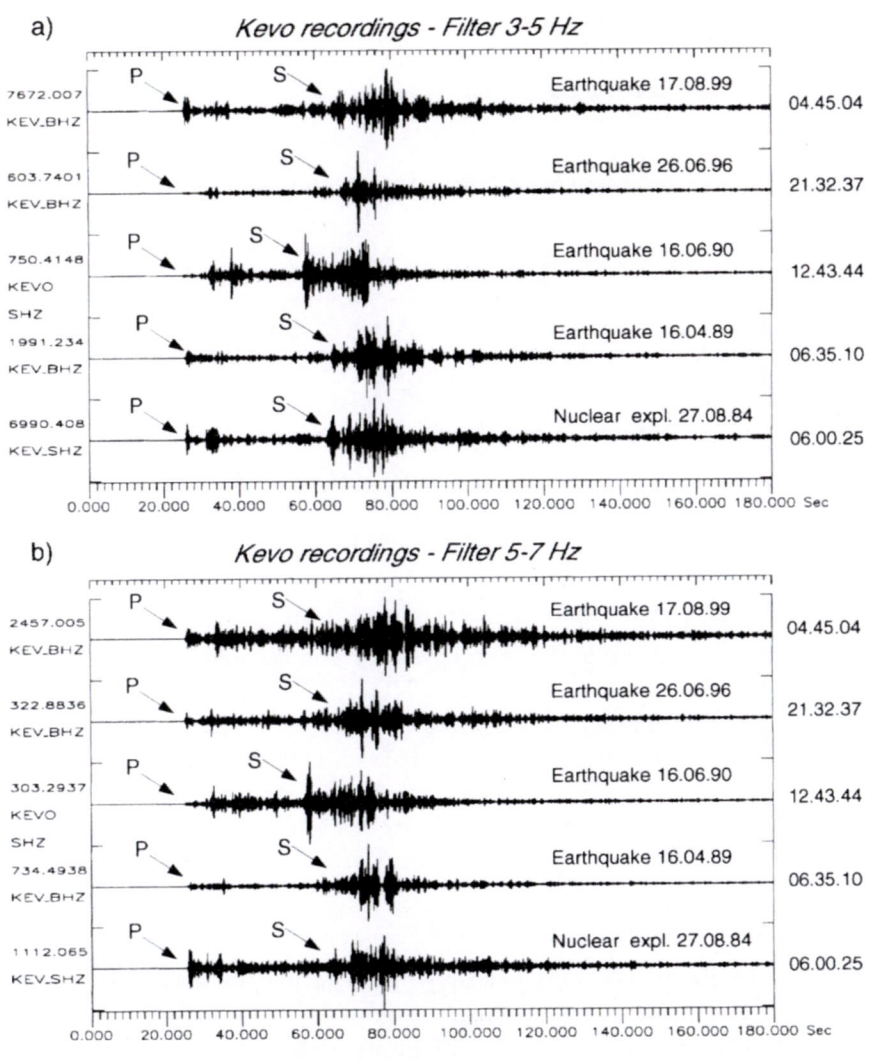

Figure 9

Recordings by the Kevo station in Finland (vertical component) for four Kola earthquakes and the 1984 Kola nuclear explosion. The epicentral distance is 2–3 degrees, and the data have been filtered in three different passbands.

caution must be exercised when applying it in narrow filter bands on the basis of one or a few stations.

We further comment on the events used in the comparisons in this section. The two nuclear explosions (1972 and 1984) were carried out inside an abandoned mine in the Khibiny Massif, for the purpose of testing ore crushing technology (MIKHAILOV et al., 1996). The largest earthquake (REVDA, 1999) is currently being actively

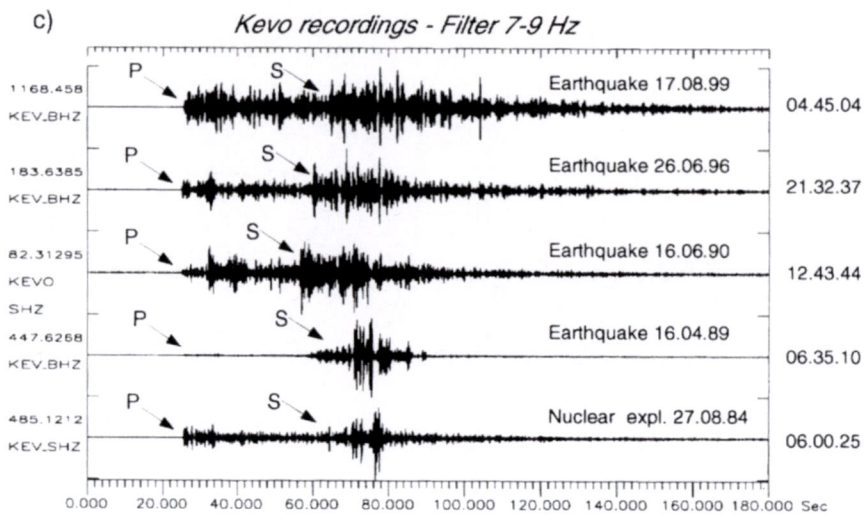

Figure 9c

studied. This earthquake, which was associated with a large mine collapse in the Lovozero Massif, was preceded by numerous foreshocks several months in advance, and was followed by several aftershocks. The 1989 Khibiny earthquake caused a surface rupture with a maximum measured displacement of 15–20 cm inside a mine, and was apparently triggered by a mining explosion (KREMENETSKAYA and ASMING, 1995). The Imandra earthquake of 1996 occurred about 30 km from the closest Khibiny mine, with an estimated depth of 15 km. The 1990 earthquake near Murmansk was not in a known mining area.

Kara Sea / Northern Urals Events

On 16 August 1997, the CTBT prototype International Data Center in Arlington, VA reported a small seismic disturbance located in the Kara Sea, near the Russian nuclear test site on Novaya Zemlya. The event caused considerable interest, since initial analysis indicated that the seismic signals had characteristics similar to those of an explosion.

NORSAR and KRSC collaborated on locating this event, each carrying out independent analysis (RINGDAL *et al.*, 1997). Since some phase onsets were very difficult to read, this was quite useful, and the results were very consistent. We were very quickly able to confirm beyond doubt that the 16 August 1997 event was located in the Kara Sea, at least 100 km from the Novaya Zemlya nuclear test site.

Figure 10 depicts Kevo recordings comparing the 16 August 1997 event and a nuclear explosion at Novaya Zemlya . The difference in P/S ratios is striking, however we note that there is a difference in magnitude of two full units between the events.

The Kevo recordings have been filtered in a considerably higher frequency band than the explosions shown in Figure 6, and we have no data to determine whether a source scaling as observed in the 1–3 Hz band might in fact also be present at these higher frequencies. We note that we have seen no examples of Novaya Zemlya nuclear explosions with observed with *P* and *S* waves of nearly the same amplitude at frequencies above 5 Hz. Unfortunately, digital recordings at regional distances are not available for any of the smaller nuclear explosions at Novaya Zemlya.

Perhaps the best indication of an earthquake source would be the presence of possible aftershocks. We have carried out a detailed search for aftershocks of the 16 August 1997 event, using both Spitsbergen array data and data that later have become available at KRSC from the Amderma station south of Novaya Zemlya.

Our search of Spitsbergen data, which was conducted by detailed visual inspection of the array beam, enabled us to find a second (smaller) event from the same site occurring about 4 hours after the main event. This second event had Richter magnitude 2.6, and could be quite clearly seen to originate from the same source area (Fig. 11).

This conclusion was supported when Amderma data became available at KRSC some weeks later. In spite of very careful analysis of both Spitsbergen and Amderma data, we have been unable to identify additional aftershocks during the two weeks following the main event.

Amderma Recordings

Figure 12 shows Amderma vertical component recordings of five seismic events at a similar epicentral distance from the station (about 300 km). The data have been filtered in the 4–8 Hz and 8–16 Hz bands. The five events are the two Kara Sea events on 16 August 1997, two mining explosions in Vorkuta south of the station, and a small event at the coast of Novaya Zemlya in 1995 (KREMENETSKAYA *et al.*, 1999).

The recordings are quite instructive. As can be seen by the scaling factor in front of the traces, the events vary in size by about an order of magnitude. It is noteworthy that the two Vorkuta explosions have very different *P/S* ratios, and encompass the range of *P/S* ratios for the other three events in the 4–8 Hz band. In the 8–16 Hz band the picture is somewhat different, with the two events of 16 August 1997, registering the lowest *P/S* ratios. Unfortunately, we have no confirmed earthquake recordings at Amderma at a similar epicentral distance.

We also note the high signal-to-noise ratio for these small events as recorded at Amderma. By a straightforward scaling procedure, we have estimated a detection threshold of about $m_b = 1.8$ for the Amderma station at this epicentral distance (300 km). For comparison, the estimated detection threshold at the Spitsbergen

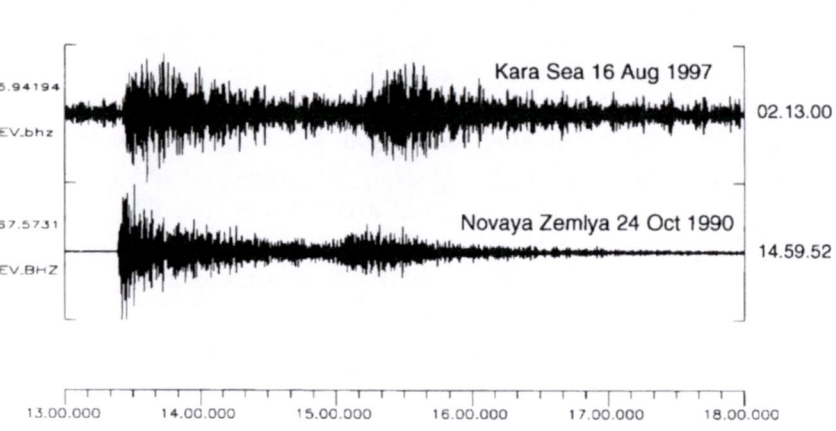

Figure 10
Kevo vertical component recordings filtered in the 7–10 Hz band. The top trace shows the 16 August 1997 seismic event (m_b = 3.5), whereas the bottom trace shows the 24 October 1990 Novaya Zemlya nuclear explosion (m_b = 5.6).

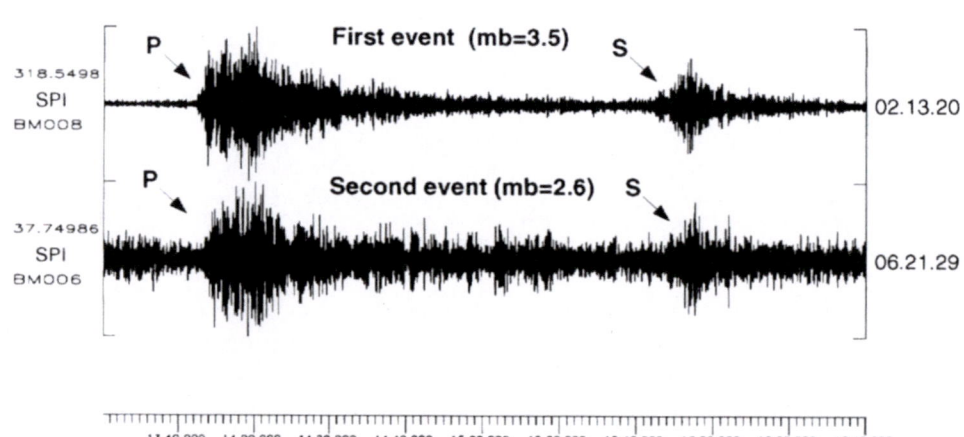

Figure 11
Recordings by the Spitsbergen array of the two events on 16 August 1997. The traces are array beams steered towards the epicenter, and with an *S*-type apparent velocity in order to enhance the *S* phase. The traces are filtered in the 4–8 Hz band. Note that the traces are very similar, although not identical. The scaling factor in front of each trace is indicative of the relative size of the two events.

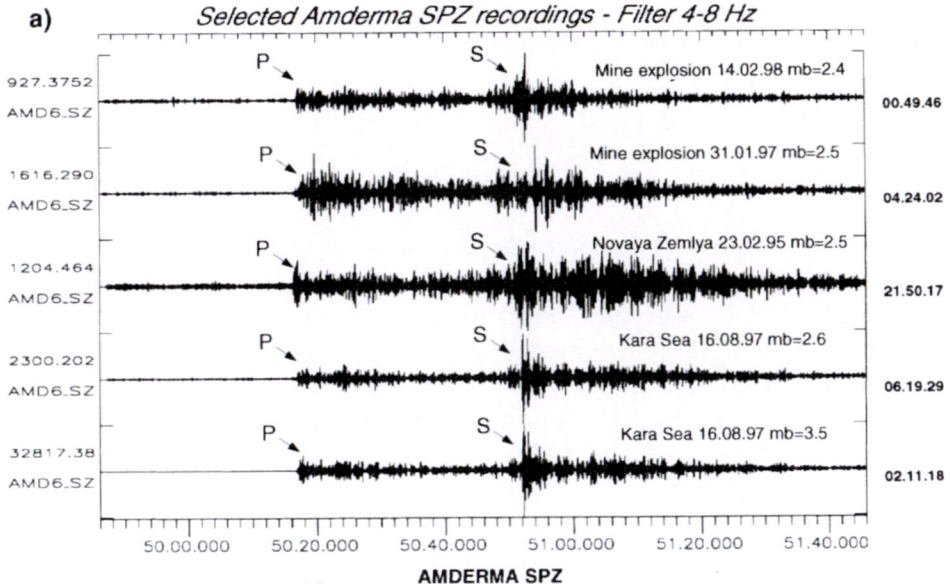

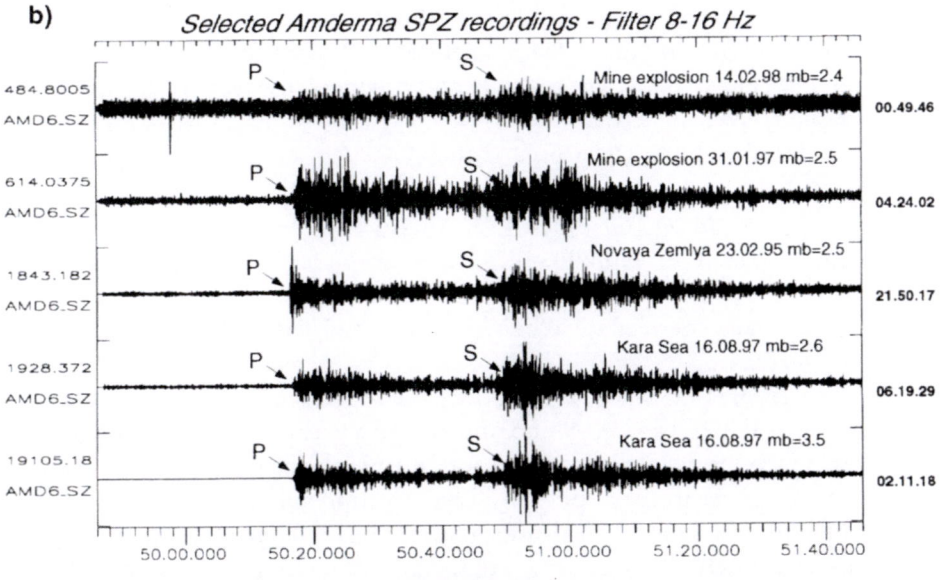

Figure 12

Amderma vertical component recordings of five seismic events at a similar epicentral distance from the station (about 300 km). The data have been filtered in the 4–8 and 8–16 Hz bands. The five events are the two Kara Sea events on 16 August 1997, two mining explosions in Vorkuta south of the station, and a small event at the coast of Novaya Zemlya in 1995. The scaling factor in front of each trace is indicative of the relative size of the events.

array is about 2.5 for seismic events near the Kara Sea site, and somewhat higher (about 3.0) for events in the Vorkuta region.

Conclusions

Case studies for the Barents/Kara Sea region, some of which are discussed briefly in this paper, have demonstrated that the P/S ratio discriminant is ineffective at low frequencies (1–3 Hz), however it shows promise at higher frequencies (e.g., 6–8 Hz) in this region. The discriminant should be applied with considerable caution when attempting to identify the source type of seismic events in the European Arctic. Future application of the P/S discriminant in this region will require extensive regional calibration and detailed station-source corrections. Future research should also focus on combining the P/S ratio with other short-period discriminants, such as complexity and spectral ratios.

We show that in the 1–3 Hz filter band the P/S ratio of Novaya Zemlya explosions recorded at the NORSAR array scale with magnitude in such a way that the larger explosions have a relatively high P/S ratio. Such an effect would make a reliable comparison difficult between P/S ratios of small and large events, and indicate a need for caution and further research when applying the P/S discriminant. We note that the S-wave signal-to-noise ratios for this set of events are not sufficient to determine if there is such a scaling relationship above 3 Hz.

The Kara Sea event on 16 August 1997 provides an interesting case study for the Novaya Zemlya region. It highlights the fact that even for this well-calibrated region, where numerous well-recorded underground nuclear explosions have been conducted, it is a difficult process to reliably locate and classify a seismic event of approximate m_b 3.5.

It is clear from this study that more research is needed on regional signal characteristics in the European Arctic and the application of discriminants such as the P/S ratio at regional distances. An interesting approach, which is beyond the scope of this paper, is to carry out a systematic study of the variability of the P/S ratio in different frequency bands and at different epicentral distances, using the comprehensive NORSAR large array database of regional explosions and earthquakes, supplemented with other available digital data. It would be a particularly useful exercise to carry out a small chemical calibration explosion, in order to improve the seismic calibration of Barents/Kara Sea region. Such an explosion, even if not recorded teleseismically, would provide valuable additional information for future studies.

REFERENCES

BAUMGARDT, D. R. (1990), *Investigation of Teleseismic Lg Blockage and Scattering Using Regional Arrays*, Bull. Seismol. Soc. Am. *86* (4), 1042–1053.

BUNGUM, H., HUSEBYE, E. S., and RINGDAL, F. (1971), *The NORSAR Array and Preliminary Results of Data Analysis*, Geophys. J.R. Astr. Soc. *25*, 15–26.

HARTSE, H. E. (1998), *The 16 August 1997 Novaya Zemlya Seismic Event as Viewed from GSN Stations KEV and KBS*, Seism. Res. Lett. *69* (3), 206–215.

HARTSE, H. E., TAYLOR, S. R., PHILLIPS, W. S., and RANDALL, G. E. (1997), *A Preliminary Study of Regional Seismic Discrimination in Central Asia with Emphasis on Western China*, Bull. Seismol. Soc. Am. *87*, 551–568.

KREMENETSKAYA, E. and ASMING, V. (1995), *Induced Seismicity in the Khibiny Massif (Kola Peninsula)*, Pure appl. geophys., *145* (1), 29–37.

KREMENETSKAYA, E., ASMING, V., and RINGDAL, F. (1999), *Seismic Location Calibration of the Barents Region*, Semiannual Technical Summary 1 October 1998 - 31 March 1999, NORSAR Sci. Rep. 2-98/99, Kjeller, Norway.

MARSHALL, P. D., STEWART, R. C., and LILWALL, R. C. (1989), *The Seismic Disturbance on 1986 August 1 near Novaya Zemlya: A Source of Concern?* Geophys. J. *98*, 565–573.

MIKHAILOV, V. N. *et al.* (1996), *USSR Nuclear Weapons Tests and Peaceful Nuclear Explosions, 1949 through 1990*, RFNC - VNIIEF, Sarov, 1996, 63 pp.

RICHARDS, P. G. and WON-YOUNG KIM (1997), *Test-ban Treaty Monitoring Tested*, Nature *389*, 781–782.

RINGDAL, F. (1990), *P-wave Focusing Effects at NORSAR for Novaya Zemlya Explosions*, Report GL-TR-90-0330, Hanscom AFB, Massachusetts.

RINGDAL, F. (1997), *Study of Low-magnitude Seismic Events near the Novaya Zemlya Nuclear Test Site*, Bull. Seismol. Soc. Am. *87* (6), 1563–1575.

RINGDAL, F., KVÆRNA, T., KREMENETSKAYA, E., and ASMING, V. (1997), *The Seismic Event Near Novaya Zemlya on 16 August 1997*, Semiannual Technical Summary 1 April - 30 September 1997, NORSAR Sci. Rep. 1-97/98, Kjeller, Norway.

SERENO, T., BRATT, S., and BACHE, T. (1988), *Simultaneous Inversion of Regional Wave Spectra for Attenuation and Seismic Moment in Scandinavia*, J. Geophys. Res. *93*, 2019–2035.

TAYLOR, S. R.(1996), *Analysis of High-frequency Pg/Lg Ratios from NTS Explosions and Western U.S. Earthquakes*, Bull. Seismol. Soc. Am. *86* (4), 1042–1053.

WALTER, W. R., MYEDA, K. M., and PATTON, H. J. (1995), *Phase and Spectral Ratio Discrimination Between NTS Earthquakes and Explosions. Part I: Empirical Observations*, Bull. Seismol. Soc. Am. *85*, 1050–1067.

(Received June 25, 1999, revised July 11, 2000, accepted July 18, 2000)

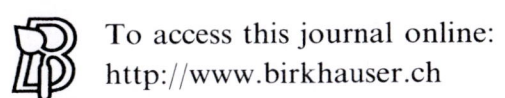

To access this journal online:
http://www.birkhauser.ch

Pure appl. geophys. 159 (2002) 721–733
0033–4553/02/040721–13 $ 1.50 + 0.20/0

❙ Pure and Applied Geophysics

Study of Regional Surface Waves and Frequency-dependent M_s:m_b Discrimination in the European Arctic

ELENA KREMENETSKAYA,[1] VLADIMIR ASMING,[1]
ZINA JEVTJUGINA[1] and FRODE RINGDAL[2]

Abstract—Accurate discrimination of seismic events with a regional network requires detailed knowledge of the propagation characteristics of seismic waves in the region. At present, such propagation characteristics are reasonably well known for P and S waves in the European Arctic, however much work remains to be done regarding surface wave propagation and magnitude estimation.

Regional long-period or broadband seismic data in digital form has been available in the European Arctic for only a few years. In order to assess regional surface wave propagation, and in particular to evaluate the M_s:m_b discriminant at regional distances, it is therefore necessary to take advantage of the historic analog recordings. The station APA in Apatity forms a unique source of such data, with high-quality long-period seismic recordings of regional earthquakes and nuclear explosions dating back about 30 years.

This paper presents initial results from a project to digitize APA surface waves of selected regional events. The recordings for recent years have been compared to a colocated broadband Guralp three-component seismometer in order to verify the response characteristics and the quality of the digitization process. It turns out that the quality of the digitized records is excellent, and can be used over a spectral band ranging from 5 seconds to at least 30 seconds period.

We demonstrate the capabilities of the APA surface wave recordings to provide a promising separation of earthquakes and explosions in the European Arctic over a range of frequencies using the M_s:m_b discriminant, although we note that additional work is required in regionalization of the propagation paths to take into account the major tectonic features in the region. We also note that the body-wave magnitudes provided by international agencies are not always reliable for events in this region, and must be reassessed in order to make full use of the earthquake-explosion discrimination potential.

Key words: Surface waves, earthquake-explosion discrimination, regional phases.

Introduction

As part of a project aimed at improving seismic monitoring capabilities for the Barents/Kara Sea region, NORSAR and Kola Regional Seismological Centre (KRSC) are conducting a comprehensive study of seismicity, seismic wave propagation and seismic event characterization in the European Arctic. This work is particularly relevant to the development of event screening criteria, which is one of

[1] Kola Regional Seismological Centre (KRSC), Apatity, Russia.
[2] NORSAR, Kjeller, Norway. E-mail: frode@norsar.no

the main tasks of the expert work conducted by Working Group B of the Preparatory Commission for the Comprehensive Nuclear-Test-Ban Treaty (CTBT).

The purpose of event screening is to "screen out" events that are thought to be consistent with natural causes (such as earthquakes), so that a detailed analysis can be focused on those events that are truly of interest for monitoring purposes. The current seismic screening procedure employed at the International Data Centre (IDC) focuses on two criteria: event focal depth and $M_s:m_b$. These are considered to be the most robust criteria currently available, but have the disadvantage that they are difficult to apply to small events or events recorded only by a few stations. Other criteria, such as the high-frequency P/S ratio, hold the promise of being applicable at considerably lower event magnitudes, and this is currently an area of active research (see e.g., RINGDAL *et al.*, 2002).

By focusing on regional recordings of surface waves, it might be possible to apply the $M_s:m_b$ discriminant to low magnitude events, perhaps approaching $m_b = 3.0–3.5$. This is the motivation for the present study. As is well known, accurate discrimination of seismic events with a regional network requires detailed knowledge of the propagation characteristics of seismic waves in the region. At present, these propagation characteristics are reasonably well known for short-period P and S waves in the European Arctic, (see e.g., SERENO *et al.*, 1988), but substantial work remains to be done regarding surface-wave propagation and magnitude estimation. In the following we describe certain initial results obtained for this region.

Database

The regional seismic network operated by the Kola Regional Seismological Center currently comprises a combination of digital and analog stations. Several stations of the analog type have been in operation for many years (see Fig. 1), whereas the digital stations in this network have only a few years of available recordings (ASMING *et al.*, 1998).

In order to assess surface wave propagation, and in particular to evaluate the $M_s:m_b$ discriminant, it is necessary to take advantage of the historic analog recordings. The station APA in Apatity forms a unique source of such data. This station has had high-quality LP recordings since 1969, and thus a database is available of regional earthquakes and nuclear explosions dating back about 30 years. The LP seismometer is a three-component system, with analog recording at a constant amplification of 1000 relative to ground displacement in the band 5–25 seconds. It is supplemented by a low-gain vertical channel (amplification 100), which is used for the largest seismic events.

We have initiated a project to digitize surface waves of selected regional events in the APA database of LP recordings. The digitization method is based on a semi-automated algorithm. The original seismograms are amplified by photocopying and

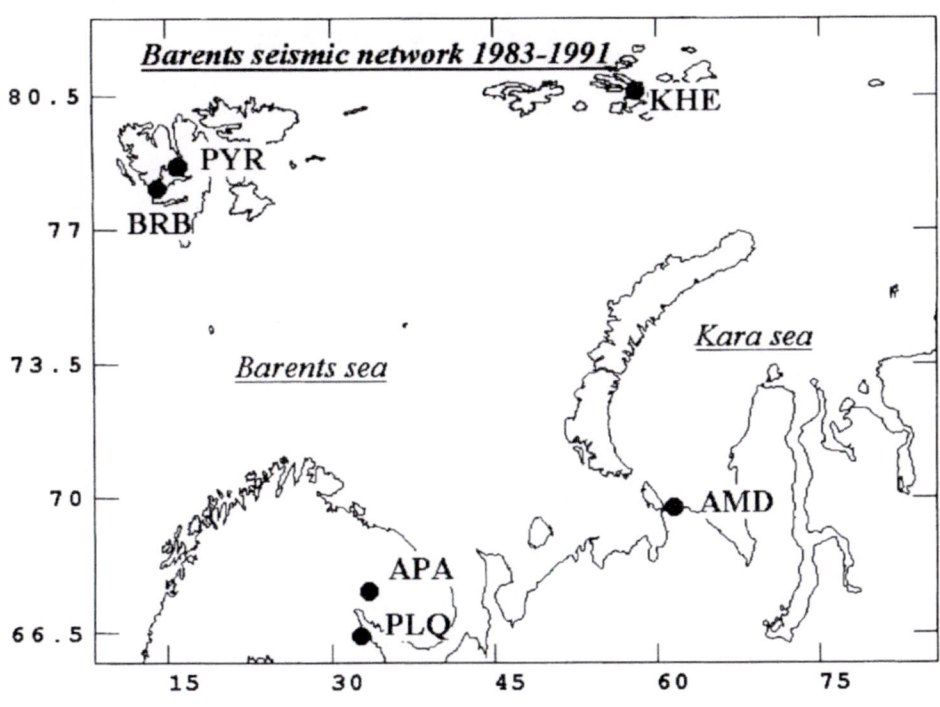

Figure 1

Stations in the Barents seismic network operated by KRSC. The station APA, which has both three-component SP and LP seismometers, is the longest in operation, from 1969 until present. APA has in addition a Guralp CMG-3T broad-band digital seismometer, which has been operational since 1991.

scanned into an image on a PC. An automatic algorithm calculates the midpoint of each trace for a given time interval, and thus creates an initial digital record. The analyst can interactively verify the output and make corrections as necessary (for example when lines on the seismogram cross each other). Finally, the record is resampled to a uniform sampling rate.

We have checked the performance of this method by comparing digitized analog LP recordings to the digital recordings of a colocated broadband station in order to verify the response characteristics and the quality of the digitization process. This comparison can only be done for the most recent years, during which a colocated broadband Guralp three-component seismometer has been in operation in Apatity. An illustration of such a comparison for an earthquake in 1998 near Spitsbergen is shown in Figures 2 and 3. As seen from these illustrations, the quality of the digitized records is excellent, and can be used over a spectral band ranging from 5 seconds to about 30 seconds period. In fact, the recordings in the various filter bands are nearly identical, except that for the lowest filter band (0.03–0.04 Hz or 25–33 seconds) the broadband recordings have slightly more ringing of the signal than the digitized LP recordings. The relative amplitudes in the different frequency bands likewise show

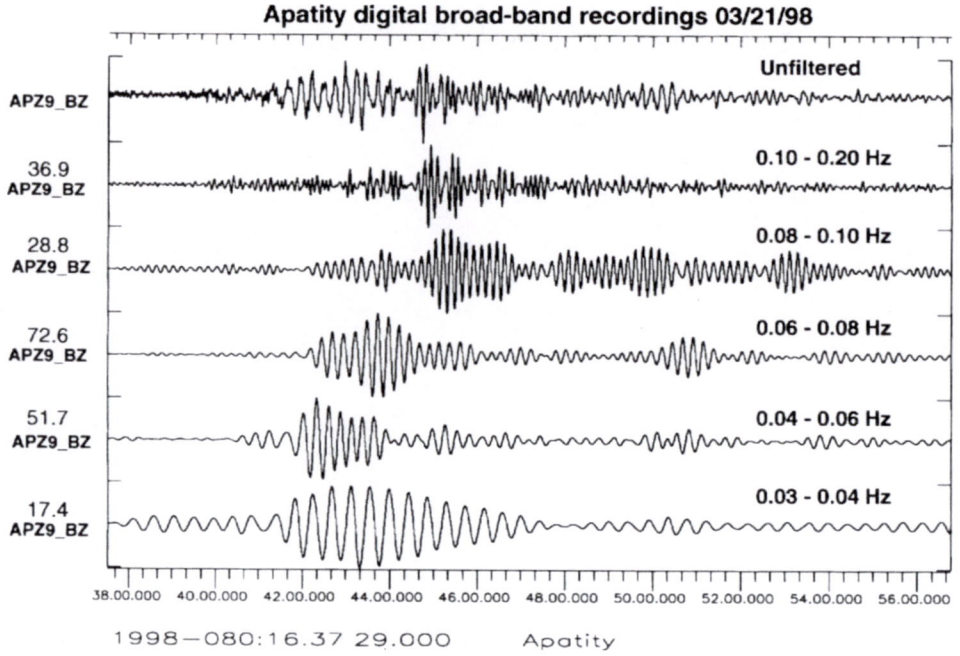

Figure 2

Digital recording by the broadband Guralp vertical seismometer in Apatity for an earthquake near Spitsbergen on 21 March, 1998. The unfiltered data are shown in the top trace, with the other traces showing a suite of narrow-band filters applied to the recording. Numbers in front of each filtered trace represent ground motion (maximum amplitude) in microns.

good agreement, again with a reservation for the lowest frequency band. We attribute the differences noted above to uncertainties in the nominal response characteristics of the seismometers and recording systems at the lowest frequencies.

Data Analysis

We have initially applied this digitization process to 28 seismic events at regional distances and various azimuths from the APA station (see Table 1). Eleven are nuclear explosions, as listed by MIKHAILOV *et al.* (1996), mostly from the Novaya Zemlya test site. The remainder are intermediate and low magnitude earthquakes (typical magnitude range 4.0–5.0). All of the earthquakes have continental propagation paths. While the earthquakes (by necessity) are at azimuths different from the majority of the explosions, we consider that the variations in azimuths and propagation paths are sufficient to provide a representative sample of the characteristics of the seismic source and propagation effects in the region being considered.

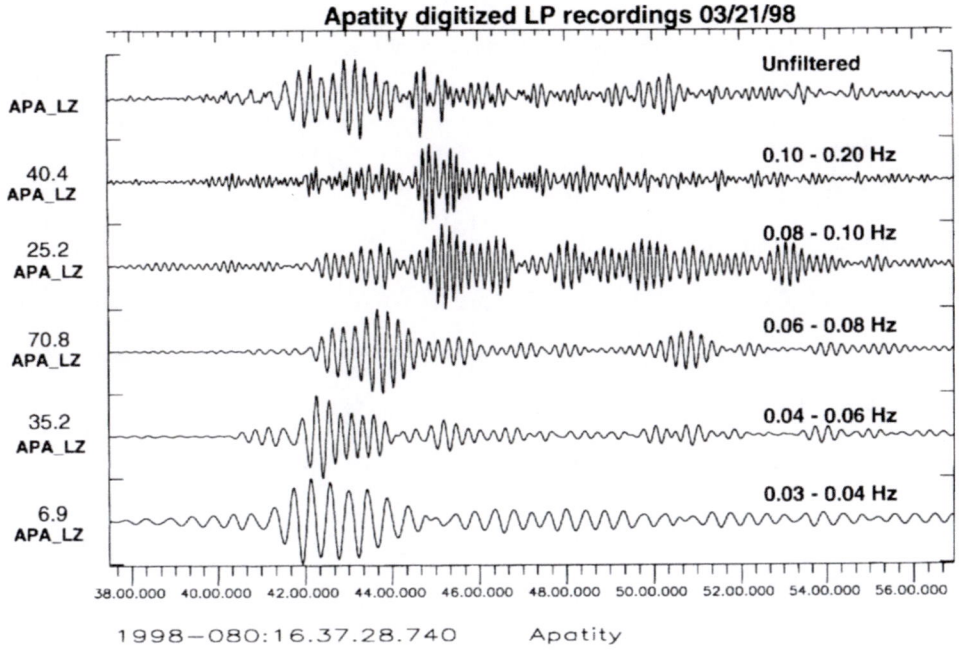

Figure 3

Digitized recordings based on the APA LP vertical component colocated with the Guralp broadband seismometer for the same event shown in Figure 2. Numbers in front of each filtered trace represent ground motion (maximum amplitude) in microns. Note the close correspondence of the traces shown in the two figures.

The lack of reliable m_b estimates by international agencies for events in this region has been a source of concern. As an example, the ISC m_b can on occasion be biased high by one full magnitude unit, e.g., when only one or two high-amplitude teleseismic stations have detected a given event. While most of the Novaya Zemlya nuclear explosions have a reasonably accurate magnitude estimate (RINGDAL, 1997), corresponding reliable estimates are not generally available for the earthquakes in our database. For this reason we have chosen to recompute m_b values for all the processed events, using the maximum-likelihood method of RINGDAL (1986). These values are listed in Table 1 as m_b (MLE).

An example of digitized data for one of the nuclear explosions is shown in Figure 4. We note that the LP signals are very clear in all the frequency bands considered. In particular, it is interesting to note the strong signals even at the highest frequency band in the figure (0.1–0.2 Hz or 5–10 seconds period).

Figure 5 shows a map of the propagation paths (top) and a comparison of the normalized surface-wave spectra (bottom). These normalized spectra have been computed by adjusting for distance as well as body-wave magnitude. The distance adjustment makes use of the standard formula for M_s computation (VANEK et al., 1962):

Table 1

List of seismic events used in this study

No.	DATE	Time	Lat. (N)	Lon. (E)	m_b (MLE)	Reference
Confirmed nuclear explosions						
1	1973/09/30	05.00.00	51.7	54.6	5.1	SULTANOV *et al.* (1999)
2	1974/08/29	10.00.00	73.4	55.1	6.5	RINGDAL (1997)
3	1974/08/29	15.00.00	67.1	62.6	4.9	SULTANOV *et al.* (1999)
4	1975/10/21	12.00.00	73.4	55.1	6.6	RINGDAL (1997)
5	1976/09/29	03.00.00	73.4	54.8	5.8	RINGDAL (1997)
6	1976/10/20	08.00.00	73.4	54.6	4.9	RINGDAL (1997)
7	1978/08/10	08.00.00	73.3	54.8	6.0	RINGDAL (1997)
8	1984/08/11	19.00.00	65.1	55.1	5.3	SULTANOV *et al.* (1999)
9	1985/07/18	21.15.00	66.0	41.0	5.1	SULTANOV *et al.* (1999)
10	1988/12/04	05.20.00	73.4	55.0	5.8	RINGDAL (1997)
11	1990/10/24	14.58.00	73.4	54.7	5.6	RINGDAL (1997)
Presumed earthquakes						
12	1971/12/16	18.35.45	77.8	18.1	4.9	ISC
13	1972/01/16	06.31.04	77.6	18.0	3.9	ISC
14	1975/01/20	10.47.29	71.7	14.2	5.0	ISC
15	1976/01/18	04.46.21	77.8	18.3	5.5	ISC
16	1976/09/09	09.27.45	77.8	7.9	5.2	ISC
17	1976/10/25	08.39.45	59.2	23.6	4.3	ISC
18	1977/07/17	09.22.24	77.8	18.6	4.6	ISC
19	1981/09/03	18.39.42	69.3	14.2	4.8	ISC
20	1986/08/01	13.56.38	73.0	56.7	4.3	MARSHALL *et al.* (1989)
21	1986/10/26	11.34.38	61.7	3.3	4.4	ISC
22	1987/05/26	02.44.48	76.6	25.7	3.6	ISC
23	1988/08/08	19.59.32	63.6	2.3	5.6	ISC
24	1990/05/28	00.35.48	55.2	58.6	4.3	LOMAKIN and YUNUSOV (1993)
25	1990/05/28	02.41.28	55.2	58.7	4.4	LOMAKIN and YUNUSOV (1993)
26	1993/09/13	05.25.10	66.3	5.8	3.9	ISC
27	1997/12/20	21.40.48	67.6	10.9	3.6	ISC
28	1998/03/21	16.33.11	79.9	1.9	5.9	NEIC

$$M_s = \log \frac{A}{T} + 1.66 \log \Delta + 3.3$$

where Δ is the epicentral distance in degrees, A is zero-to-peak amplitude in microns and T is the corresponding signal period. This formula is currently used by the International Seismological Centre (ISC), and as noted by VANEK *et al.* (1962), the formula is considered valid in the 10–60 second range and for distances from 5–160 degrees (depth less than 60 km). There are several other proposed formulas for

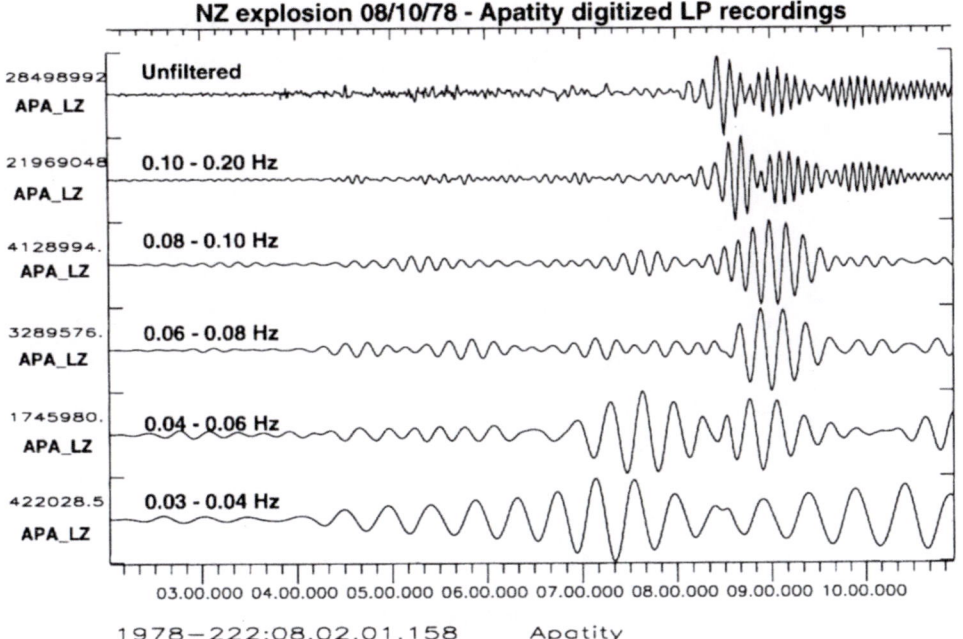

Figure 4

Digitized recordings based on the APA LP vertical component seismometer for the nuclear explosion at Novaya Zemlya on 10 August, 1978. Numbers in front of each trace are maximum amplitudes (not converted to ground motion). Note the high SNR in all the filter bands.

surface-wave magnitudes (notably MARSHALL and BASHAM, 1972), and in many such formulas the distance coefficient is significantly less than 1.66 as used above. Since most events discussed in this study are at a similar distance from the APA station (typically about 10 degrees), the distance coefficient is not critical, and we have therefore chosen to use the standard M_s formula in this analysis.

The adjustment for body-wave magnitude makes use of the current event screening criterion for M_s:m_b employed at the prototype International Data Centre (MURPHY et al., 1998), which is of the form:

$$M_s = 1.25m_b - 2.20 \quad .$$

The important term in our context is the slope (1.25) of this relationship. We use the two above formulas to normalize the distance-corrected spectral log amplitude values by calculating, at each frequency, the quantity:

$$\log \frac{A}{T} + 1.66 \log \Delta - 1.25m_b \quad .$$

In this way we obtain measurements of log A/T, shown in the figure, that have been corrected for distance and body-wave magnitude. In all cases, the signal-to-

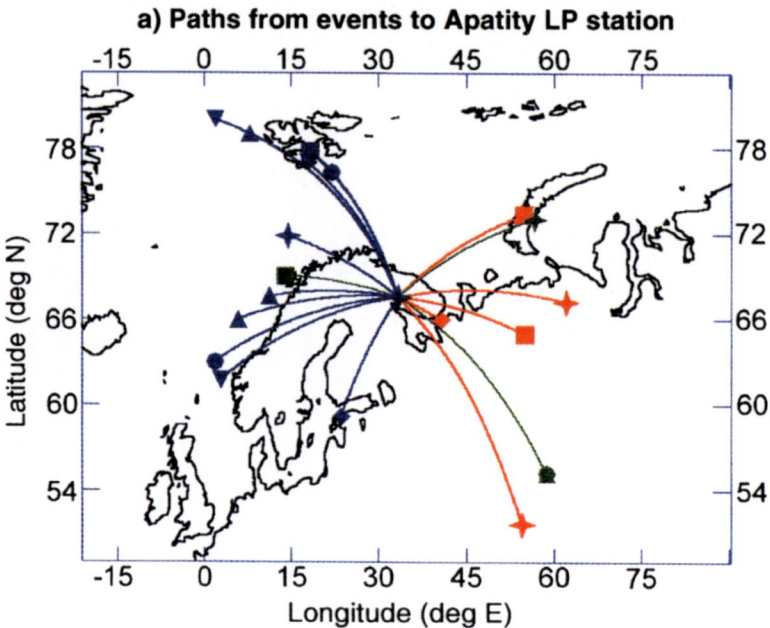

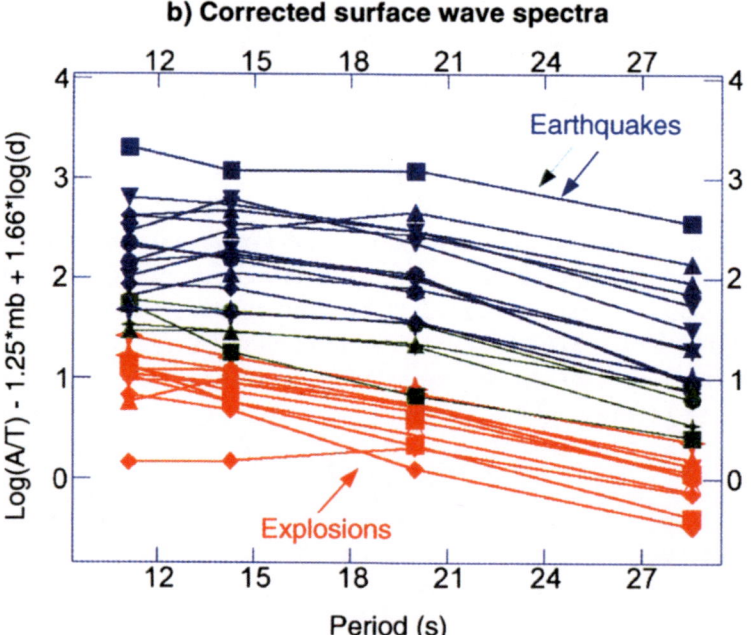

Figure 5

Spectral plot for a database of earthquakes and nuclear explosions with continental travel paths to the APA LP station. The top part shows the events and the travel paths to APA, whereas the bottom part shows surface-wave spectral levels ranging from 10 to 30 second periods. The spectral levels have been corrected for distance and body-wave magnitude (see text for details). The four "borderline" events marked in green are discussed separately in the text.

noise ratios of the recordings are sufficient to ensure that the amplitudes are true measurements of the signal, and not of the microseismic noise. Although somewhat simplified, the diagram can be seen as a frequency-dependent $M_s:m_b$ plot, and the separation between the earthquakes and explosions is similar over the entire frequency range. Four "borderline" cases (marked in green color on the plot) are discussed briefly below.

The first borderline case is an earthquake off the coast of Norway, very close to the oceanic/continental margin (event no. 19 in Table 1). The table shows that the location of this earthquake is on the oceanic side, and this could account for the anomalously small surface waves (the event is in fact inseparable from the explosion population). To study this topic further we digitized surface-wave recordings from several additional earthquakes in the oceanic part of the Norwegian Sea close to the oceanic/continental margin. These earthquakes were consistently close to the explosion population (they are not shown in the plot). It is well known that for earthquakes on an oceanic path the Airy phase is shifted to higher frequencies and thus these earthquakes exhibit lower amplitudes in the 5–20 seconds range (see e.g., MARSHALL and BASHAM, 1972). However, in the cases studied here, most of the propagation paths are continental, and this suggests that the low surface-wave amplitudes could, at least in part, be due to attenuation when crossing the oceanic/continental margin.

The next borderline cases are two colocated seismic events in the Ural Mountains (event nos. 24 and 25 in Table 1). According to LOMAKIN and YUNOSOV (1993) these events were rockbursts in the "Kurgazakskaya" bauxite mine. We have searched the ISC bulletins for other events in the central and northern Ural Mountains to be used in this study, however those events which we have found have all been identified as mine rockbursts or collapses. We have therefore been unable to include any tectonic earthquakes from this region in our study.

The fourth and final borderline case in Figure 5 is the 1 August 1986 event near Novaya Zemlya (event no. 20 in Table 1). This event is generally presumed to have been an earthquake, based upon the depth estimate (19 km) and focal mechanism provided by MARSHALL et al. (1989). We note that the event is separated from the explosion population, nevertheless it is rather close. In any case, we agree with MARSHALL et al. (1989) that caution should be exercised in interpreting $M_s:m_b$ data unless a network of stations at a variety of azimuths is available, and Figure 5 should not be interpreted as an attempt to identify this event on the basis of the $M_s:m_b$ criterion.

Interfering Surface Waves

We have shown that the measurement of surface wave magnitudes at regional distances holds significant promise to lower the limit for applying the $M_s:m_b$

criterion, and would be of particular importance for the event screening currently being implemented at the IDC. Furthermore, regional surface waves have a significant energy at shorter periods (down to 5–10 seconds), and this could be exploited in extending the magnitude range for useful M_s measurements.

In particular, measurement of such shorter period surface waves at regional distances could contribute to reducing the influence of coda from surface waves of large teleseismic earthquakes, which often mask ordinary surface waves from small events for hours. This is brought about because these strong surface waves generally have a dominant period of 20 or more seconds, with far less energy in the shorter period bands.

An example of this is given in Figures 6 and 7, which make use of data recorded by the large-aperture NORSAR array (Bungum *et al.*, 1971). The figures show steered array beams (phase velocity 3.5 km/s and azimuth 40 degrees) based on 7 LPZ seismometers distributed over an area 60 km in diameter. Figure 6 is recordings of a large nuclear explosion ($m_b = 5.8$) at Novaya Zemlya on 25 October, 1984. An unfiltered array beam is displayed together with the beam filtered in the "standard" 17–25 seconds band and a "high-frequency" 8–10 seconds band. Note the high SNR of this regional recording (distance = 20 degrees) even at the higher frequencies. Figure 7 shows a similar plot of NORSAR LPZ array beam recordings of a small

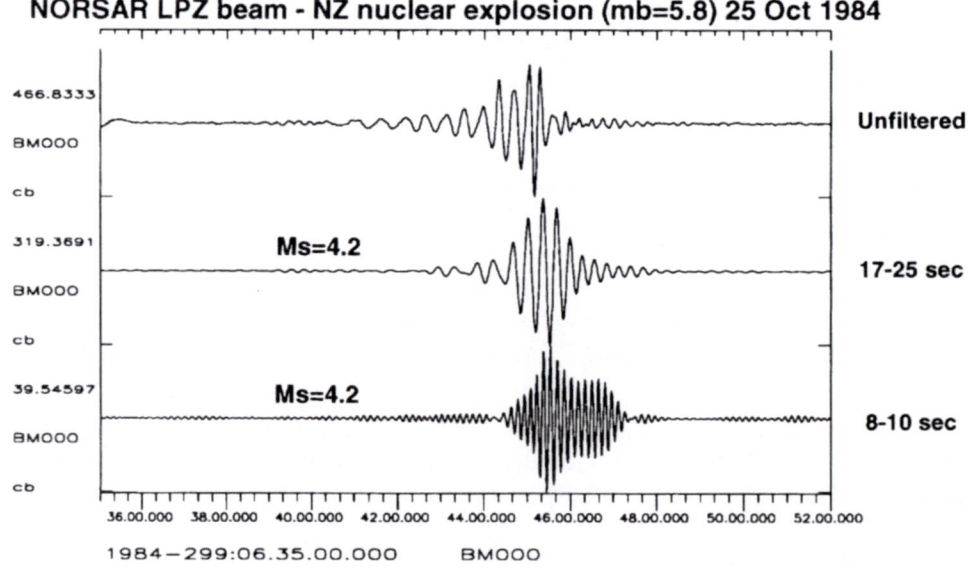

Figure 6

NORSAR LPZ steered array beam recordings of a large nuclear explosion ($m_b = 5.8$) at Novaya Zemlya on 25 October, 1984. An unfiltered beam is shown together with the beam filtered in the "standard" 17–25 seconds band and a "high-frequency" 8–10 seconds band. Note the high SNR of this regional recording (distance = 20 degrees) even at the higher frequencies.

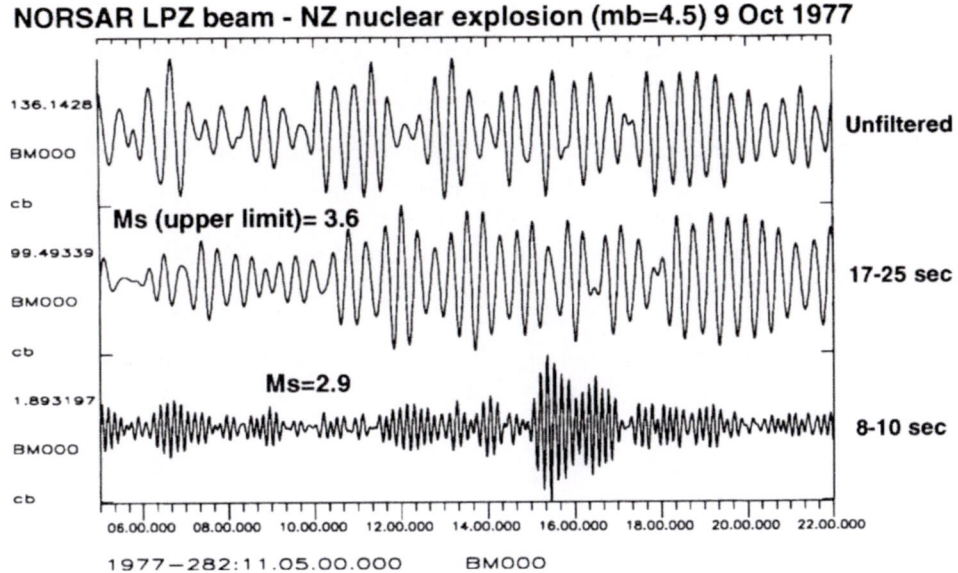

Figure 7

NORSAR LPZ steered array beam recordings of a small nuclear explosion ($m_b = 4.5$) at Novaya Zemlya on 9 October, 1977. The traces correspond to those shown in Figure 6. Note that an interfering earthquake masks the explosion surface waves in the 17–25 seconds band, whereas the explosion signal is clearly seen in the 8–10 seconds band.

nuclear explosion ($m_b = 4.5$) at Novaya Zemlya on 9 October, 1977. A few minutes before this latter explosion, a large earthquake occurred in the Philippine Islands, causing significant interfering surface waves at NORSAR during the expected time of surface wave arrival for the explosion. From Figure 7, we can see how the interfering event masks the explosion surface waves in the 17–25 seconds band, and it is only possible to estimate an upper bound on the M_s in this band. On the other hand, the explosion signal is clearly seen in the 8–10 seconds band, and a surface wave magnitude can be calculated by direct comparison with the explosion in Figure 6.

There are several other techniques, such as matched filtering using a nearby event, that could be applied to enhance the signal-to-noise ratio of surface waves in the presence of an interfering earthquake. The use of narrow band filtering of regional recordings at high frequencies as discussed above has the advantage of being applicable even if no suitable reference event is available, and will become progressively more important for the International Monitoring System as the density of the station deployment increases.

Conclusions

The station APA is situated at a regional distance from the Novaya Zemlya test site, and has a long history of surface wave recordings of nuclear explosions from there, all prior to being converted to digital operation. We have demonstrated the quality of the analog recordings at this station by comparing recordings from a modern broadband seismometer at the same place as signals digitized from the analog equipment.

We further show that the APA surface-wave recordings, normalized for distance and magnitude, provide an encouraging degree of separation between earthquakes and explosions in the European Arctic. We demonstrate that this separation can be achieved in a wide frequency band (at least 10–25 seconds period). We note that this gives promise for applying the $M_s:m_b$ discriminant down to lower magnitudes and at lower signal periods than is possible using teleseismic recordings. We also note that the shorter-period energy available in surface waves recorded at regional distances can be exploited in improving the monitoring capabilities during periods with strong interfering surface waves from large distant earthquakes.

In order to further develop the $M_s:m_b$ discriminant for event screening purposes, it will be necessary to study extensive historical recordings of nuclear explosions and earthquakes in various tectonic regions. Fortunately, many of the stations in the emerging International Monitoring System (IMS) have retained such recordings, but nevertheless the majority of IMS stations were not established at the time when most of the historic nuclear explosions were conducted. The event screening criteria must therefore be developed based to a large extent on non-IMS data, including available high-quality stations (such as APA) with a long history of analog recording. Additional work is also required in regionalization of the propagation paths to allow for the major tectonic features in calibrating the monitoring network.

Furthermore, since event magnitudes are important in most of the proposed screening criteria, the problem of computing magnitudes of historic seismic events in a way compatible with the current magnitude calculations must be addressed. We plan to develop a more general application of the maximum-likelihood method of RINGDAL (1986), and to compare the derived values to coda-based estimates (MAYEDA, 1993) and to other available magnitude estimators.

REFERENCES

ASMING, V., KREMENETSKAYA, E., and RINGDAL, F. (1998), *Monitoring Seismic Events in the Barents/Kara Sea Region*, Semiannual Technical Summary 1 October 1997–31 March 1998, NORSAR Sci. Rep. 2-97/98, Kjeller, Norway.

BUNGUM, H., HUSEBYE, E. S., and RINGDAL, F. (1971), *The NORSAR Array and Preliminary Results of Data Analysis*, Geophys. J. R. Astr. Soc. *25*, 15–26.

LOMAKIN, V. S. and YUNUSOV, Ph. (1993), *Effective methods for seismological observations in mines*. In: *Prediction and Prevention of Rockbursts in Ore Deposit*. Apatity, Kola Science Centre, Mining Institute, pp. 73–76 (in Russian).

MARSHALL, P. D. and BASHAM, P. W. (1972), *Discrimination Between Earthquakes and Underground Explosions Employing an Improved M_s scale*, Geophys. J. R. Astr. Soc. *28*, 431–458.

MARSHALL, P. D., STEWART, R. C., and LILWALL, R. C. (1989), *The Seismic Disturbance on 1986 August 1 near Novaya Zemlya: A Source of Concern?* Geophys. J., *98*, 565–573.

MAYEDA, K. M. (1993), m_b (Lg Coda): *A Stable Single-sensor Estimator of Magnitude*, Bull. Seismol. Soc. Am. *83*, 851–861.

MIKHAILOV, V. N. *et al.* (1996), *USSR Nuclear Weapons Tests and Peaceful Nuclear Explosions, 1949 through 1990*, RFNC – VNIIEF, Sarov, 1996, 63 pp.

MURPHY, J. R., STEVENS, J. L., BENNETT, T. J., BARKER, B. W., and COOK, R. W. (1998), *Development of improved magnitude measures for the prototype International Data Center (PIDC)*, Proc. 20th Annual Seismic Research Symposium on CTBT Monitoring, pp. 369–378.

RINGDAL, F. (1986), *Study of Magnitudes, Seismicity and Earthquake Detectability Using a Global Network*, Bull. Seismol. Soc. Am. *76*, 1641–1659.

RINGDAL, F. (1997), *Study of Low-magnitude Seismic Events near the Novaya Zemlya Nuclear Test Site*, Bull. Seismol. Soc. Am. *87*(6), 1563–1575.

RINGDAL, F., KREMENETSKAYA, E., and ASMING, V. (2002), *Observed Characteristics of Regional Seismic Phases and Implications for P/S Discrimination in the European Arctic*, Pure appl. geophys. *159*(4).

SERENO, T., BRATT, S., and BACHE, T. (1988), *Simultaneous Inversion of Regional Wave Spectra for Attenuation and Seismic Moment in Scandinavia*, J. Geophys. Res. *93*, 2019–2035.

SULTANOV, D. D., MURPHY, J. R., and RUBINSTEIN, Kh. D. (1999), *A Seismic Source Summary for Soviet Peaceful Nuclear Explosions*, Bull. Seismol. Soc. Am. *89*(3), 640–647.

VANEK, J., ZATOPEK, A., KARNIK, V., KONDORSKAYA, N. V., RIZNICHENKO, Y. V., SAVARENSKY, Y. F., SOLOV'EV, S. L., and SHEBALIN, N. V. (1962), *Standardization of Magnitude Scales*, Bull. (Izvest.) Acad. Sci. USSR, Geophys. Ser., *2*, 108.

(Received June 23, 1999, revised July 11, 2000, accepted July 18, 2000)

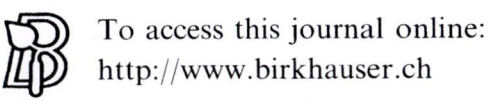

To access this journal online:
http://www.birkhauser.ch

Pure appl. geophys. 159 (2002) 735–757
0033–4553/02/040735–23 $ 1.50 + 0.20/0

|Pure and Applied Geophysics

Identification of Earthquakes and Explosions Using Amplitude Ratios: The Vogtland Area Revisited

KARL KOCH[1] and DONAT FÄH[2]

Abstract — Identification of seismic events is a major scientific issue in the framework of verification of the Comprehensive Nuclear-Test-Ban Treaty (CTBT). Of special interest in this context is the identification of the numerous low-yield mining or blasting events, especially those occurring in the same area as earthquakes, such as the Vogtland area in the border region of Germany and the Czech Republic. Seismic events in this area were investigated by WÜSTER (1993, 1995), who achieved complete discrimination using measures of spectral decay and spectral variance at the GERESS array and a quadratic discrimination function.

A subset of these events, for which ground-truth information is available, has been analyzed in this study using multivariate statistical analysis. Various parameters based on measurements from seismic waveforms of the broadband stations of the German Regional Seismic Network (GRSN) and short-period elements of the GERESS array are tested for statistical significance in a linear regression analysis, in particular spectral amplitude ratios for the L_g phase and P_g/S_g amplitude ratios. The subset includes a total of 35 explosions and 24 earthquakes. The results of our study argue that identification based on spectral L_g and high-frequency P_g/S_g ratios is promising. However, discrimination success is strongly varying from station to station; thus, weighting according to station success rates could improve the overall identification capability.

Key words: CTBT verification, seismic identification, earthquakes, explosions, amplitude ratio.

1. Introduction

Identification of seismic events poses a considerable problem in the framework of verification of the Comprehensive Nuclear-Test-Ban Treaty (CTBT). Therefore, the CTBT contains provisions to perform event screening of the numerous events detected by the International Monitoring System (IMS), although most of the procedures are and will be geared towards identification of the overwhelming number of large events (i.e., magnitude $m_b = 4$ and above) at teleseismic distances, such as, for example, by the m_b:M_s criterion, since the IMS is a sparse teleseismic network. Low-magnitude events, i.e., events closely above the detection threshold of the IMS,

[1] Federal Institute for Geosciences and Natural Resources, Stilleweg 2, D-30655 Hannover, Germany.
E-mail: koch@sdac.hannover.bgr.de
[2] Swiss Seismological Service, ETH-Hoenggerberg, CH-8093 Zurich, Switzerland.
E-mail: faeh@seismo.ifg.ethz.ch

will be recorded only by a few stations at regional distances, for which several discriminants are still under closer scrutiny (WALTER *et al.*, 1994; GOLDSTEIN, 1995; PATTON and TAYLOR, 1995; HARTSE *et al.*, 1997; TAYLOR and HARTSE, 1998). Suspicious events being of major concern to CTBT verification (e.g., VAN DER VINK and WALLACE, 1996; RINGDAL, 1997; HARTSE, 1998) are expected to fall into or even below this low-magnitude range.

Of special interest within the CTBT context, however, could be the identification of the numerous low-yield mining or blasting events, which may be the focus of allegations and debates (VAN DER VINK and WALLACE, 1996; HENGER *et al.*, 1997; RICHARDS and KIM, 1997; HEDLIN, 1998). Particular interest may be directed to explosions occurring in the very same area as earthquakes, such as the Vogtland area in the border region of Germany and the Czech Republic, since in case of available ground-truth information new identification approaches can be tested on related seismic or acoustic data. Seismic event identification in this area was extensively studied by WÜSTER (1993, 1995), who tested a number of identification parameters such as P/S amplitude ratios, spectral cross-correlation and time-independent structures in sonograms, autoregressive-moving average (ARMA) models of L_g spectra, as well as measures of spectral decay and roughness, called spectral variance hereafter, determined from waveform data of the GERESS array. In some of his work (WÜSTER, 1995), when using variance and decay calculated from stacked P- and S-wave spectra of GERESS recordings, discrimination of Vogtland events was completely achieved (Fig. 1a). These events form the basis of the analysis presented herein, because they provide detailed ground-truth information, which is not readily available for events in other parts of Germany.

A subset of these events (Fig. 1b) has been analyzed in this study using multivariate statistical analysis in order to evaluate the statistical significance and identification capability of various identification parameters based on measurements from seismic waveforms of the broadband seismic stations of the German Regional Seismic Network (GRSN) and the short-period GERESS array. In particular, these identification parameters are spectral amplitude ratios for the S_g phase, called synonymously 'spectral L_g ratios' throughout the text, and P_g/S_g amplitude ratios determined for a number of frequency bands. The subset of events is determined by the start of operations of the GRSN in mid-1991. Between July 1991 and September 1993, a total of 35 explosions and 24 earthquakes, with local magnitudes between $M_L = 1.0$ and $M_L = 2.5$, were recorded with sufficient signal-to-noise ratio. Similar waveform-based parameters were determined for a representative set of GERESS elements (GEA1, GEA2, GEC2, GEC5, GEC7, GED1, GED4, GED7) to provide comparisons with results from the GRSN stations.

This paper first focuses on the statistical aspects of the success of WÜSTER's identification parameters by means of the same multivariate statistical analysis which is later applied to spectral amplitude ratios. The Vogtland events from our data set are then investigated by spectral L_g ratios (KOCH, 2002) and, in addition, by P_g/S_g

Vogtland Events (Wüster,1995)

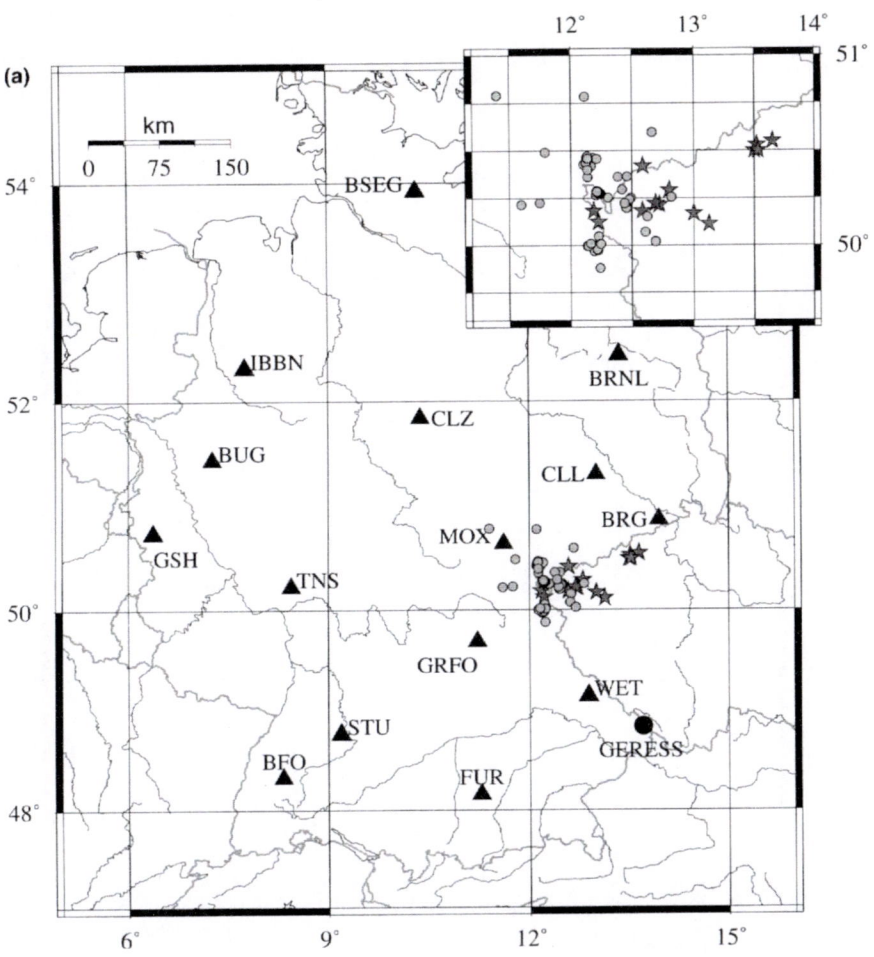

Figure 1

(a) Location of events used by WÜSTER (1995) for discrimination between earthquakes (circles) and explosions (stars) in the Vogtland region. A total of 77 earthquakes, frequently clustered due to earthquake swarm activity, and 69 explosions that are confined to about a dozen quarry and mining locations were investigated in his study. (b) Location of the subset of events available for this study – due to onset of GRSN operations not before mid-1991, only 35 explosions and 24 earthquakes from WÜSTER's data set were available for this investigation.

ratios using a scoring scheme as outlined in the related work. The following multivariate statistical analysis deals in particular with the separation of earthquake and explosion populations as expressed by a quantity called standard distance, which

Vogtland Events (This Study)

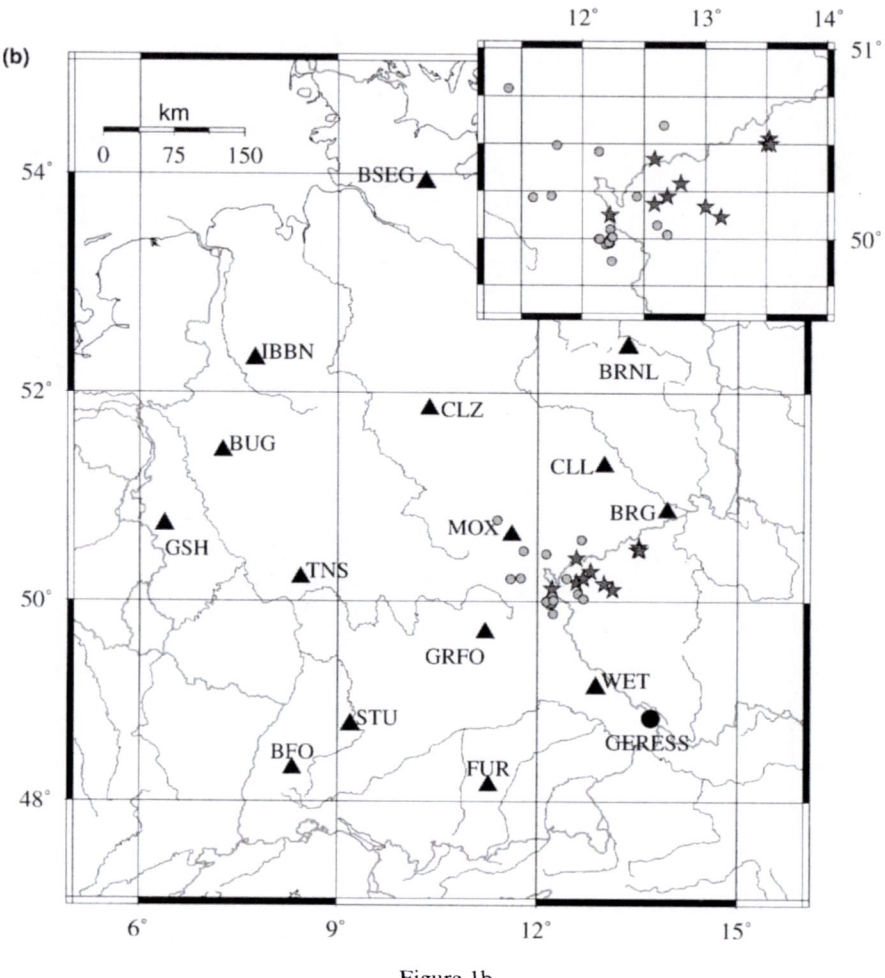

Figure 1b

is a measure of the group means of each population scaled to their standard deviations.

2. The Success of Parameters Derived from Stacked Amplitude Spectra in Identifying Vogtland Events

The problem of identification of explosions and earthquakes in the Vogtland region was studied by WÜSTER (1995), and complete discrimination was achieved by determining a quadratic discrimination function based on four parameters measured in stacked amplitude spectra at the GERESS array: spectral decay and spectral variance of *P*- and *S*-wave spectra. Stacking was introduced to reduce the effects of spectral roughness usually obtained when calculating simple FFT spectra from single records. Spectral decay is determined as the slope of a regression line fit to the spectra (Figs. 2 and 3), while spectral variance is a measure of the misfit or, equivalently, the area between the stacked spectrum and this regression line (black areas). WÜSTER (1995) argued that the smaller spectral decay as well as the reduced spectral variance of earthquakes compared to explosions are the key factors to obtain complete separation of the two populations, based on the experience gained from a training data set of 21 explosions and 30 earthquakes. Figures 2a,b provide examples of each of the two cases, with earthquakes manifesting smoother spectra, while explosions indicate more pronounced peaks and troughs as, for example, was also found by HEDLIN (1998) to be typical for mining explosions in other regions of the world.

As we intend (see next section) to use spectral L_g ratios and high-frequency amplitude ratios of *P* and *S* waves for event identification at local and regional distances, it should be instructive to first investigate the multivariate significance among these four parameters, because spectral L_g ratios are a gross measurement of spectral decay of *S*-wave spectra. Hence, we have determined multivariate as well as univariate standard distances for the four parameters used by WÜSTER (1995) according to multivariate statistical analysis (FLURY and RIEDWYL, 1988). The standard distance between two populations is also called Mahalanobis distance when squared. Defined as the difference between group means divided by the pooled standard deviation of both groups (FLURY and RIEDWYL, 1988, p. 15), a value of the standard distance larger than 4 indicates, as a rule of thumb, statistically significant separation of two populations. The corresponding results for WÜSTER's identification data are summarized in Table 1. The standard distance of about 4, including all four discrimination parameters, reflects the almost complete discrimination of earthquakes and explosions in the Vogtland area when using a linear discrimination function. To achieve complete discrimination a quadratic discrimination function was thus introduced by WÜSTER (1995). On the other hand, when the univariate case for the individual parameters is examined, it becomes evident that the spectral variance of the *S*-wave spectra is the parameter most promising for population separation, as the standard distance is rather close to the standard distance obtained in the multivariate case. The spectral variance of *P* is the next most significant parameter, while standard distances for the decay of *P*- and *S*-wave spectra must be considered poor identification parameters. However, it must be argued that propagation path effects dominate these two parameters since no distance correction was applied by WÜSTER (1995) to remove the effects of anelastic attenuation. According to KOCH (2002) spectral L_g ratios from GRSN data show a clear distance scaling, thus *P*- and *S*-wave spectra in the frequency band between 3 and 15 Hz, as

used by WÜSTER (1995) in his analysis, may be severely affected; and this effect will be
similar for earthquakes and explosions.

For the multivariate case, we further investigated the significance of all
parameters by forward and backward elimination of parameters in the regression

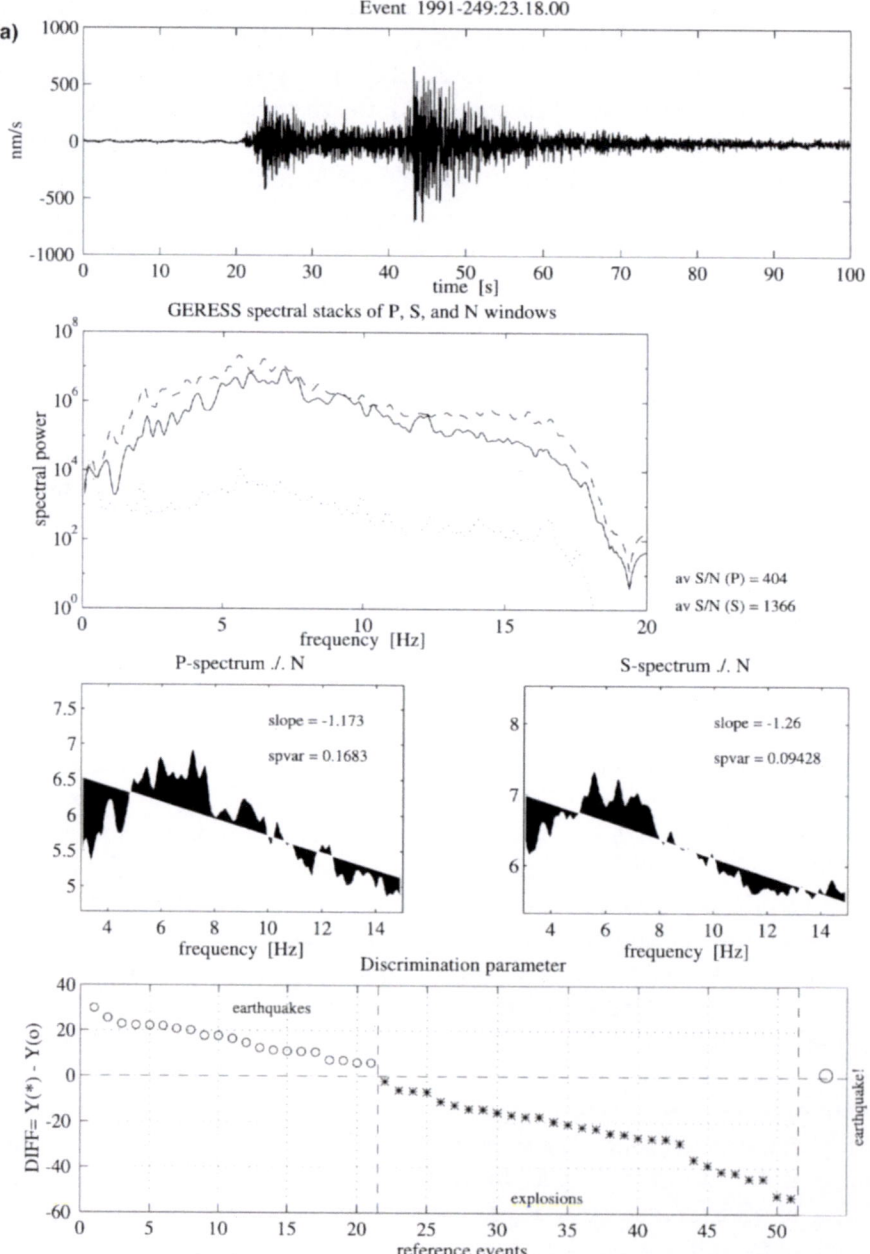

analysis. We found that the spectral variance of the *P*-wave spectra is a redundant parameter, as it is only eliminated when the spectral variance of *S* waves is used in the regression. Conversely, from forward elimination, i.e., when parameters are added according to statistical significance, it followed that the parameters of spectral decay are rather weak, as is reflected in the small standard distances in the univariate case discussed earlier, and that the spectral variance of the *S*-wave spectrum is the parameter with the strongest identification power. The implication from this result can easily be demonstrated by analyzing an earthquake that is incorrectly classified by WÜSTER's method. In Figure 3 it is obvious that the strong modulation in the *S*-wave spectrum, most likely caused by a corner frequency effect, results mainly in a larger spectral variance of *S*, which is typical for a mining explosion (e.g., HEDLIN, 1998) and therefore leads to the misclassification of this particular earthquake.

3. Spectral L_g Ratios and P_g/S_g Amplitude Ratios

In related work on seismic event identification, KOCH (2002) has investigated the effectiveness of spectral L_g ratios for identification of seismic events in Germany. Spectral phase ratios have also been investigated by MURPHY and BENNETT (1984) and TAYLOR (1996) for discrimination between nuclear explosions and earthquakes in the western U.S. Spectral L_g ratios are calculated in these studies from either maximum amplitudes or the averaged spectral level in different frequency bands. In this study we follow the procedure of KOCH (2002) and calculate the spectral L_g ratio from the (spectral) amplitude in the 1–2 Hz frequency band divided by this amplitude in a higher frequency band, where we used both the 6–8 Hz and 7–9 Hz frequency bands, with the latter even containing the antialiasing corner frequency of 8.5 Hz applied at GRSN stations. The second frequency band was used to investigate whether a tendency can be observed that data with frequencies higher than 6–8 Hz provide better identification capability. To determine the spectral amplitudes, the velocity waveform data were bandpass-filtered according to the frequency bands and (spectral) amplitude was measured as the maximum amplitude in a 5 sec window embracing the L_g arrival, which was picked during analyst review. Because a standard bandpass of 1–8 Hz is used during routine analysis, and all phases were

◄

Figure 2

Typical earthquake (a) and explosion (b) recordings (at GEC2) and stacked (velocity) spectra used in WÜSTER's (1995) discrimination method. The second subfigures from the top show the stacked spectra of *P* (solid), *S* (dashed), and the pre-*P* noise (dotted) from 12.8 sec time windows each for the frequency range from 0 Hz to the Nyquist frequency. The subfigures below show the noise-corrected *P* and *S* spectra for the frequency band from 3 to 15 Hz, for which the estimate of spectral slope and spectral variance was determined to avoid complications due to the instrument response. The bottom figures show the difference in Mahalanobis distances for the particular event to the earthquake (o) and explosion (*) populations. A negative Mahalanobis difference implies an explosion, while a positive value indicates an earthquake.

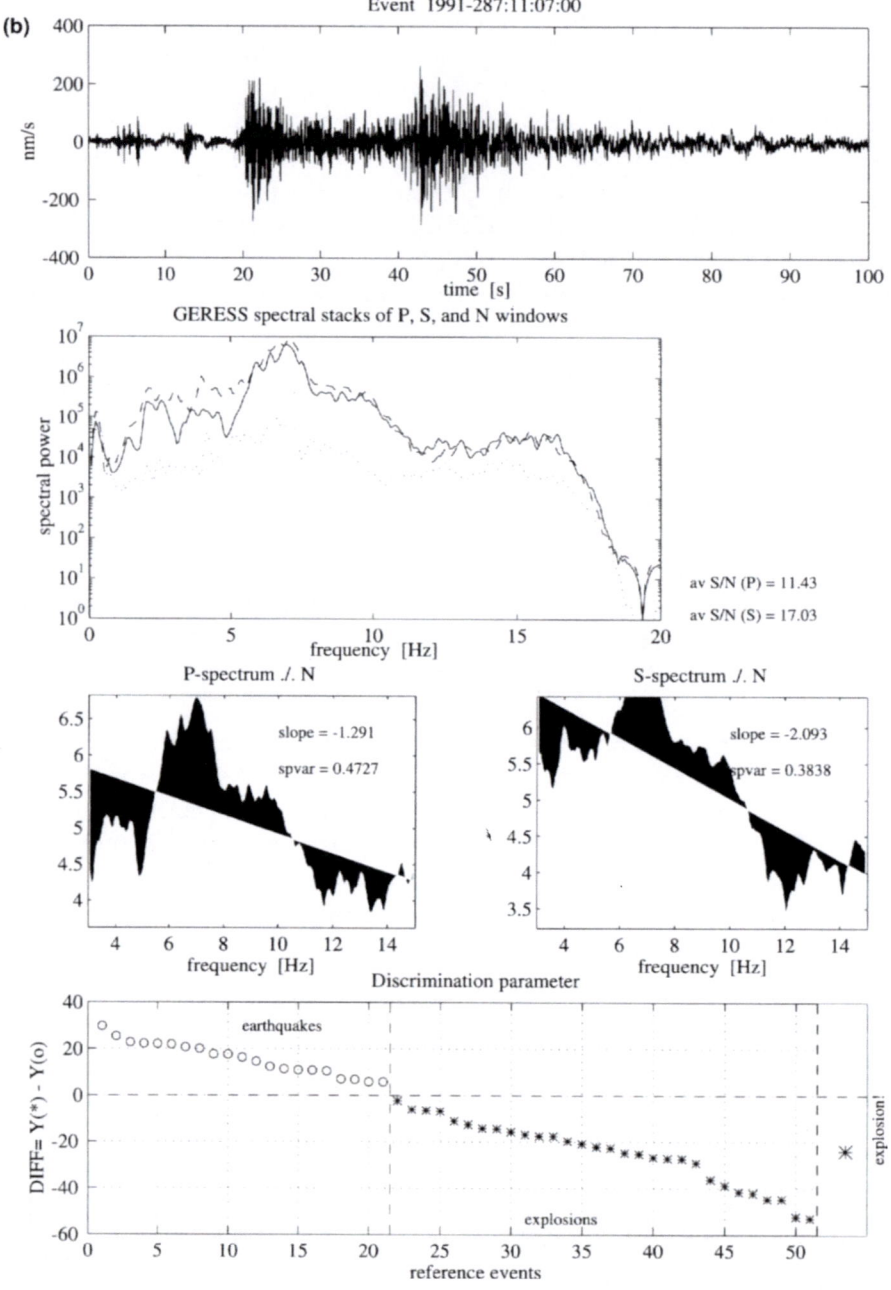

Figure 2b

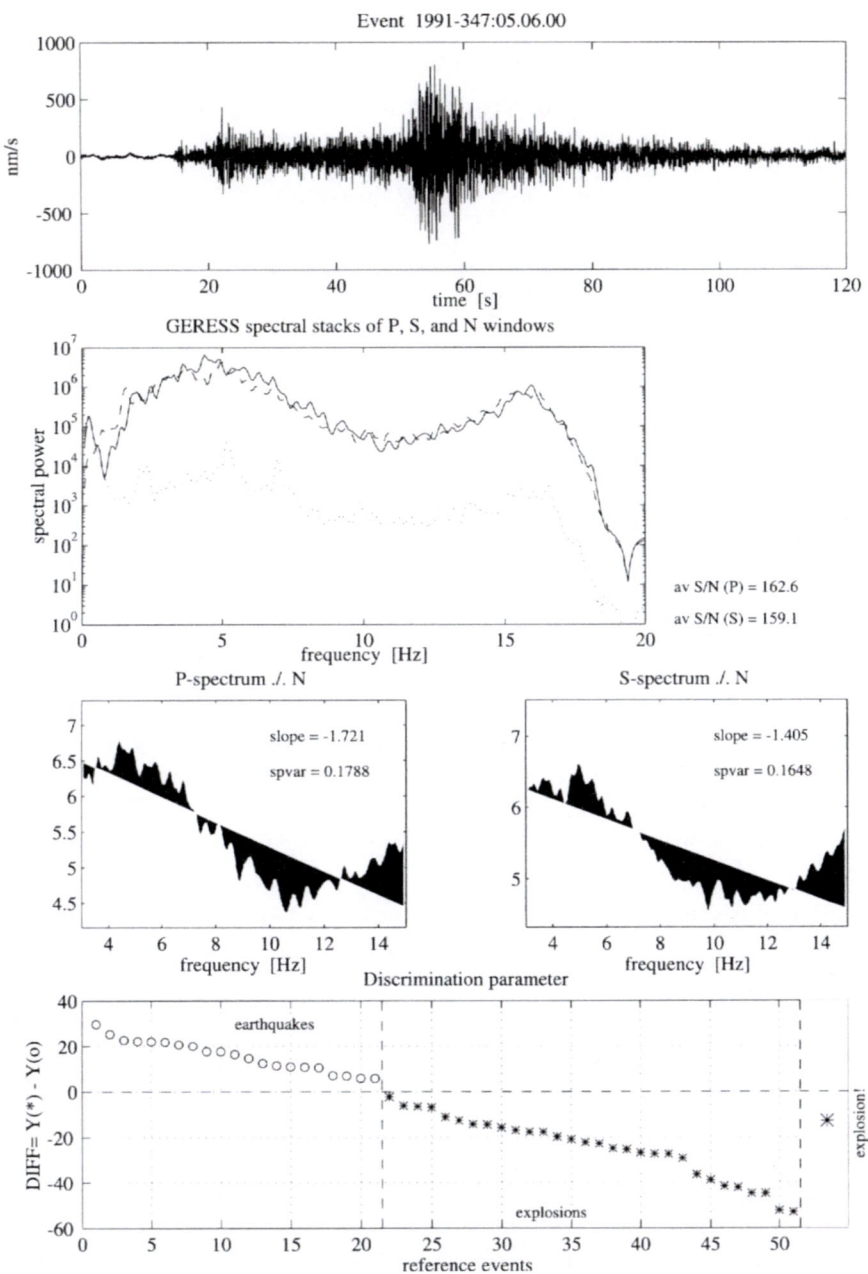

Figure 3

Atypical earthquake record showing strong modulations of the signal spectra most likely caused by corner frequency and propagation path effects. Due to the larger spectral variance for S and the relative importance of this parameter in the classification method, the event is incorrectly classified as an explosion. (For further explanations, see Figure 2).

Table 1

Performance of the identification parameters used by WÜSTER (1995) for identification of earthquakes and mining explosions in the Vogtland region, investigated for univariate and multivariate cases based on the training data set from 21 earthquakes and 30 explosions. Checks indicate the parameter used in each case. LDF indicates the overall discrimination rate (success) when a linear discriminant function is used, while QDF represents the case of a quadratic discriminant function. Numbers in parentheses give the success rate for explosions: earthquakes.

Case	Spectral decay of P waves	Spectral decay of S waves	Spectral variance of P waves	Spectral variance of S waves	Standard distance	LDF	QDF
1	✔	✔	✔	✔	3.93	98% (100:97)	100% (100:100)
2	✔	✔		✔	3.92	98% (100:97)	100% (100:100)
3		✔		✔	3.44	98% (100:97)	98% (100:97)
4	✔			✔	3.35	96% (100:93)	98% (100:97)
5				✔	3.34	96% (100:93)	96% (100:93)
6			✔		2.47	90% (100:83)	86% (90:83)
7		✔			1.25	76% (81:73)	80% (90:73)
8	✔				0.65	63% (71:57)	69% (86:57)

picked manually, a sufficiently high signal-to-noise ratio was assumed for all measurements and no further signal-to-noise criteria were imposed.

We have applied this L_g ratio method to two different data sets for the events of the Vogtland region. One data set comprised data from the GRSN stations, while the other data set was obtained from the three-component elements of the GERESS array. Various measurements of the spectral L_g ratio were taken, as displayed in Figures 4a–b, including single component ratios as well as three-component estimates to identify the most prospective of these parameters by the multivariate statistical analysis which follows below. From Figure 4, where it is important to note for Figure 4a that data at certain distances are related to certain stations due to the relatively small size of the Vogtland area, it can be concluded that none of the parameters is superior for separating the two populations. However, there does seem to be a different level of discrimination capability for different stations.

For the purpose of testing the identification method of KOCH (2002) on the Vogtland ground-truth events, however, we used only the 6–8 Hz ratio from the vertical component. KOCH (2002) found this identification parameter to be, after proper distance correction to compensate for the effects from attenuation, a prospective candidate for screening out many of the explosion and blasting events detected in Germany and adjacent areas from the earthquake population. This identification method uses station-dependent thresholds to classify the station data as earthquake, explosion, or "unknown." The application of a scoring scheme to all station classifications yields then a network-averaged classification of an event.

The results of this identification method, when applied to the events of the Vogtland region for both data sets from the GRSN network and the GERESS array

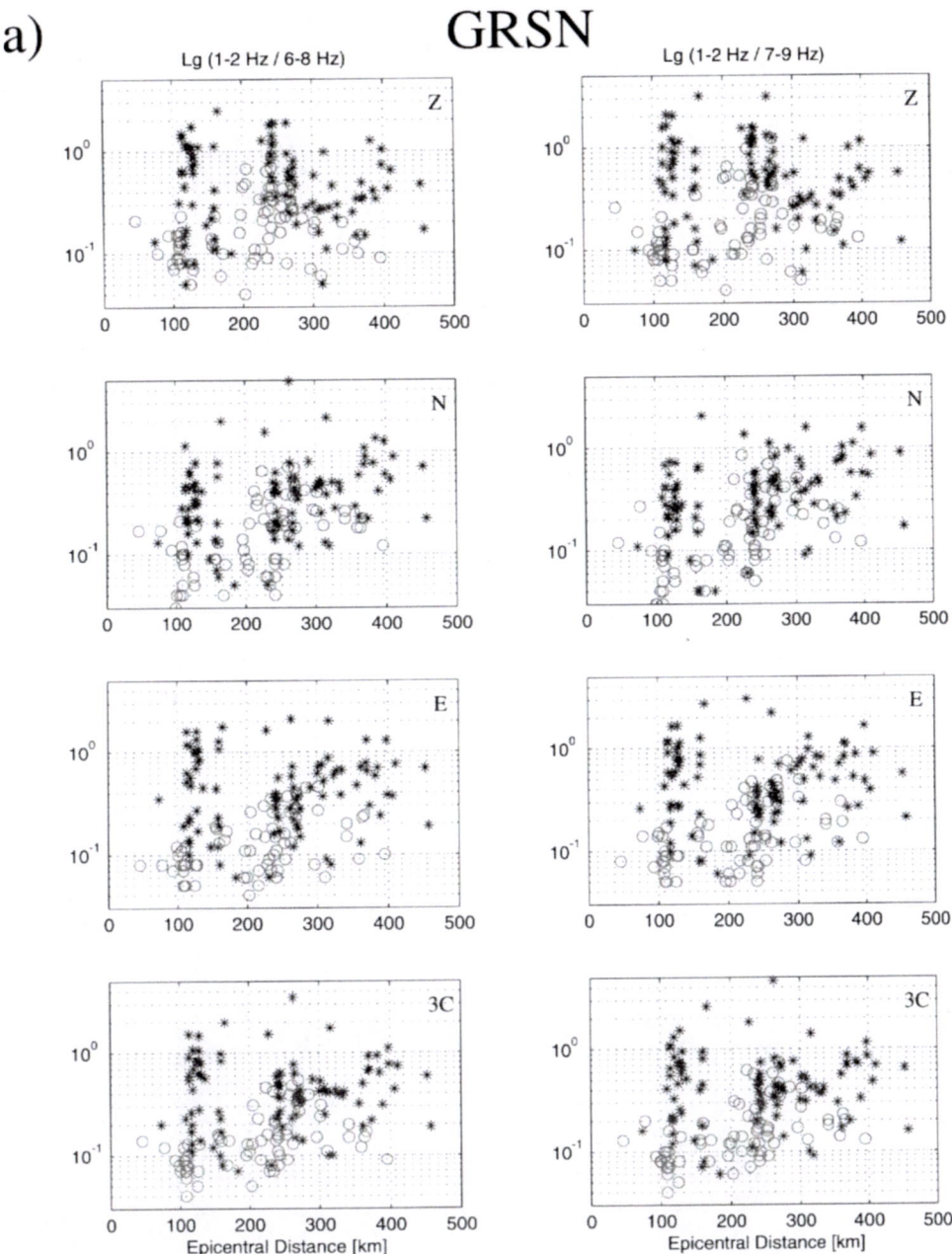

Figure 4

Spectral L_g ratios estimated for different components (top to bottom: Z – vertical; N – north/south; E – east/west; 3C – vectorially summed three-component measurement) and frequency bands (left: L_g ratio (1–2 Hz/6–8 Hz); right: L_g ratio (1–2 Hz/7–9 Hz)) at GRSN (a) and GERESS (b) stations. Data from earthquakes are shown by circles and data from explosions are represented by asterisks.

b) GERESS

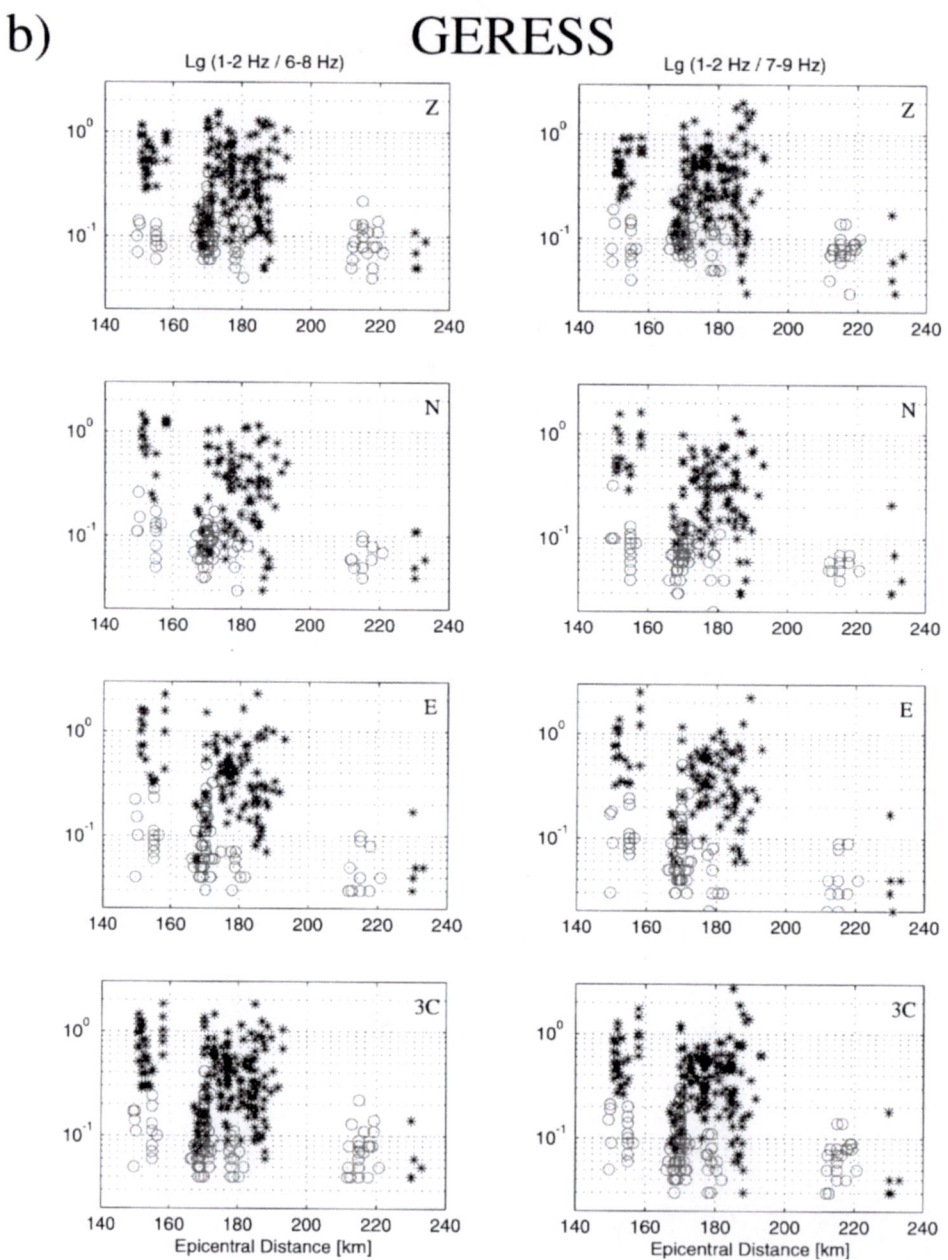

Figure 4b

stations, are shown in Figures 5a–b. All of the earthquakes except for two ambiguous cases are identified by the GRSN data, while all these events were correctly classified with GERESS L_g data. Furthermore, many of the explosions are correctly classified

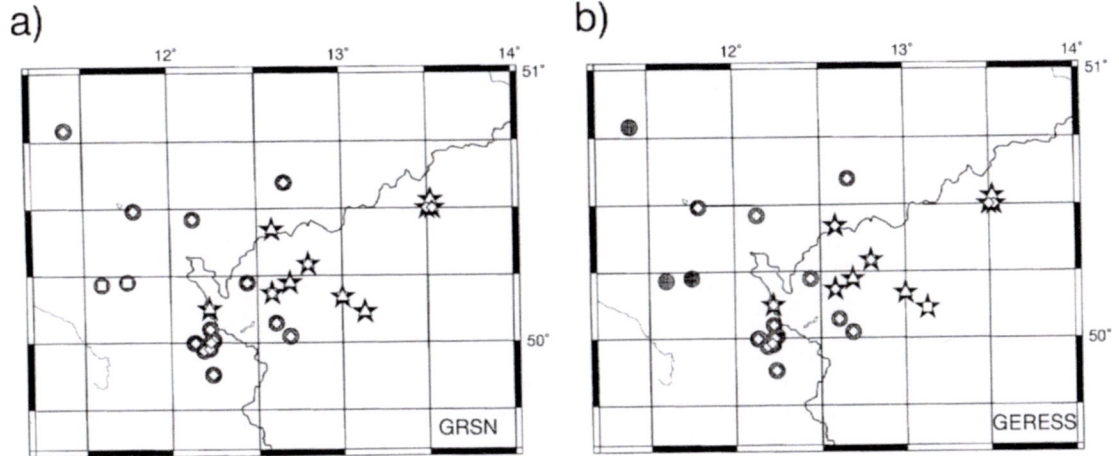

Figure 5

Identification analysis of the Vogtland events using the L_g ratio method and GRSN (a) or GERESS (b) data – most earthquakes (solid circles) are identified correctly, however, a considerable number of explosions (stars) (and in particular those from specific blasting sites) are incorrectly classified. Classifications by the L_g ratio method are as follows: earthquakes (◊), explosions (Δ), and undecided (unknown) cases (□). (Note: for graphical representation the order of plotting was chosen according to symbols (□) first and symbols (◊) last.)

by both data sets (GRSN 52% , GERESS 48%). However, it is also obvious that both data sets consistently misclassified explosion events from two mining sites in the northeastern part of the study area. In addition, spectral L_g ratio data from GERESS stations for events from two explosion sites in the central part have no capability of identifying these explosions. Combining the two data sets improves these results only slightly.

As an extension of the spectral L_g ratio method, we have further investigated the use of high-frequency amplitude ratios of regional P and S waves. Due to the geographical locations of events and stations, the most useful amplitude ratio was identified to be the P_g/S_g ratio, as many GRSN and all GERESS stations show considerable P_g energy, while P_n is often a weak and noisy arrival due to the magnitude range ($1.0 < M_L < 2.5$) of the events studied. Amplitude ratios studied were the broadband P_g/S_g, where the data were filtered with a high-pass filter with corner frequency at 0.5 Hz, and two high-frequency ratios determined in the 6–8 Hz and 7–9 Hz frequency bands. The measurements were taken accordingly to the L_g ratios, in that we determined maximum amplitudes of the bandpass-filtered waveforms in a 5 sec window including the P_g or S_g arrivals. The corresponding parameters are shown in Figures 6a–b separately for GRSN and GERESS data. For these data, no distance correction was applied, as, contrary to the case of L_g ratios, no obvious distance dependency is recognizable, such as a systematic increase or decrease of earthquake and explosion data with distance. It seems obvious from these

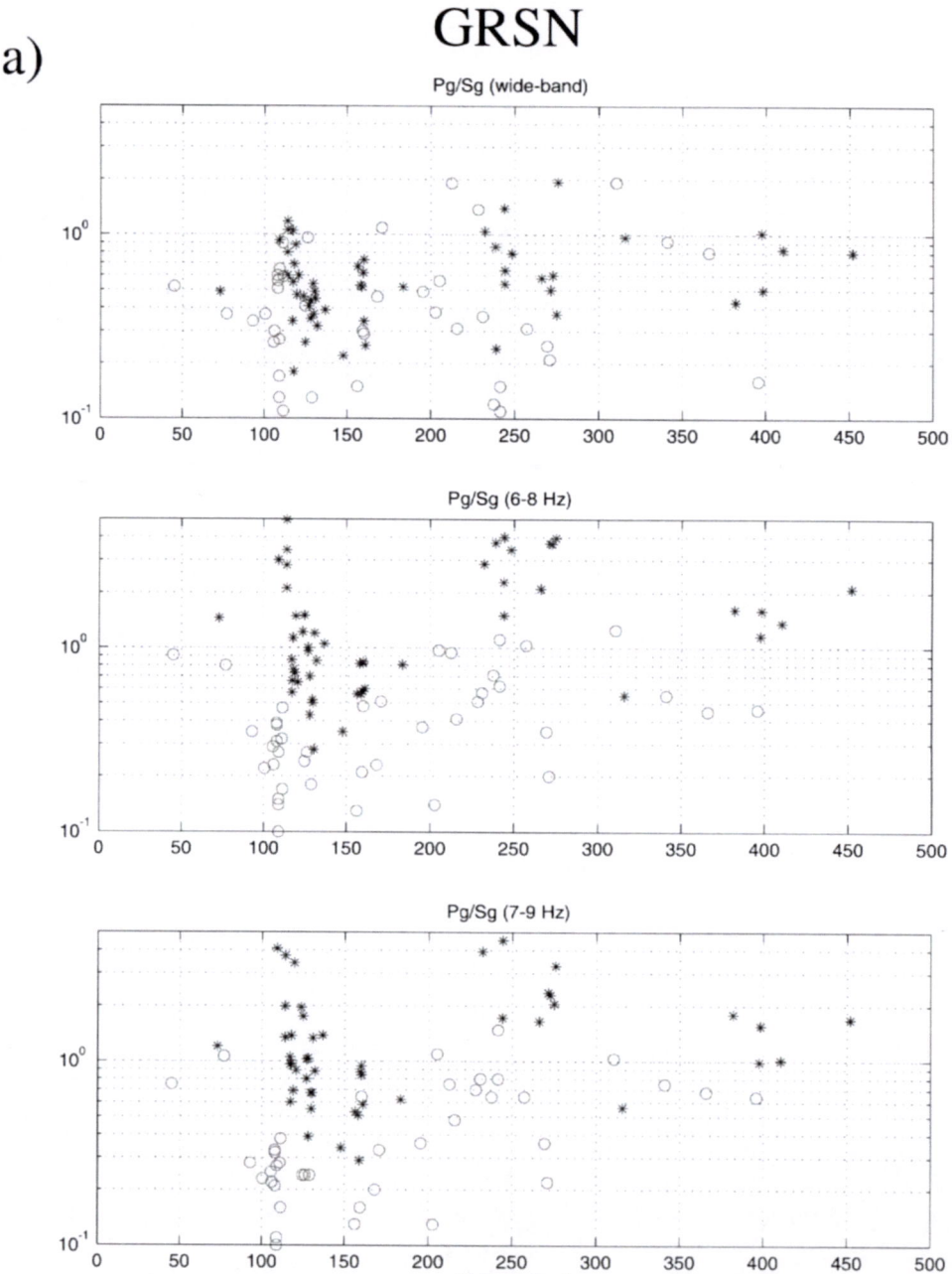

Figure 6

P_g/S_g ratios estimated in a wide frequency band as well as in the frequency bands 6–8 Hz and 7–9 Hz at GRSN (a) and GERES (b) stations. While earthquake (circles) and explosion (asterisks) populations are not well separated in the wide-band case, populations are more clearly separated in the high-frequency bands.

GERESS

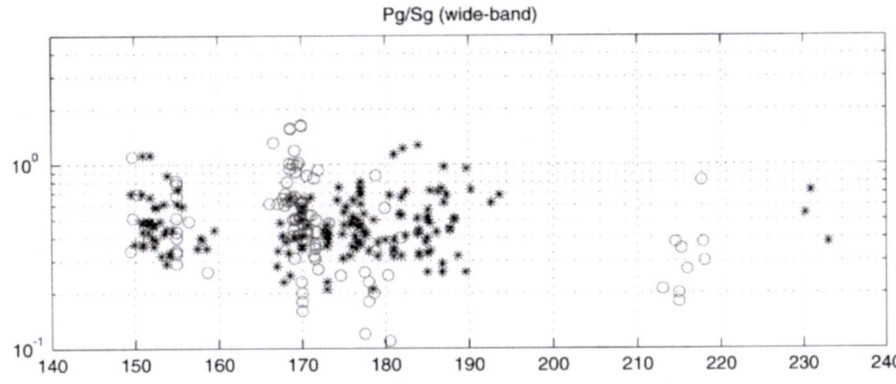

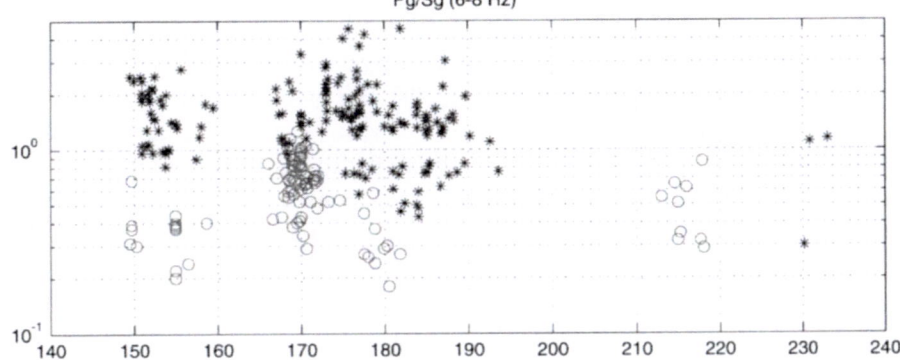

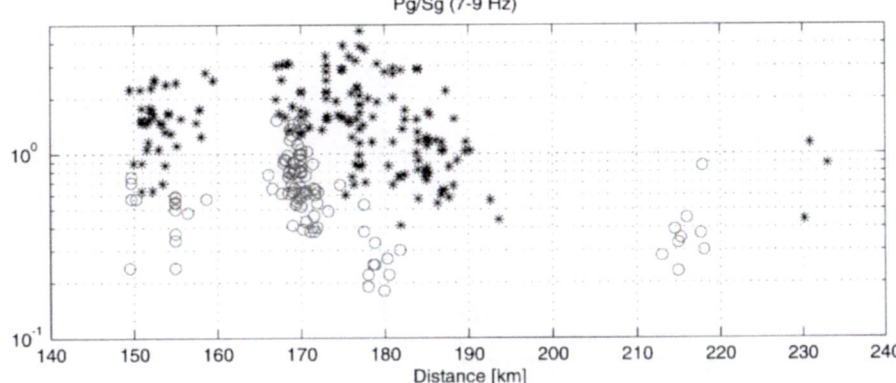

Figure 6b

figures that the broadband ratio may have no discrimination capability, as populations do not separate at all, which is consistent with WÜSTER's (1995, p. 77) findings that Vogtland events are not discriminated well with this identification parameter (Incidentally, WÜSTER used the same cut-off frequency as we did for his P/S ratios). However, looking at the high-frequency P_g/S_g ratios, the two populations indicate a fairly large amount of separation, with P_g/S_g ratios larger than 1 indicative for explosion population and ratios below 0.5 indicating earthquakes. Considering a possible distance dependency of the P_g/S_g ratios, it must be pointed out that apparent trends in Figure 6a may be related to individual stations which contribute the data at certain ranges and discriminate between the two populations at different thresholds, thus implying the use of station-dependent classification thresholds. Since GERESS stations show only slight variations of this feature, and as the number of Vogtland events is small, determination of station-dependent thresholds for GRSN station is left to future study.

Using a similar scoring scheme as applied in the spectral L_g method, applying the classification thresholds given above for all stations, we obtained consistent identification of earthquakes with the GRSN data and equally good results for GERESS data (Figs. 7a–b). Concerning the explosions, we have a reasonably good identification success for the GERESS data, with most events identified as explosions except for one site on the German/Czech border in the northern part of the Vogtland region. The same applies for the GRSN data, however to a lesser degree, as events from two explosion sites in the northeast are consistently classified as earthquakes. As the overwhelming number of explosion data is from only one station, WET, this

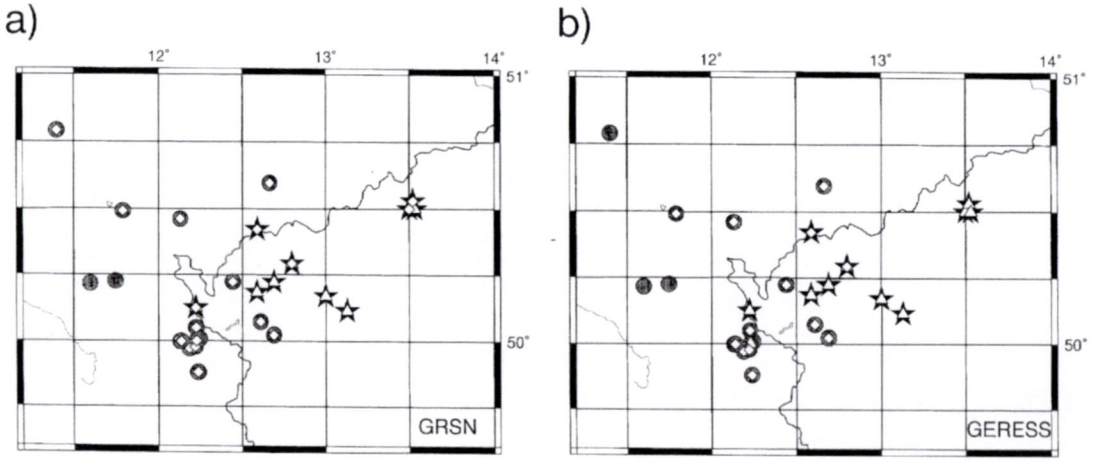

Figure 7

Identification analysis with the P_g/S_g ratio method using GRSN (a) and GERESS (b) data in the 6–8 Hz band – results are fairly similar as those obtained by the spectral L_g ratio method, in that almost all earthquakes are identified correctly and that there is a larger number of incorrectly classified explosions. Symbols and order of plotting as in Figure 5.

feature could be related to the source-receiver geometry with P-wave radiation to WET reduced or strong amplification of S waves. Similar arguments may apply for the one explosion site with respect to GERESS data, although the events from this explosion site are also consistently misclassified by the GRSN data.

4. Multivariate Statistical Analysis of Candidate Identification Parameters

In order to evaluate the significance of the various parameters shown in Figures 4 and 6, we have used a standard statistical analysis package to perform a multivariate regression analysis. In the case of a binary variable in the regression analysis (or a two-sample problem), the resulting multivariate linear regression model yields an equivalent linear discrimination function (FLURY and RIEDWYL, 1988).

A method of identification of redundant and uncorrelated regressors in the multivariate analysis is achieved by successive backward elimination (Figs. 8a–b). Starting from the full regression model, including all identification parameters to be tested, partial F-statistics provides for a test on the redundancy or lack of correlation of a particular regressor, successively eliminating the parameters with the smallest partial F-statistics in each step. Partial F-statistics is a measure for the increase of the variance, or equivalently, for the decrease in the standard distance, when eliminating a regressor. For example, when the change in the standard distance after eliminating a regressor is negligible (see Fig. 8), then this regressor is either redundant or uncorrelated, and, hence, not contributing significantly to the discrimination function.

This procedure was applied separately to the selected spectral L_g ratios and to the P_g/S_g ratios, because the database was substantially larger for the former data set, related to the fact that L_g is often the strongest phase in regional seismograms (e.g., BAUMGARDT, 1990; HANSEN et al., 1990). For spectral L_g ratios, the standard distances for both the GRSN and GERES data sets are nearly unaffected by four of the nine parameters tested. Elimination of four additional parameters results in a decrease of only about 20% in the squared multiple regression coefficient, translating into a reduction of the standard distance, which is about 1.5 for all parameters, by about 0.2. This reduction of the already small initial standard distance does not seem to be significant, especially as both data sets produce a different parameter in the final elimination step.

The results for the P_g/S_g ratios differ in several ways: the ratio determined for the 6–8 Hz frequency band is not eliminated in the regression analysis until the final step for both data sets, while the other regressors are eliminated in the same order, with the broadband P_g/S_g ratio eliminated first and the high-frequency ratio (7–9 Hz) eliminated last. For the P_g/S_g ratios, there is also a marked difference in the standard distance between the two data sets. While the standard distance is fairly small (1.25) for the GRSN data set, the standard distance is about 50% higher (1.8) for the

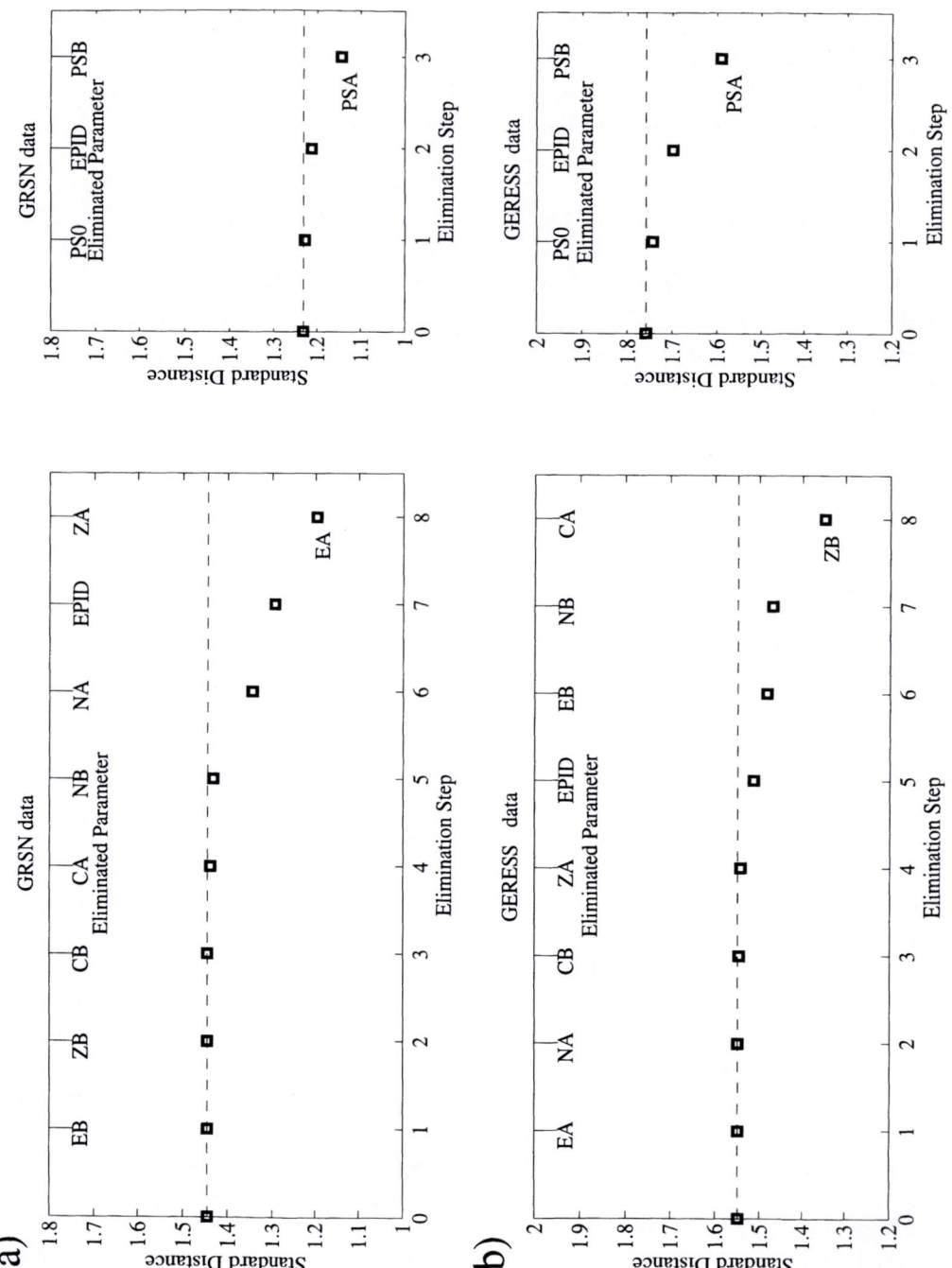

GERESS data, which explains the larger identification success of GERESS shown in Figure 6b.

It should further be noted that the epicentral distance is eliminated rather early in the elimination process, although the epicentral distance was considered a significant regressor by FÄH (1997) for shorter distances (< 100 km). This may be related to the fact that we have corrected the spectral L_g ratios for propagation path effects due to attenuation and have no clear indication for strong propagation path effects for P_g/S_g, when inspecting its distance dependency. In conclusion, the multivariate regression analysis suggests that for both spectral L_g ratios and high-frequency P_g/S_g amplitude ratios, one parameter performs nearly as well as any other; however, there seems to be some minor benefit of using more than one of the parameters simultaneously, as this coincidence measure may provide additional constraints to the identification of earthquakes and explosions in the Vogtland area.

5. Discussion

Motivated by the pattern found in Figures 8 that there is moderate variability in the standard distance of the P_g/S_g ratios for GRSN and GERESS data, we investigated individual station performance for the spectral L_g ratio and the high-frequency P_g/S_g amplitude ratio. For the spectral L_g ratio we selected the parameters as determined on the vertical component in the 6–8 Hz frequency band, which was also chosen for the P_g/S_g data.

For the GERESS data (Fig. 9b), the results show a slight variation for the L_g ratio standard distance for the different array elements. On the other hand, the P_g/S_g results show considerable variability between standard distances of 1.4 and 2.3. This indicates that GEC2 and GEC5 will perform considerably better than, for example, GEA2 or GED7. As GERESS has only an aperture of about 4 km, differences in radiation pattern are very unlikely to be responsible for this difference at epicentral distances exceeding 100 km. Serious factors to be considered for this effect are, therefore, the local site response and local noise conditions, which could cause a significant difference in the observed P_g and S_g amplitudes at these higher frequencies.

◄

Figure 8

Results of multivariate regression analysis with the identification parameters as shown in Figures 4 and 6 for the step-wise elimination procedure, step 0 indicating all tested parameters contributing to the regression. First letter in the left part of figures stands for component (i.e., Z, E , N, C for 3-C) and second letter (in both subfigures) for frequency band (i.e., A for 1–2 Hz/6–8 Hz and B for 1–2 Hz/7–9 Hz). PS in right figures indicates P_g/S_g ratio. EPID is the epicentral distance and PS 0 represents the wide-band P_g/S_g ratio. In each step, the labeled parameter (top axis of graph) is eliminated. The label at the last symbol represents the parameter not eliminated. (a) GRSN data set – (b) GERESS data set.

a)

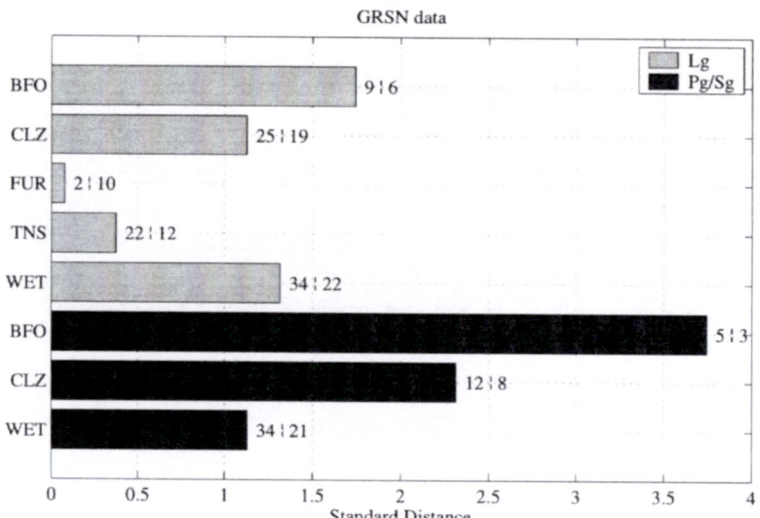

b)

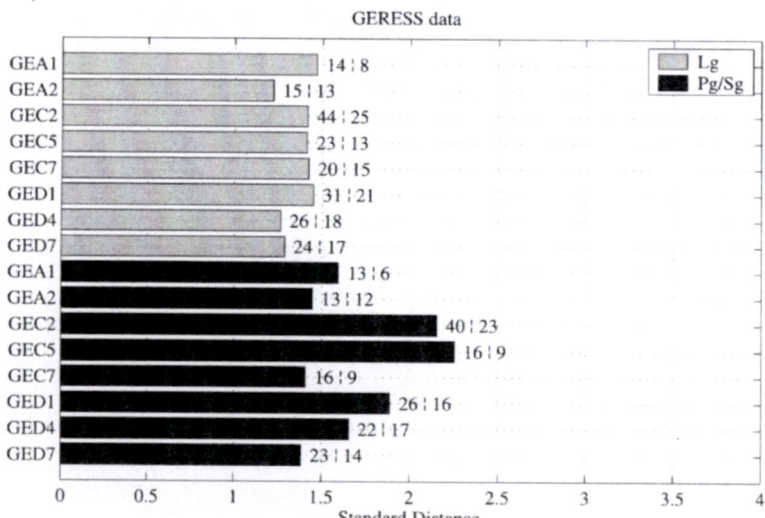

Figure 9

Station specific standard distances of spectral L_g ratios (1–2 Hz /6–8 Hz) and high-frequency P_g/S_g amplitude ratios for all GERESS (a) and GRSN (b) stations. Even across the small aperture GERESS array, differences of 20–50% occur. The numbers to the right of each bar indicate the number of data at a station based on the event type (i.e., explosions | earthquakes). Note: FUR and TNS did not contribute P_g/S_g ratios due to the lack of useful P_g observations.

For GRSN stations, the results shown in Figure 9a are contrary, as for both L_g parameters and P_g/S_g ratios the standard distance strongly varies from station to station. While BFO, CLZ, and WET perform at about the same level with respect to spectral L_g ratios as all GRSN stations combined (Fig. 8), FUR and TNS show very little discrimination capability, neglecting the fact that the statistical basis for FUR seems inadequate. This result appears to be reflected in Figure 4a, where populations between 200 and 300 km epicentral distance (i.e., the distances to FUR and TNS) are not well separated. As FUR is a fairly noisy station for frequencies beyond a few Hertz, much of this inability to discriminate earthquakes and explosions could be related to the local noise conditions. For the P_g/S_g ratio, the difference between stations seems even more intriguing, neglecting the very narrow database for BFO, which indicates that discrimination from BFO data could be complete, while WET performs rather poorly.

6. Conclusions

Seismic events in the Vogtland area were successfully discriminated by WÜSTER (1995) using a quadratic discrimination function based on four spectral parameters calculated from stacked spectra at the GERESS array, and these events provide a reference due to the available ground-truth information. By our multivariate analysis we have statistically re-examined the parameters of spectral decay and variance, and have found that spectral variance of the S-wave spectrum plays a major role, and that the remaining parameters are useful in making the discrimination success perfect. Thus, if spectral variance is atypical for the proper population, then the discrimination method most likely will fail.

Considering spectral amplitude ratios for L_g, which can be considered a coarse approximation to spectral decay of S waves, this work indicates that they provide more discrimination power than was found for WÜSTER's univariate S-decay data, and which may be related to the fact that WÜSTER (1993) did not correct for propagation path effects. Path corrections, besides correction for source effects, have been shown by PHILLIPS (1999) and TAYLOR and HARTSE (1998) to be essential for improved regional seismic discrimination. In addition, while broadband P_g/S_g ratios provide scant discrimination capability, high-frequency P_g/S_g ratios at frequencies higher than 6 Hz show some potential in discrimination of earthquakes and explosions in the Vogtland area, similar to the results obtained by TAYLOR (1996) for earthquakes and nuclear explosions in the western U.S.

Results obtained in this study indicate that identification success strongly varies from station to station. Further emphasis should thus be given to weighting according to station success rates in order to improve the overall identification capability of the GRSN (and to a lesser extent to GERESS) stations. Although this method may not be directly applicable to CTBT monitoring, owing to the magnitude range of typical Vogtland events, the results could be useful for the assessment of suspicious small-magnitude events (e.g., RINGDAL, 1997; RICHARDS and KIM, 1997) detected and recorded by regional and local networks, and which could be used for clarification with respect to CTBT compliance.

Acknowledgements

We would like to thank J. WÜSTER for helpful discussions and for providing his analysis program for the discrimination of earthquakes and explosions in the Vogtland area, and J. Schlittenhardt for reviewing the manuscript. We appreciate the valuable comments of T. J. Bennett and associate editor H. Hartse, as well as an anonymous reviewer, which helped to improve the manuscript. The multivariate statistical analysis was carried out using the SYSTAT® for Windows® software package.

REFERENCES

BAUMGARDT, D. R. (1990), *Investigation of Teleseismic L_g Blockage and Scattering Using Regional Arrays*, Bull. Seismol. Soc. Am. *80*, 2261–2281.

FÄH, D. (1997), *Discrimination Between Earthquakes and Chemical Explosions by Multivariate Statistical Analysis–A Case Study for Switzerland* (abstract), 29th General Assembly IASPEI, Thessaloniki, Greece, Aug. 18–29, 1997.

FLURY, B. and RIEDWYL, H., *Multivariate Statistics: A Practical Approach* (Chapman & Hall, London, 1998) 296 pp.

GOLDSTEIN, P. (1995), *Slopes of P- to S-wave Spectral Ratios – A Broadband Regional Seismic Discriminant and a Physical Model*, Geophys. Res. Lett. *22*, 3147–3150.

HANSEN, R. A., RINGDAL, F., and RICHARDS, P. G. (1990), *The Stability of RMS L_g Measurements and their Potential for Accurate Estimation of the Yields of Soviet Underground Nuclear Explosions*, Bull. Seismol. Soc. Am. *80*, 2106–2126.

HARTSE, H. (1998), *The 16 August 1997 Novaya Zemlya Seismic Event as Viewed from GSN Stations KEV and KBS*, Seism. Res. Lett. *69*, 206–215.

HARTSE, H., TAYLOR, S., PHILLIPS, W., and RANDALL, G. (1997), *A Preliminary Study of Regional Seismic Discrimination in Central Asia with Emphasis on Western China*, Bull. Seismol. Soc. Am. *87*, 551–568.

HARTSE, H., FLORES, R., and JOHNSON, P. (1998), *Correcting Regional Seismic Discriminants for Path Effects*, Bull. Seismol. Soc. Am. *88*, 596–608.

HEDLIN, M. A. H. (1998), *A Global Test of a Time-Frequency Small-Event Discriminant*, Bull. Seismol. Soc. Am. *88*, 973–988.

HENGER, M., KOCH, K., RUUD, B. O., and HUSEBYE, E. S. (1997), *Comments on "The Political Sensitivity of Earthquake Locations by van der Vink and Wallace"*, IRIS Newsletter *16*, 20–22.

KOCH, K. (2002), *Seismic Event Identification of Earthquakes and Explosions in Germany Using Spectral L_g Ratios*, Pure appl. geophys. (this volume).

PATTON, H. and Taylor, S. (1995), *Analysis of L_g Spectral Ratios from NTS Explosions: Implications for the Source Mechanisms of Spall and the Generation of L_g Waves*, Bull. Seismol. Soc. Am. *85*, 220–236.

PHILLIPS, W. S. (1999), *Empirical Path Corrections for Regional Phase Amplitudes*, Bull. Seismol. Soc. Am. *89*, 384–393.

RICHARDS, P. G. and Kim, W.-Y. (1997), *Testing the Nuclear Test-Ban Treaty*, Nature *389*, 781–782.

RINGDAL, F. (1997), *Study of Low-Magnitude Seismic Events near the Novaya Zemlya Test Site*, Bull. Seismol. Soc. Am. *87*, 1563–1575.

TAYLOR, S. (1996a), *Analysis of High-Frequency P_g/L_g Ratios from NTS Explosions and Western U.S. Earthquakes*, Bull. Seismol. Soc. Am. *86*, 1042–1053.

TAYLOR, S., *A Review of Broadband Regional Discrimination Studies of NTS Explosions and Western U.S. earthquakes*. In *Monitoring a Comprehensive Nuclear Test-Ban Treaty* (eds. Husebye E. S., and Dainty A. M.) (Kluwer Academic Publishers, Dordrecht, The Netherlands 1996b).

TAYLOR, S. and HARTSE, H. (1998), *A Procedure for Estimation of Source and Propagation Amplitude Corrections for Regional Seismic Discriminants*, J. Geophys. Res. *103*, 2781–2789.

VAN DER VINK, G. E. and Wallace, T. C. (1996), *The Political Sensitivity of Earthquake Locations*, IRIS Newsletter *15*, 20–23.

WALTER, W., MAYEDA, K., and PATTON, H. (1994), *Phase and Spectral Ratio Discrimination Between NTS Earthquakes and Explosions, Part 1: Empirical Observations*, Bull. Seismol. Soc. Am. *85*, 1050–1067.

WÜSTER, J. (1993), *Discrimination of Earthquakes and Explosions in Central Europe – A Case Study*, Bull. Seismol. Soc. Am. *83*, 1184–1212.

WÜSTER, J. (1995), *Discrimination of Earthquakes and Industrial Explosions in the Vogtland Region and NW-Bohemia*, Ph.D. Thesis, Ruhr-University Bochum, Series A, No. 42.

(Received June 7, 1999, revised July 6, 2000, accepted July 19, 2000)

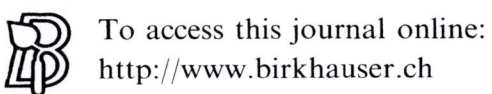

To access this journal online:
http://www.birkhauser.ch

Pure appl. geophys. 159 (2002) 759–778
0033–4553/02/040759–20 $ 1.50 + 0.20/0

▌ **Pure and Applied Geophysics**

Seismic Event Identification of Earthquakes and Explosions in Germany Using Spectral *Lg* Ratios

KARL KOCH[1]

Abstract — At the German NDC initial work on seismic event identification has focused on the application of spectral amplitude ratios for *Lg* in order to discriminate naturally occurring seismic events from other events associated with mining and quarry activities. Only about 10% of all seismic events occurring in Germany and adjacent areas are due to natural seismicity and are mostly constrained to the Alpine regions and areas along the Rhinegraben, Rhenish massif, Swabian Jura, and the Bohemian massif (Vogtland region). Using data from the broadband GRSN network, spectral amplitude ratios are calculated from maximum trace amplitudes in the 1–2 Hz and 6–8 Hz frequency bands, which are within the passbands of the deployed STS-2 instruments and the recorded 20 Hz data streams. These amplitude ratios then must be corrected with an appropriate attenuation model in order to remove propagation paths effects. For event identification, a scoring scheme is applied across the GRSN network, based on station-dependent scoring thresholds. In a case study aimed at testing the identification scheme, events are investigated from a quarry in southern Germany that provided ground-truth information for six events in 1997 to demonstrate the suitability of this identification approach. Except for one event with a rather strong earthquake signature, i.e., a low spectral *Lg* ratio, these events could be screened out from the earthquake population by their large *Lg* ratios. In a second step, aimed at applying the identification scheme, all events in Germany and neighboring areas that occurred in 1995 were processed, with approximately 800 out of more than 1200 events showing explosion-type *Lg* ratios, while only 10% remain in the earthquake population. However, specific mining areas appear to consistently produce earthquake-type spectral ratios indicative of particular blasting practices.

Key words: Earthquake, explosion, blast, mining-induced event, seismological bulletin, identification, spectral amplitude ratio.

1. Introduction

Each year more than a thousand seismic events are detected by the German Regional Seismic Network (GRSN) for which an epicenter location can be carried out. From this number of events only approximately 10%, or on the order of 100–200 events annually (e.g., HENGER and LEYDECKER, 1998), are related to natural seismicity, which is observed in various regions: Along the Rhinegraben, where rifting occurs as the response to the regionally imposed stress regime of the Alpine orogenesis; within the Rhenish massif and lower Rhine embayment, where active graben formation is taking place; in the Swabian Jura as the result of intraplate stress

---

[1] Federal Institute for Geosciences and Natural Resources, Stilleweg 2, D-30655 Hannover, Germany.

relaxation, and in the Bohemian massif (Vogtland region), where microearthquake swarm activity is the dominant seismicity pattern. Besides, many more events are located within the Alps and to its south as the expression of the collision process involving the Eurasian and African plates (Fig. 1). In addition, a few areas with induced seismicity can be identified, such as the Ruhr and Saar coal mining districts.

German Seismicity 1975 - 1992

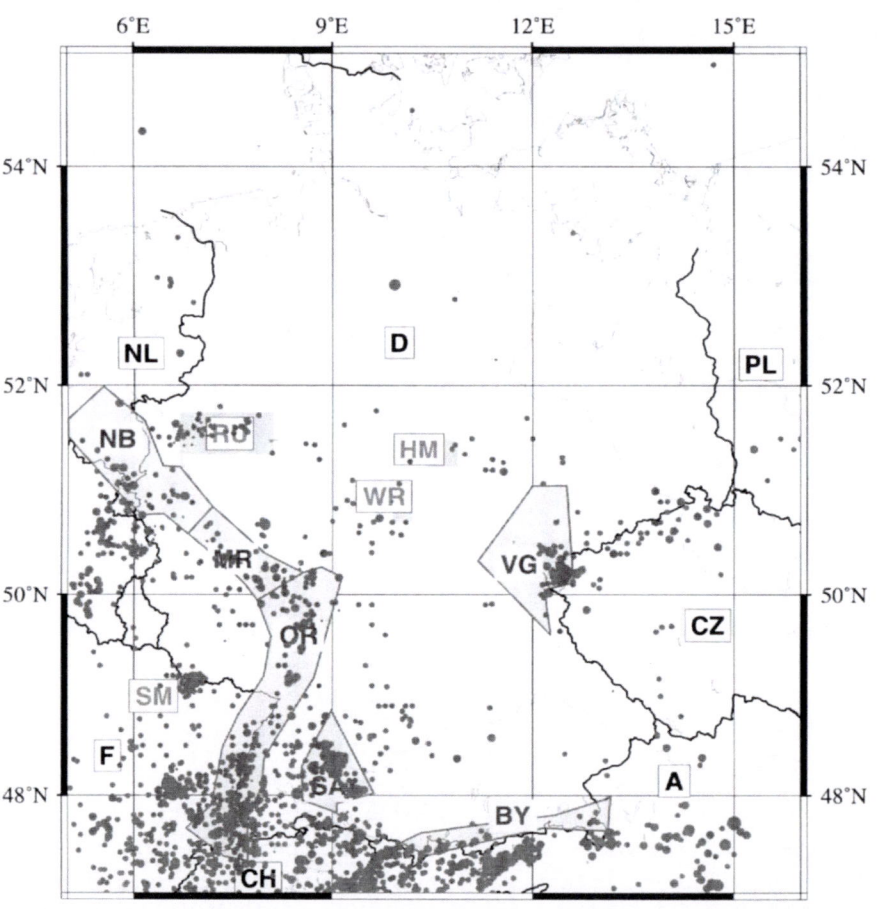

Figure 1
German seismicity between 1974 and 1992 taken from the German earthquake catalogue, with seismogeographic regions according to LEYDECKER and AICHELE (1998) outlining the seismically active regions in Germany along the Rhine River (OR – Upper Rhine Graben, MR – Middle Rhine area in the center of the Rhenish Massif, NB – Lower Rhine area/embayment), the Swabian Jura (SA), the Vogtland region (VG) and the Alps (including and south of BY – Bavarian Alps). Other regions marked are mining districts (HM – Harz, RU – Ruhr, SM – Saar, and WR – Werra mining district). Annotations are also given to identify the countries in Central Europe.

However, for establishing seismic bulletins that are frequently used in seismic hazard assessments of certain regions, it is desirable to identify any event not related to natural seismicity in order to remove these unwanted events from earthquake catalogues. For example, during GSETT-3 (Group of Scientific Experts Technical Test 3) most of the events located within Germany originated from mining activity. In particular, the Ruhr and Saar mining districts, northern Bohemia, and the east Saxony region produce many of these events (Fig. 2).

German Seismicity 1995 - 1997

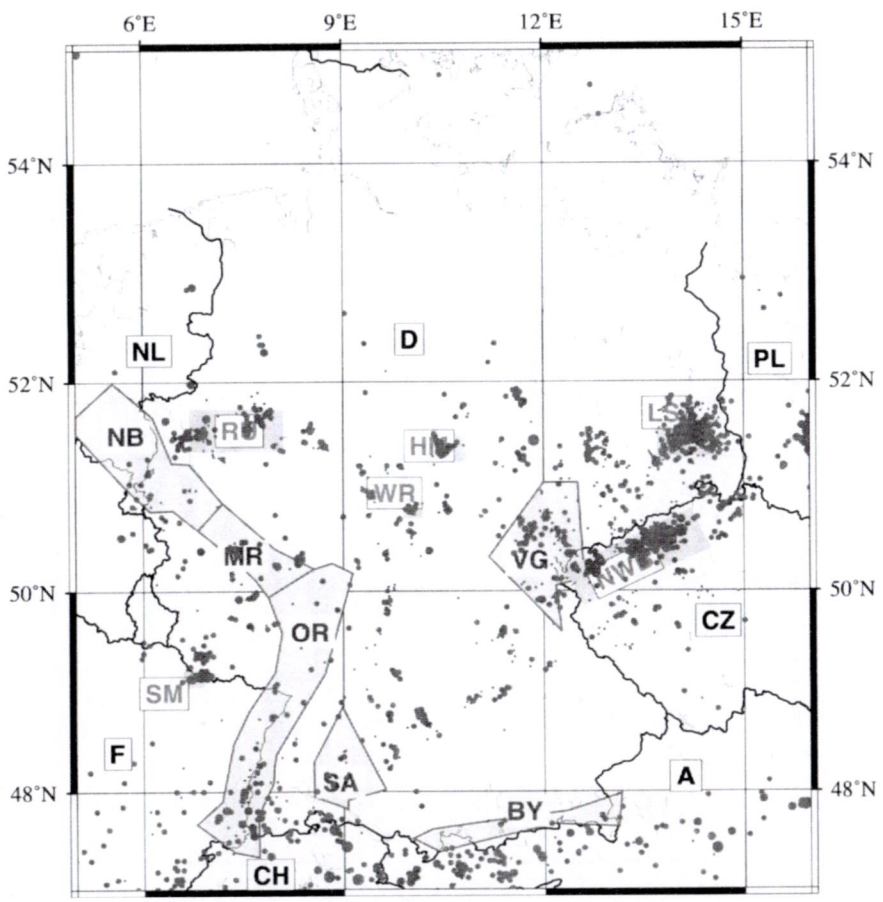

Figure 2

Seismic events detected and located in Germany during the first three years of GSETT-3, where the distribution of seismic events is more even and highlighting areas of intensive mining such as northern Bohemia and eastern Saxony. In addition to the relevant seismogeographic regions (see Fig. 1), the Lausitz (LS) and NW Bohemia (NWB) quarry regions are sketched.

By applying a search algorithm, it can be verified that many events shown in Figure 2 can be assigned to quarry locations. For this purpose a rectangular area of about 10 by 10 km was selected around the location of active quarries and mines, and the number of seismic events falling into this search region was determined. For the three years 1995–1997, some 70 quarries and mines were identified where more than 10 events were located within the search area (Fig. 3). In particular, the Hoyerswerda-Lausitz region is outstanding, with one particular quarry registering

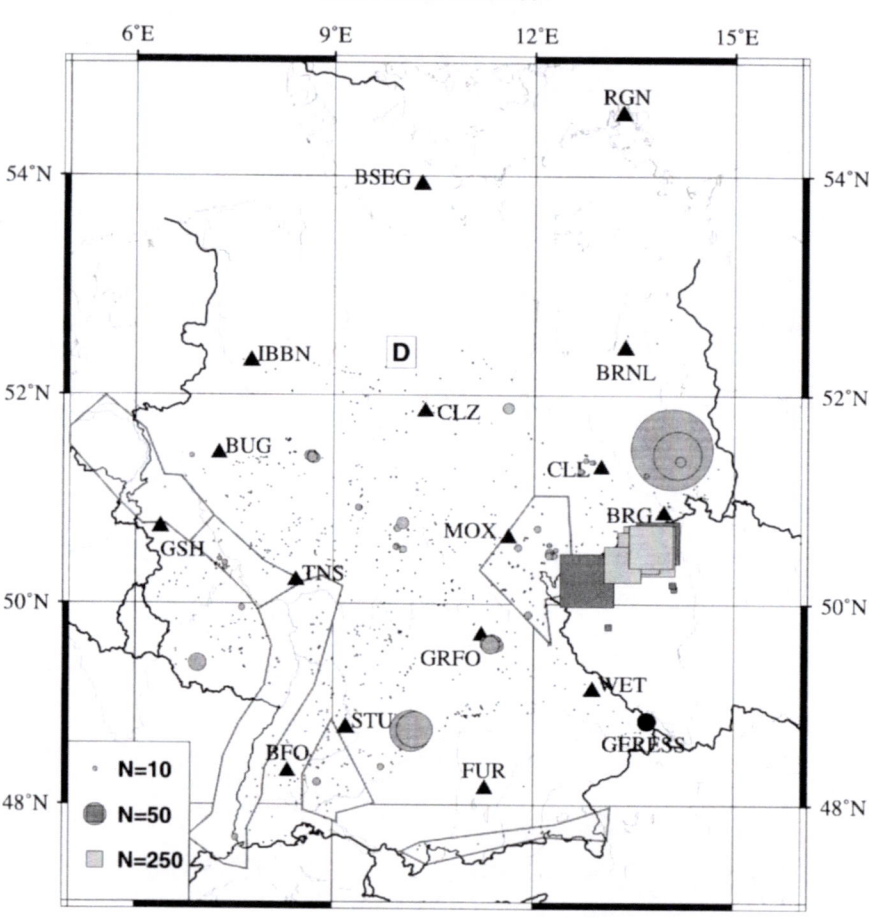

Figure 3
Location of all quarries in Germany (dots) and those (symbols) with at least 10 seismic events located within a distance of about 10 km for the period Jan. 1995 and Dec. 1997. Each symbol (circle for events in Germany, squares for events in northern Bohemia) represents a quarry location with its size proportional to the number of seismic events near that location (note the different shading for the squares representing different scaling).

almost 200 events close-by. Another area within Germany with a rather large number of seismic events close to active quarries is the eastern part of the Swabian Jura near Heidenheim. Quarries or mines with the largest number of associated events, however, are located in northwestern Bohemia, where one particular quarry had nearly 800 events associated with the selection box, i.e., about one event per day on average. This area is also of considerable importance in terms of natural seismicity, as the Vogtland area, where episodic earthquake swarm activity occurs, is located within this region (e.g., KOCH and FÄH, 2002).

So as to identify those seismic events that are associated with mining activities, seismological methods must be developed that are capable of separating natural from man-made events. A prospective approach is the use of amplitude ratios, in which the spectral ratios of the *Lg* phase seem to be reasonably promising (MURPHY and BENNETT, 1982; BENNETT and MURPHY, 1986; WÜSTER, 1993, 1995; WALTER *et al.*, 1994; GOLDSTEIN, 1995; PATTON and TAYLOR, 1995; TAYLOR, 1996; KOCH and FÄH, 2002). WÜSTER (1995) has shown that events in the Vogtland area can be identified reliably as explosions or earthquakes, based on spectral slope and variance of stacked waveform spectra calculated from the waveform data of the sensors of the GERESS array.

WALTER *et al.* (1994) and PATTON and TAYLOR (1995) have investigated the use of a number of different amplitude ratios for discrimination of earthquakes and explosions in and near the Nevada Test Site (NTS). They found that there is a varying degree of capability for these ratios for success in identification. WALTER *et al.* (1994) argued for the combination of various parameters to enhance discrimination success. Some of these studies are based on the original work by MURPHY and BENNETT (1982) and BENNETT and MURPHY (1986), which found significant capability of spectral *Lg* ratios for discrimination between earthquakes and underground nuclear explosions in the Western United States from observations at regional distances.

In this paper I will solely focus on the use of spectral amplitude ratios of the *Lg* phase, and throughout this study *Sg* and *Lg* will be used as synonymous terms. *Lg* was chosen as it is probably the most important phase for regional and local distances due to the fact that it is the highest amplitude phase in most observations and is observed to long distances (cf., LILWALL, 1988). With three-component seismograms from the broadband GRSN-network it is possible to further enhance the *Lg* feature by vectorial summation or station-averaging for all components, to determine a network discriminant based on a station-dependent scoring scheme.

2. Data Examples from Earthquakes and Explosions and Data Processing Scheme

The problem at hand can be demonstrated by the velocity waveforms shown in Figure 4a, which were recorded from events at similar ranges at the broadband

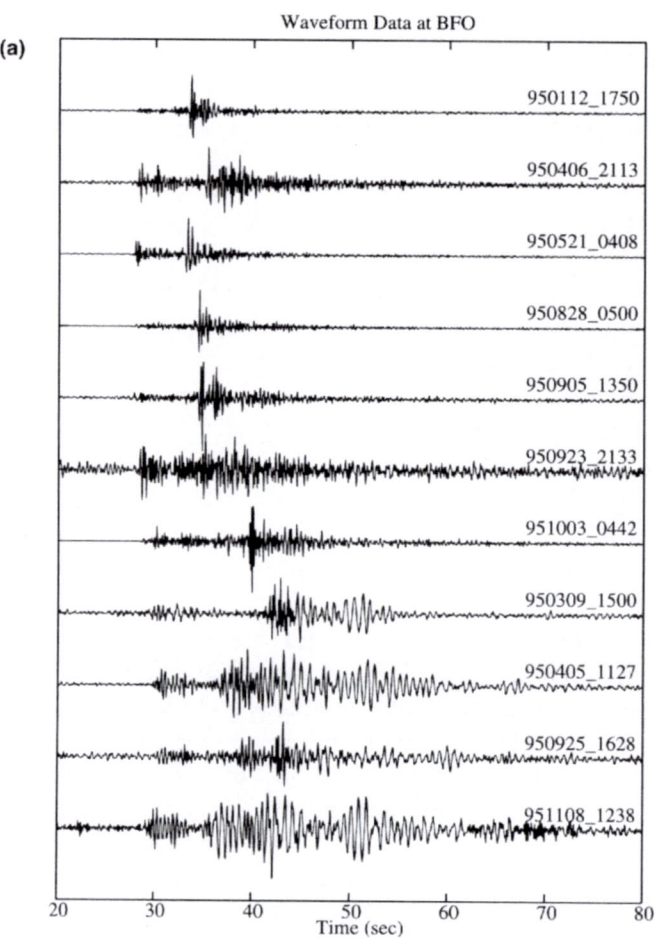

Figure 4

Waveforms (a) and vertical and radial component *Lg* spectra (b) for events at comparable distance ranges (40–60 km) from recording station BFO. Earthquake signals (thick lines) show more impulsive and higher frequency *Lg* signals which is consistent with positive slopes in the spectra for earthquakes and negative slopes for explosions (thin lines) for the velocity data. Earthquake spectra are scaled by a factor of 100 for plotting purposes. The shaded patches indicate the used frequency bands 1–2 Hz and 6–8 Hz.

station BFO. The seven traces on the top are from earthquakes, with the lower four traces from explosions. Clearly, the earthquake records show considerably more impulsive *Sg* (or *Lg*) waves, while the explosion data are more emergent and exhibit lower frequency content. When calculating spectra from the traces, which are shown in Figure 4b, the visual impression is confirmed, consistent with numerous observations, that earthquakes show rather flat spectra to higher frequencies compared to explosions, which often show spectra that are enhanced at lower frequencies (STUMP *et al.*, 1994). This pattern is expressed in Figure 4b for

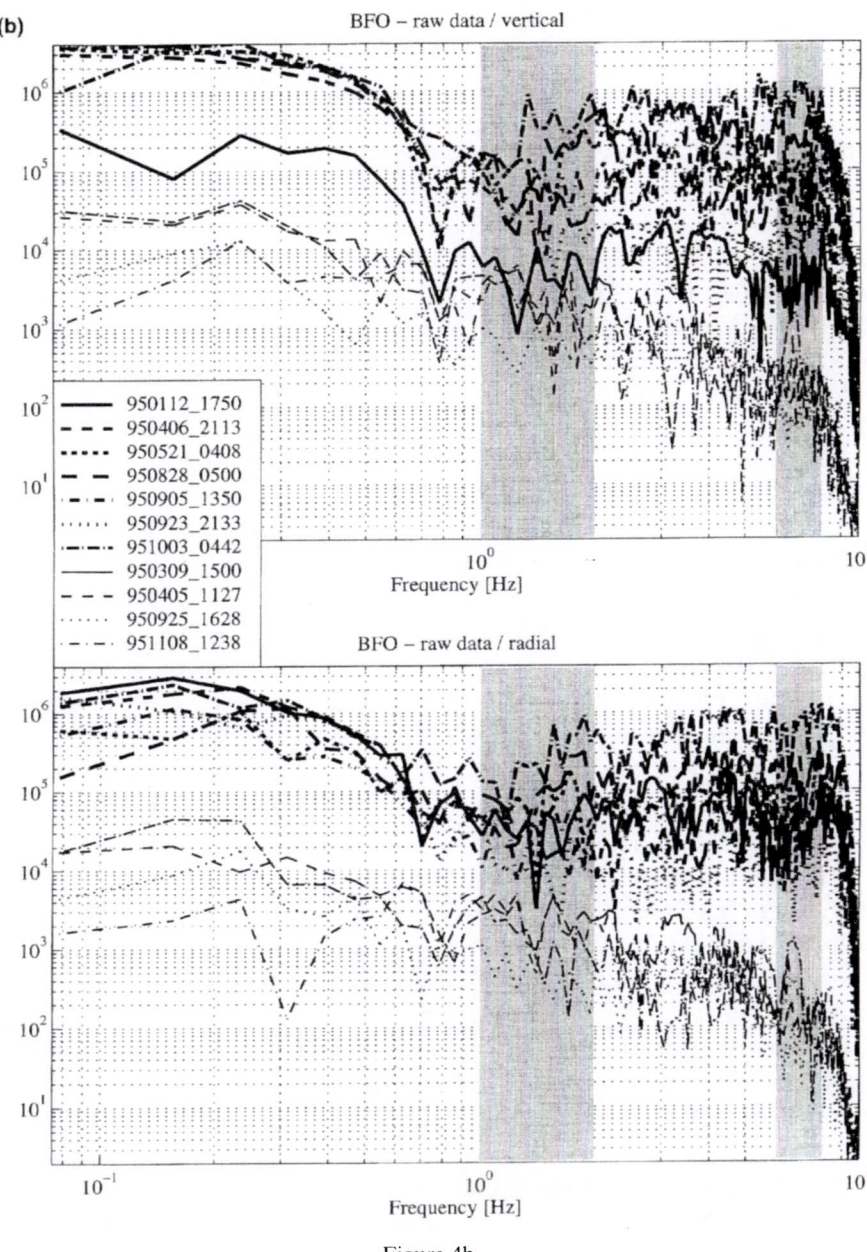

Figure 4b

earthquakes by the positive slope for the velocity spectra at BFO, while the explosion data show negative slopes. This result is also consistent with the discrimination work done by WÜSTER (1995) who argued for more flat and smoother stacked spectra for earthquakes than for explosions. A similar pattern was found by BENNETT and

MURPHY (1986) for seismic spectra from earthquakes compared to nuclear explosions in the western U.S. However, the depletion of higher frequencies in the spectral characterization of explosions, also found by STUMP *et al.* (1994), is in conflict with the theoretical study by LILWALL (1988) arguing for high-frequency contributions from shallow ($h \to 0$) explosions.

The data processing is carried out as outlined in Figures 5a and b. The unfiltered data are passed through Butterworth filters for the 1–2 Hz and the 6–8 Hz band and the maximum amplitude of the resulting waveforms in a window of 5 sec around the *Lg* onset is determined. The earthquake records show considerable energy in the higher passband, while the explosion data show negligible amplitudes for the upper band. Specifically, the spectral amplitude ratio of *Lg* used throughout this study is defined as the ratio of the maximum amplitude at 1–2 Hz to the maximum amplitude at 6–8 Hz. It represents a gross estimate of the slope of the *Lg* spectrum between these two frequency bands. No specific noise measurements on the filtered traces were taken, although the fact that all *Lg* phases analyzed were picked during analyst review should provide for sufficient signal-to-noise ratio with respect to the selected frequency bands.

The next required step is related to a distance correction for the observed amplitude ratio that has proven to be essential for regional discrimination work in many different geological settings (HARTSE *et al.*, 1998; TAYLOR and HARTSE, 1998; RODGERS *et al.*, 1999). As geometrical spreading can be regarded as independent of the frequency considered, at least to first order when only considering far-field terms, the only effect that must be taken into account is the attenuation of the seismic waves along the propagation path. To accomplish the correction for attenuation, I have chosen the attenuation model of JIA (1996) for Central and Southern Europe, which was established based on data from earthquakes at regional distances in Germany and the Mediterranean region and from a hydraulic fracturing experiment at the German continental deep drilling site KTB (cf., EMMERMANN and LAUTERJUNG, 1997). These models are consistent between about 1 and 50 Hz. Applying the *Q*-law determined by JIA (1996), a sequence of curves of amplitude-distance scaling can be found describing the effects due to attenuation for all possible frequency pairs (Fig. 6). However, if a constant frequency spacing is used, appropriate for cases when the maximum amplitudes are governed by the corner frequencies in each band, i.e., when the source spectrum either rises as in the case of earthquake data, or decreases as for explosion data, then the amplitude distance relation is nearly constant, matching the average curve of all amplitude-

▶

Figure 5

Seismograms are filtered in the 1–2 Hz and 6–8 Hz frequency bands and the maximum amplitude is taken as spectral amplitude in each frequency band. (a) Sample seismograms for an earthquake source, (b) seismograms for an explosion source. *Pg* and *Lg* phase windows are sketched. To enhance visual phase identification, high-pass filtered data are also shown.

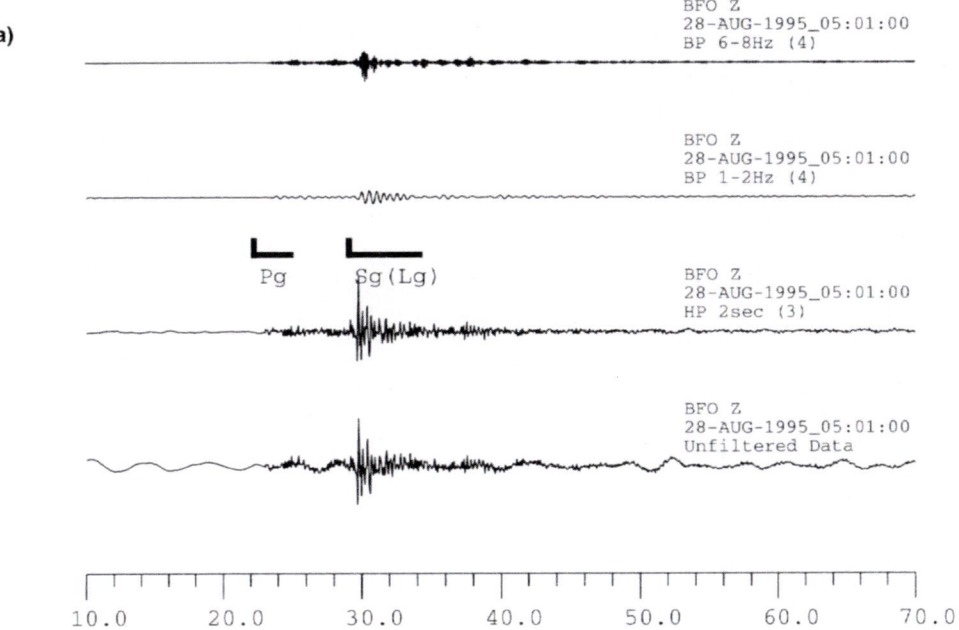

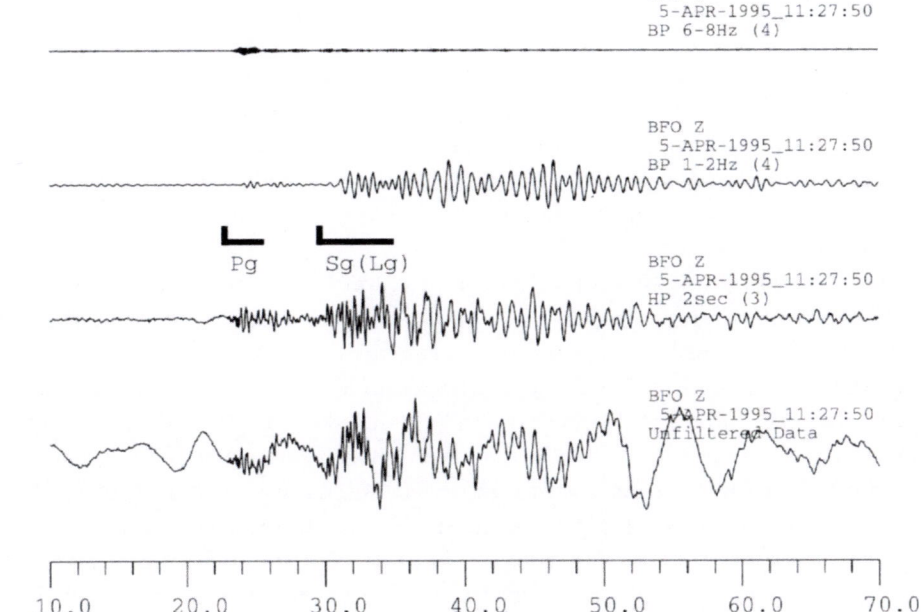

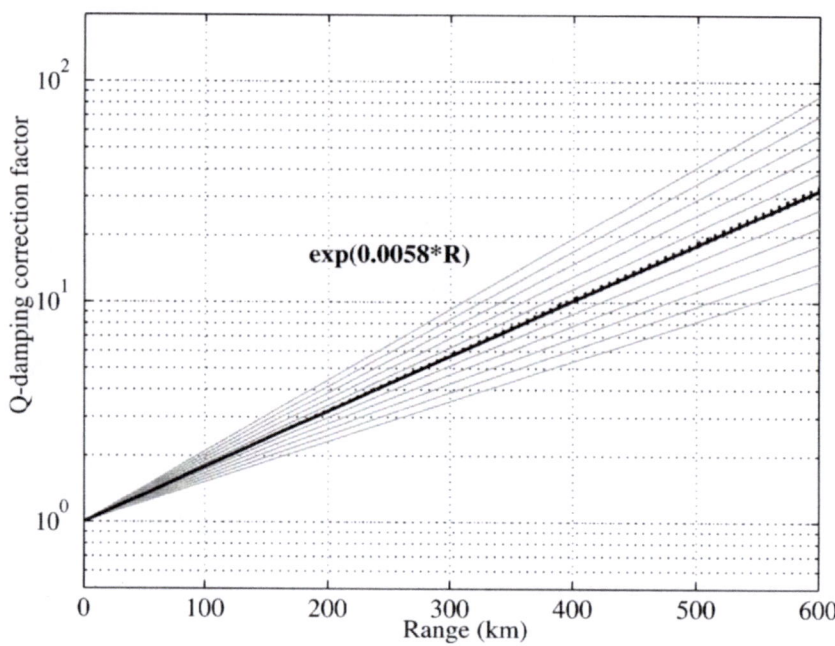

Figure 6

Amplitude-distance correction term due to apparent attenuation following the Q model of JIA (1996) for Central Europe. Gray curves are for various frequency pairs from the two passbands including the extreme cases, while the thick dotted curve is the average with the form $e^{0.0058*R}$, where R is hypocentral distance.

distance scaling curves. The observed spectral amplitude ratios then must be corrected accordingly with the distance correction term $e^{0.0058\ R}$ (Fig. 6). The appropriateness of this amplitude distance correction can be checked through the observations in Figure 7. In the upper part the observed uncorrected Lg ratios are displayed, which clearly show an increase for increasing distance, as the higher frequencies are more strongly attenuated than the lower frequencies. When correcting with the proposed distance correction the spectral ratio-distance dependency is no longer observed.

The propagation path corrected spectral amplitude ratios were then used to define station-dependent spectral ratio thresholds for classification of the spectral characteristic (see Table 1) of each measurement as explosion- or earthquake-like. These thresholds were defined for each station by examining the data from the stations independently for events that can be assumed as earthquakes with very high-probability, i.e., taken from the German earthquake catalogue. This, however, does not strictly constitute ground-truth information, as was available for WÜSTER's (1995) work and which is also used in the study carried out by KOCH and FÄH (2002) for Vogtland events. The earthquake threshold for a station was established such that the majority of data from these earthquakes fell below it. The explosion threshold was set by these data at a level that was exceeded only by outliers.

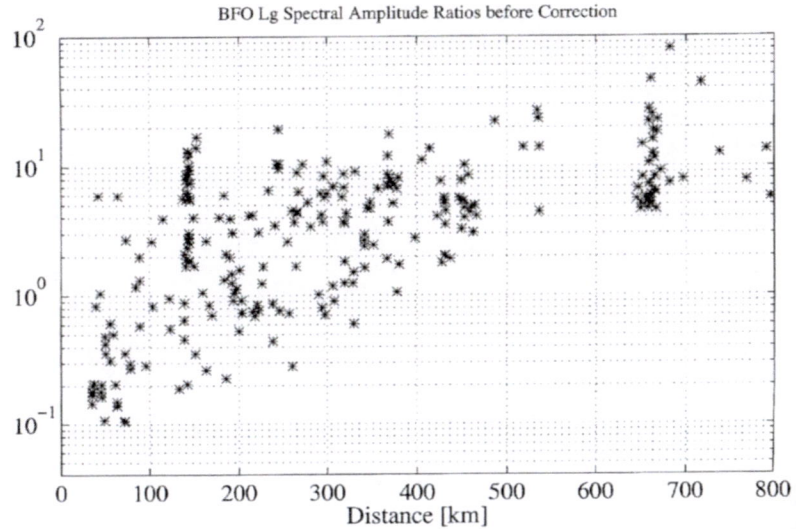

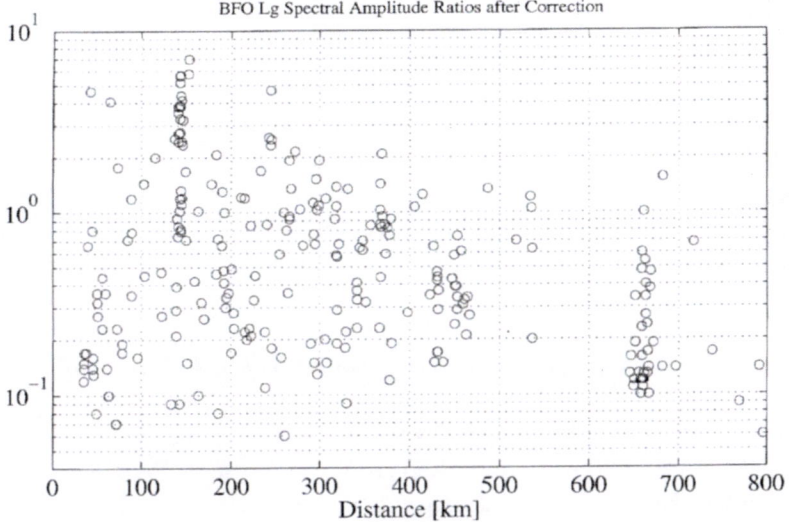

Figure 7

Observed spectral *Lg* amplitude ratios at a GRSN station before (upper figure) and after (lower figure) amplitude distance correction. The trend in the uncorrected spectral ratio data clearly follows the average curve from Figure 6, which is presented by the dashed line normalized to 0.3 at 0 km. After correction, the constant level follows the trend of the data, although low magnitude explosions may dominate the distribution at short distances.

In a second step, the station-dependent classifications for an event are submitted to a scoring scheme, where station classifications are weighted with +3 for earthquake types, −3 for explosion types, and +1 otherwise, summed, and

Table 1

Station-dependent thresholds of Lg amplitude ratios used in event classification

Station	EQ THRESH.	EX THRESH.
BFO	0.25	0.4
BRG	0.1	0.25
BUG	0.25	0.4
CLL	0.1	0.3
CLZ	0.15	0.4
FUR	0.2	0.4
GRFO	0.3	0.5
GSH	0.25	0.4
MOX	0.1	0.3
TNS	0.3	0.6
WET	0.1	0.25

normalized by the number of data. The rationale for a score of +1 instead of no score for spectral ratios lying between the two established classification thresholds (Table 1), is the desire to retain as many events as possible with ambiguous data in the earthquake population. If all stations classify the spectral ratios for an event as earthquake-type, then the resulting score will be +3.0 and, accordingly, if all data show explosion type, the overall score will be −3.0. Therefore, this procedure provides a network classification of events under study. As KIM *et al.* (1997) have shown, network averaged data provide for an increased discrimination potential than single station data. This scoring technique is intended to use different types of data, such as spectral *Lg* ratios and phase amplitude ratios (e.g., *Pg/Sg*), in a consistent manner for network averaging, without the need for weighting each measurement *a priori*, however with increased flexibility to introduce additional discriminants.

3. A Case Study Based on "Ground-truth" Information

In spring/summer of 1997 six events in southern Germany were detected at the German NDC (Table 2) that were located close to the site of the tallest concrete-steel bridge in Central Europe, thus drawing particular attention based on the potential for induced seismicity. However, during initial inquiry more events were found that had been detected in this region and, from the similar origin times, it soon became obvious that these events were explosions from a quarry operated close-by. Hence, it was decided to test the spectral *Lg* ratio method on these six explosions and compare the results with the ground-truth information obtained from the quarry operator. Besides confirmation of origin time, detailed additional information was made available on the blasting method, including number of holes, charge per hole, firing procedure, and type of explosives. Typically, the explosions in this quarry consist of ripple-fired blasts including 2–5 holes with an explosive yield of 80–100 kg each and

Table 2

Seismic source parameters of quarry events and ground truth information such as total yield, number of blasting holes, delay firing (dly for delayed, sim for simultaneous firing) and type of explosives

Date	Time	Lat.	Long.	Mag.	Yield	Holes	Explosives
19970401	050609.7	49.220	9.730	2.0	200	2 (dly)	A + B
19970505	044026.9	49.225	9.752	1.7	240	3 (dly)	A + B
19970609	045301.4	49.170	9.790	2.1	320	3 (dly)	A + B
19970610	044541.5	49.190	9.760	2.1	555	5 (dly)	A + B
19970613	043841.8	49.220	9.770	1.8	245	8 (sim)	B
19970923	095532.6	49.210	9.740	1.9	240	3 (dly)	A + B

A: Ammonsilit (Eurodyn), B: Wallonit

using delay caps. However, this information further stated that explosions were occasionally carried out with smaller charge sizes without using any delay caps or in different geological settings on benches or in the quarry bottom.

Spectral *Lg* amplitude ratios of these quarry events (Table 3) reflect mostly high values larger than 0.5, or even above 1, thus indicating explosion-type characteristics. Only one event shows pronounced smaller spectral ratios. This event is associated with a blast in which the holes were detonated simultaneously in the quarry bottom. Whether the smaller charge sizes per hole or if the difference in the local geology played the dominant role in the stronger excitation of the high-frequency contributions to the *Lg* phase could not be resolved. Applying the voting scheme described above for identification of the quarry events examined (Table 3), all of the events of this case study except for one show pronounced explosion-type voting scores (i.e., below or close to a value of −2). For the earthquake-like event, the small amplitude ratios lead to a small positive score, indicating that mining practice may result in the possible misclassification of this type of event.

4. Results for Seismic Events in Germany in 1995

As the previously discussed case study has shown, there is strong evidence that spectral *Lg* ratios are useful for identification of quarry and/or mining explosions. Hence, such a procedure could be used to investigate all events occurring in Germany in order to obtain an objective measure for event characterization. Such an objective tool is desirable to screen out mining events from earthquake data catalogues which are published by national seismological centers such as, for example, the Federal Institute for Geosciences and Natural Resources, which publishes annual seismological bulletins for Germany (e.g., HENGER and LEYDECKER, 1998). On the other hand, activities related to the Group of Scientific Experts Technical Test 3 (GSETT-3) within the United Nations Conference of Disarmament included the detection and

Table 3

Distance-corrected spectral Lg amplitude ratios for the quarry events given in Table 2, associated scoring classification, and final score based on voting scheme assigning + 3 for classification as earthquake (EQ), −3 for explosion (EX) and + 1 for unknown (UK)

DATE - OT	MAG	STN	Z		N		E		SCORE
19970401-05:06	2.0	BFO	0.64	EX	1.00	EX	0.82	EX	−1.9
		FUR	0.12	EQ	0.25	UK	0.20	EQ	
		GRFO	1.60	EX	0.91	EX	1.64	EX	
		MOX	0.44	EX	0.71	EX	0.39	EX	
		TNS	1.38	EX	1.08	EX	1.12	EX	
19970505-04:40	1.7	BFO	1.20	EX	2.03	EX	1.71	EX	−2.3
		GRFO	0.54	EX	0.36	UK	0.59	EX	
		MOX	0.37	EX	0.43	EX	0.26	UK	
		TNS	0.82	EX	0.91	EX	0.82	EX	
19970609-04:53	2.1	GRFO	0.86	EX	0.41	UK	0.59	EX	−1.7
		MOX	0.25	UK	0.35	EX	0.24	UK	
		TNS	1.98	EX	0.79	EX	1.08	EX	
19970610-04:45	2.1	BFO	1.19	EX	1.82	EX	1.03	EX	−3.0
		GRFO	2.89	EX	0.71	EX	1.52	EX	
		TNS	2.00	EX	1.79	EX	1.42	EX	
19970613-04:38	1.8	BFO	0.56	EX	2.09	EX	1.41	EX	0.9
		GRFO	0.35	UK	0.18	EQ	0.31	UK	
		MOX	0.24	UK	0.18	UK	0.14	UK	
		TNS	0.24	EQ	0.27	EQ	0.19	EQ	
19970923-09:55	1.9	WET	0.12	UK	0.14	UK	0.09	EQ	−3.0
		BFO	0.95	EX	1.93	EX	1.59	EX	
		GRFO	0.69	EX	0.66	EX	1.08	EX	
		TNS	1.24	EX	0.81	EX	0.85	EX	

location of all seismic activity on a global as well as a regional scale. Within this framework, a comprehensive set of all event types was located by the German NDC since 1995. This comprehensive data set was then used for application and evaluation of the spectral *Lg* ratio method.

Performing a similar analysis as in the case study for all seismic events in our database that occurred in 1995, the results shown in Figure 8 are obtained. Events identified as earthquakes are mostly confined to the Alps and the Rhinegraben area, closely matching the pattern of German seismicity shown in Figure 1. Most events identified as explosion-type are located in the known mining areas, such as eastern Germany or Northern Bohemia. Here also a number of events certainly originating from quarries were identified as earthquakes, which reflects the known fact that some stations (e.g., MOX) often record extremely high frequency and strong *Lg* waves. This could be related to special blasting practices, as demonstrated for the events of the case study, where simultaneous firing and/or small charge size produced one event with significantly smaller spectral *Lg* ratios, typical for earthquakes. Further-

Seismic events in Germany for 1995

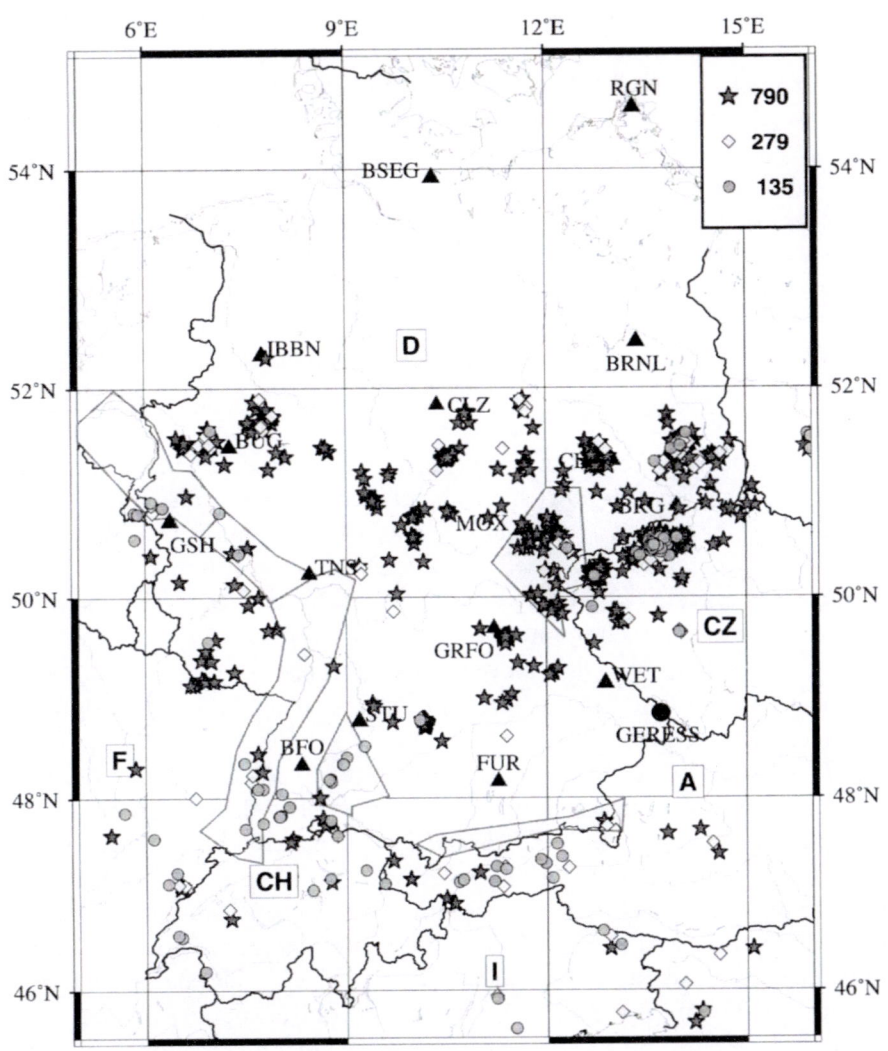

Figure 8

Identification results for the seismic events in Germany in 1995. From about 1200 events, nearly 800 were classified as explosion-like events (stars) while only 135 were identified as earthquake-like sources (circles). Diamonds represent ambiguous cases. Also, the seismicity pattern shown in Figure 1 is reflected by the distribution of the earthquake-like events (i.e., circles). For reference to map annotations, see Figure 1.

more, numerous events in central Germany, which lacks natural seismicity, are identified as explosions. Obviously, most of the induced events in the Ruhr and Saar mining districts are classified as explosions, therefore arguing for large spectral ratios indicating shallow events, as source depth of the induced events is constrained to the 1–2 km depth range.

From approximately 1200 events that occurred in 1995, about two-thirds were identified as explosion-type, while about 10% are identified as earthquake-type. Although some of these latter events are explosions in reality, the objectives of this study aim at the proper identification of explosions, which should result in some bias toward increasing the probability of misclassifying a few explosions. Therefore, the considerable number of explosion-classified events in the Alps, which must be assumed to be misclassified earthquakes due to the pattern of natural seismicity (Fig. 1), needs further consideration.

5. Discussion and Conclusions

As a significant number of larger events in the Alpine region, actually being earthquakes based on the fact that explosions generally do not exceed magnitudes of $M_L = 2.5$ and that this region is seismically active, were identified as explosions (Figs. 8 and 9), the influence of event size may play an important role on spectral Lg ratios, which follows from models of seismic source scaling (i.e., BRUNE, 1970). TAYLOR and HARTSE (1998) have argued that both source and path amplitude corrections must be applied to improve discrimination effectiveness. That source size, and hence corner frequency, has indeed a strong impact on the spectral Lg amplitude ratio which can be demonstrated by plotting this ratio versus the local magnitude (Fig. 10) for the southernmost stations of the GRSN network (i.e., BFO, FUR, WET). These stations are located closest to the regions of natural seismicity, outlined by the seismicity pattern in Figure 1, or Figure 9 for the Alpine region. Once the source strength exceeds magnitude $M_L = 3.5$, a strong trend for an increasing spectral Lg ratio is observed. Following the procedures of TAYLOR and HARTSE (1998) for source amplitude corrections, i.e., assuming a source model with an ω^2 spectral fall-off such as proposed by BRUNE (1970), and using scaling relations determined for earthquakes in northwestern Germany (ONCESÇU et al., 1994) describing the moment-magnitude relationship and linking magnitude and corner frequency, a correction term for event magnitude was determined. This correction term qualitatively matches the observed increase of the spectral Lg ratio well, in that the spectral Lg ratio starts to increase rapidly above $M_L = 3.5$ for stations BFO and WET, even for events that are mostly from the Alpine region and may not adequately be described by ONCESÇU et al.'s relation. FUR in contrast shows no tendency, which could be related to the noise conditions at this station. From the previous discussion it follows that data from events with $M_L > 3.5$ must be corrected using a correction function as shown in Figure 10, as otherwise they will most likely be classified as explosions.

Another interesting result of the identification process carried out for the events shown in Figure 8 relates to several explosion-classified events on the Swiss–German

German catalogue

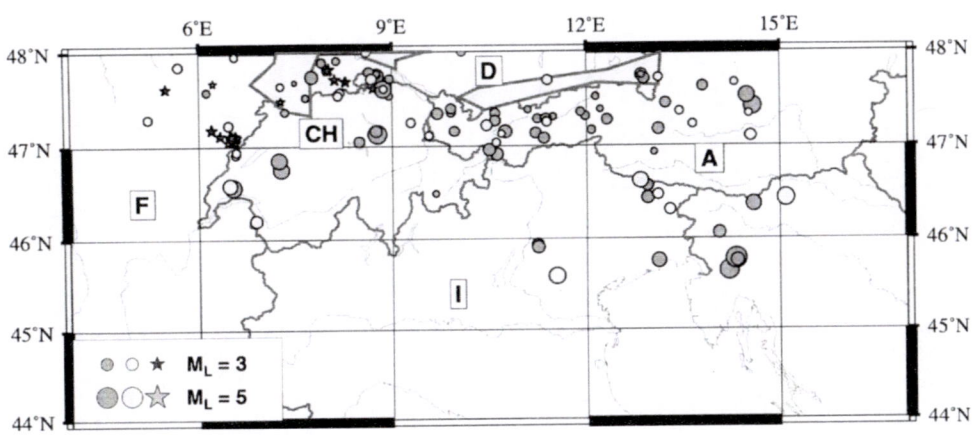

Figure 9

Distribution of seismic events in the Alpine region with symbol size according to local magnitude. Stars represent events where a reporting agency indicated an artificial source (explosion, quarry blast, mining event). Circles represent earthquakes, although not all may be verified (light circles). For reference to map annotations, see Figure 1.

border south of station BFO, at the intersection of the Rhine and Aare rivers, which is in an area where earthquakes commonly occur. Closer investigation of the location of these events by means of satellite imagery revealed a quarry site in this very same area. Therefore, the results for explosions in areas where earthquakes and explosions occur, tend to be in agreement with ground-truth information for the explosion-classified cases, when properly correcting for magnitude bias. However, events larger than magnitude 3 can confidently be excluded from the explosion population, as blasting regulations in Germany cause explosions to be constrained to magnitudes less than $M_L = 2.5$.

This study has shown the potential for using spectral Lg ratios from the 1–2 Hz and 6–8 Hz frequency bands as a measure for the identification of earthquakes and mining or quarry explosions. After correcting the observed data using an appropriate attenuation model, amplitude distance dependencies can be removed. However, special consideration must be placed on the specific mining practices, which is evident both from events in a case study as well as from the unsatisfactory identification success for specific mining areas. Also, further focus needs to be directed towards the influence of noise on the estimated spectral ratio, and the effect of region-dependent amplitude thresholds. Finally, further work is warranted to introduce other amplitude or spectral ratios in the identification approach in order to extend this method toward increasing identification capabilities for induced events, as the

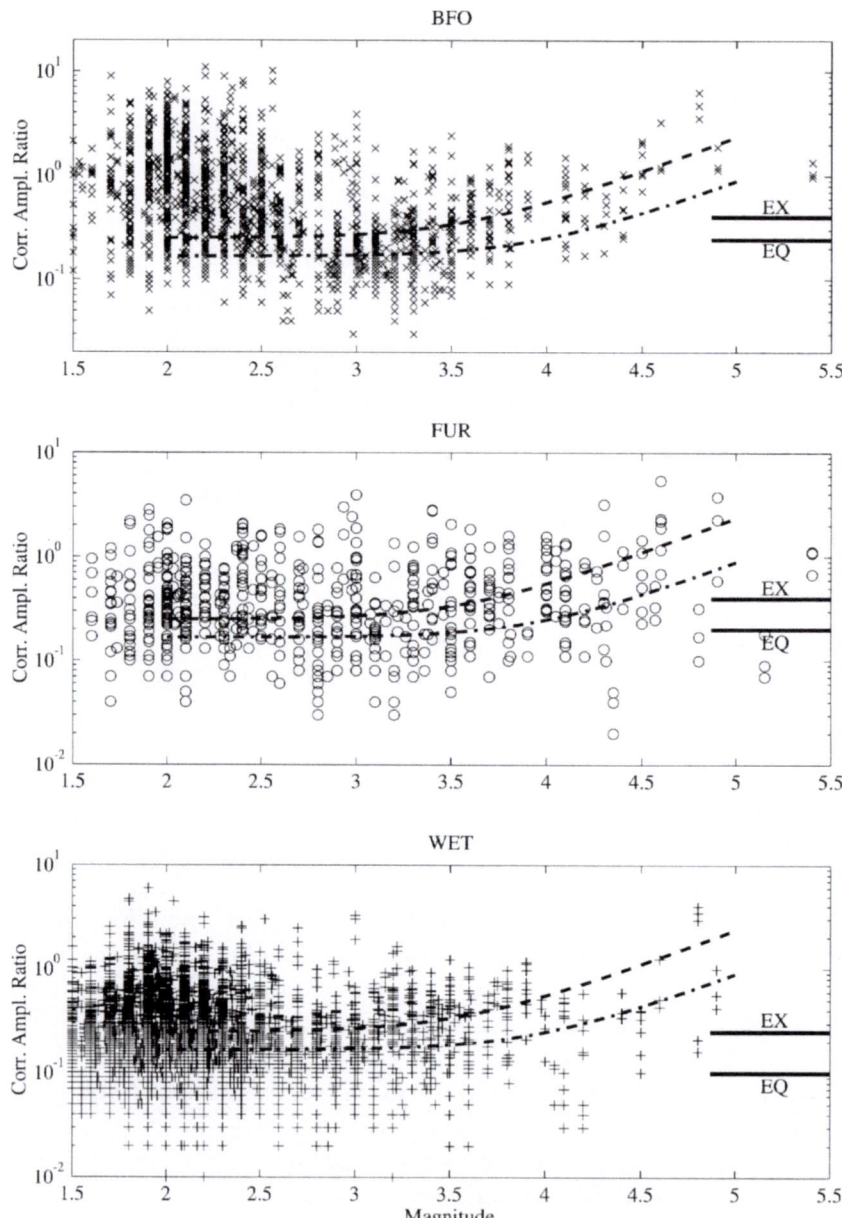

Figure 10
Path-corrected spectral amplitude ratios for the southernmost GRSN stations recording the most earthquakes, which are mostly located to the south with clear indication for BFO and also WET for magnitude dependence above $M_L = 3.5$. Dashed lines describe the effect of source magnitude on the spectral Lg ratio for a BRUNE (1970) source model with respect to frequencies at 6 Hz (dash-dotted) and 8 Hz (dashed), respectively. Explosion and earthquake thresholds for each station are also given.

current method seems to be more strongly correlated to differences in source depth than source mechanism.

REFERENCES

BENNETT, T. J., and MURPHY, J. R. (1986), *Analysis of Seismic Discrimination Capabilities Using Regional Data from Western United States Events*, Bull. Seismol. Soc. Am. *76*, 1069–1086.

BRUNE, J. (1970), *Tectonic Stress and the Spectra of Seismic Shear Waves from Earthquakes*, J. Geophys. Res. *75*, 4997–5009.

EMMERMANN, R., and LAUTERJUNG, J. (1997), *The German Continental Deep Drilling Program KTB: Overview and Major Results*, J. Geophys. Res. *107*, 18,179–18,201.

GOLDSTEIN, P. (1995), *Slopes of P- to S-spectral Ratios – A Broadband Regional Seismic Discriminant and a Physical Model*, Geophys. Res. Lett. *22*, 3147–3150.

HARTSE, H., FLORES, R., and JOHNSON, P. (1998), *Correcting Regional Seismic Discriminants for Path Effects*, Bull. Seismol. Soc. Am. *88*, 596–608.

HENGER, M., and LEYDECKER, G. (eds.) (1998), *Erdbeben in Deutschland 1993*, Bundesanstalt für Geowissenschaften und Rohstoffe, Hannover, Germany, ISBN 3-510-95808-X.

JIA, Y. (1996), *Bestimmung der scheinbaren Dämpfung seismischer Wellen in der europäischen Lithosphäre*, Ph.D. Thesis, Ruhr University Bochum, Geophysical Institute Reports, Series A, No. 47.

KIM, W.-Y., AHARONIAN, V., LERNER-LAM A. L., and RICHARDS, P. G. (1997), *Discrimination of Earthquakes and Explosions in Southern Russia Using Regional High-frequency Three-component Data from the IRIS/JSP Caucasus Network*, Bull. Seismol. Soc. Am. *87*, 569–588.

KOCH, K., and FÄH, D. (2002), *Identification of Earthquakes and Explosions Using Amplitude Ratios – The Vogtland Area Revisited*, (this volume).

LEYDECKER, G., and AICHELE, H. (1998), *The Seismogeographical Regionalisation of Germany: The Prime Example for Third-Level Regionalisation*, Geol. Jb. *E 55*, 5–24.

LILWALL, R. C. (1988), *Regional m_b : M_s, Lg/Pg Amplitude Ratios and Lg Spectral Ratios as Criteria for Distinguishing Between Earthquakes and Explosions: A Theoretical Study*, Geophys. J. *93*, 137–147.

MURPHY, J. R. and Bennett, T. J. (1986), *A Discrimination Analysis of Short-period Regional Seismic Data Recorded at Tonto Forest Observatory*, Bull. Seismol. Soc. Am. *72*, 1351–1366

ONCESÇU, M.-C., CAMELBEECK, T., and MARTIN, H. (1994), *Source Parameters for the Roermond Aftershocks of 1992 April 13 – May 2 and Spectra for P and S Waves at the Belgian Seismic Network*, Geophys. J. Int. *116*, 673–682.

PATTON, H., and TAYLOR, S. (1995), *Analysis of Lg Spectral Ratios from NTS Explosions: Implications for the Source Mechanisms of Spall and the Generation of Lg Waves*, Bull. Seismol. Soc. Am. *85*, 220–236.

RODGERS, A. J., WALTER, W. R., SCHULTZ, C. A., MYERS, S. C., and LAY, T. (1999), *A Comparison of Methodologies for Representing Path Effects on Regional P/S Discriminants*, Bull. Seismol. Soc. Am. *89*, 394–408.

STUMP, B. W., RIVIERE-BARBIER, F., CHERNOBY, I., and KOCH, K. (1994), *Monitoring a Test-ban Treaty Presents Scientific Challenges*, EOS, Trans. Am. Geophys. Union *75*, 265–273.

TAYLOR, S. R. (1996), *A Review of Broadband Regional Discrimination Studies of NTS Explosions and Western U.S. Earthquakes*, In. *Monitoring a Comprehensive Test-Ban Treaty*, (eds E.S. Husebye and A.M. Dainty,) Kluwer Acad. Publ. 1996, pp. 755–775.

TAYLOR, S., and HARTSE, H. (1998), *A Procedure for Estimation of Source and Propagation Amplitude Corrections for Regional Seismic Discriminants*, J. Geophys. Res. *103*, 2781–2789.

WALTER, W. R., MAYEDA, K. M., and PATTON, H. J. (1996), *Phase and Spectral Ratio Discrimination Between NTS Earthquakes and Explosions, Part 1: Empirical Observations*, Bull. Seismol. Soc. Am. *85*, 1050–1067.

WÜSTER, J. (1993), *Discrimination of Chemical Explosions and Earthquakes in Central Europe: A Case Study*, Bull. Seismol. Soc. Am. *83*, 1184–1212.

WÜSTER, J. (1995), *Diskrimination von Erdbeben und Sprengungen im Vogtlandgebiet und Nordwest-Böhmen*, Ph.D. Thesis, Ruhr University Bochum, Geophysical Institute Reports, Series A, No. 42.

(Received June 24, 1999, revised June 12, 2000, accepted June 19, 2000)

 To access this journal online:
http://www.birkhauser.ch

Pure appl. geophys. 159 (2002) 779–801
0033–4553/02/040779–23 $ 1.50 + 0.20/0

❘ Pure and Applied Geophysics

Signal Processing for Indian and Pakistan Nuclear Tests Recorded at IMS Stations Located in Israel

YEFIM GITTERMAN,[1] VLADIMIR PINSKY,[1]
and RAMI HOFSTETTER[1]

Abstract — In compliance with the Comprehensive Nuclear-Test-Ban-Treaty (CTBT) the International Monitoring System (IMS) was designed for detection and location of the clandestine Nuclear Tests (NT). Two auxiliary IMS seismic stations MRNI and EIL, deployed recently, were subjected to detectability, travel-time calibration and discrimination analysis. The study is based on the three recent 1998 underground nuclear explosions: one of India and two of Pakistan, which provided a ground-truth test of the existing IMS. These events, attaining magnitudes of 5.2, 4.8 and 4.6 correspondingly, were registered by many IMS and other seismic stations.

The MRNI and EIL broadband (BB) stations are located in Israel at teleseismic distances (from the explosions) of 3600, 2800 and 2700 km, respectively, where the signals from the tests are already weak. The Indian and the second Pakistan NT were not detected by the short-period Israel Seismic Network (ISN), using standard STA/LTA triggering. Therefore, for the chosen IMS stations we compare the STA/LTA response to the results of the more sensitive Murdock-Hutt (MH) and the Adaptive Statistically Optimal Detector (OD) that showed triggering for these three events. The second Pakistan NT signal arrived at the ISN and the IMS stations in the coda of a strong Afghanistan earthquake and was further disturbed by a preceding signal from a local earthquake. However, the NT signal was successfully extracted at EIL and MRNI stations using MH and OD procedures. For comparison we provide the signal analysis of the cooperating BB station JER, with considerably worse noise conditions than EIL and MRNI, and show that OD can detect events when the other algorithms fail. Using the most quiet EIL station, the most sensitive OD and different bandpass filters we tried in addition to detect the small Kazakh chemical 100-ton calibration explosion of 1998, with magnitude 3.7 at a distance approaching 4000 km. The detector response curve showed uprising in the expected signal time interval, but yet was low for a reliable decision.

After an NT is detected it should be recognized. Spectra were calculated in a 15-sec window including P and P-coda waves. The spectra for the first Pakistan NT showed a pronounced spectral null at 1.7 Hz for all three components of the EIL station. The effect was confirmed by observation of the same spectral null at the vertical component of the ISN stations. For this ground-truth explosion with a reported shallow source depth, the phenomenon can be explained in terms of the interference of P and pP phases. However, the spectral null feature, considered separately, cannot serve as a reliable identification characteristic of nuclear explosions, because not all the tests provide the nulls, whereas some earthquakes show this feature. Therefore, the multi-channel spectral discrimination analysis, based on a spectral ratio of low-to-high frequency energy (in the 0.6–1 Hz and 1–3 Hz bands), and a semblance of spectral curves (in the 0.6–2 Hz band), was conducted. Both statistics were calculated for the vertical component of the ISN stations as well for the three components of the EIL station. The statistics provided a reliable discrimination between the recent NT and several nearby earthquakes, and showed compliance with the former analysis of Soviet and Chinese NT, where nuclear tests demonstrated lower values of energy ratio and spectral semblance than earthquakes.

---

[1] Seismology Division, Geophysical Institute of Israel, P.O. Box 2286, Holon 58122, Israel.
E-mail: yefim@iprg.energy.gov.il

Accurate location of NT requires calibration of travel time for IMS stations. Using known source locations, IASPEI91 travel-time tables and NEIC origin times we calculated expected arrival time for the *P* waves to the EIL and MRNI stations and showed that the measured arrival time has a delay of about 4 sec. Similar results were obtained for the nearby Pakistan earthquakes. The analysis was complimented by the *P* travel-time measurements for the set of Semipalatinsk NT, which showed delays of about 3.7 sec to the short-period MBH station which is a surrogate station for EIL. Similar delays at different stations evidence a path- rather than site-effect. The results can be used for calibration of the IMS stations EIL and MRNI regarding Asian seismic events.

Key words: Nuclear test, adaptive detector, IMS station, travel time calibration, spectral discrimination.

Introduction

The recent nuclear test activity in southern Asia (Fig. 1), comprising the announced tests in India (May 11, 1998) and Pakistan (May 28 and 30, 1998), provided new recordings for verification of teleseismic discriminants, and for testing new detection and location procedures, as well as an opportunity for travel-time correction for new stations of the International Monitoring System (IMS). These discriminants and location procedures can serve as a tool for improving the monitoring executed by the Comprehensive Test-Ban Treaty. The explosion recordings were analyzed mainly based on the closest station, Nilore (NIL), at near-regional distance range of about 7°–9° (e.g., WALTER *et al.*, 1998). In August 1998, a 100-ton chemical explosion was conducted at the former Semipalatinsk Nuclear Test Site (STS), Kazakhstan (Fig. 1), as a calibration experiment for the CTBT verification, and was recorded by many stations worldwide (CTBTO, 1998b).

The number of active IMS stations in the Middle East is rather limited, and thus we stress the need to assess, as precisely as possible, the monitoring capabilities of each station. The IMS stations EIL and MRNI were installed in Israel in 1996 and 1998, respectively, when nuclear testing was nearing termination, therefore every chance to analyze ground-truth explosion data at these stations should be used. We analyze and interpret here the recordings of the two IMS stations, the cooperating BB station JER and short-period stations of the Israel Seismic Network (ISN; Fig. 2), of the nuclear and chemical explosions, located at a distance range of 25°–38°. We show that these measurements can be a useful tool for monitoring the CTBT and the calibration of IMS stations in the region. The case of the second Pakistan test is of special interest, because it simulates in some way the scenario of a clandestine test masked by other two events: a strong earthquake in Afghanistan (06:22, $M_S = 6.9$), and a local earthquake (06:57, $M_L = 2.5$) 20 km south of EIL station.

Data

There is no ground-truth information GT0 for the recent Indian and Pakistan nuclear tests. BARKER *et al.* (1998) presented location and origin time for these tests,

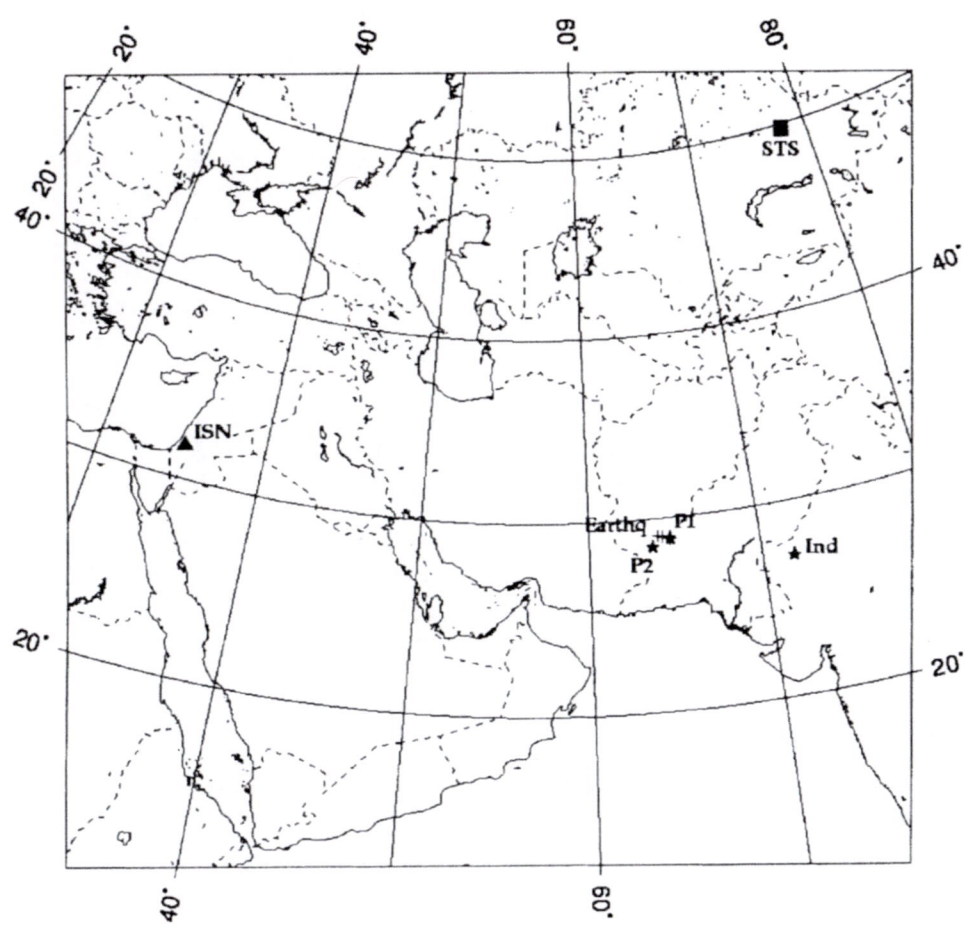

Figure 1

Relative location of the Israel network and the selected teleseismic events.

based on **PIDC REB**, Joint Epicenter Determination and Satellite Imagery. Owing to additionally using in our study several nearby Pakistan earthquakes and old Semipalatinsk nuclear tests, we address here source parameters taken from the PDE (USGS) Bulletin (see Table 1), in order to keep homogeneity of the data.

We analyzed seismograms observed by the short-period stations (1 Hz seismometer, L4C Mark Product) and broadband stations EIL, MRNI and JER (STS-2 seismometer and Quanterra data logger) of the ISN. Only the Pakistan test on May 28, 1998 triggered the short-period stations (Fig. 3a). Two tests on May 11 and 28, 1998, triggered the 80 Hz channel of the broadband station EIL. The second Pakistan nuclear test on May 30, 1998, occurred about 38 minutes after a major $M_S = 6.9$ earthquake, and was revealed in the continuous 20-Hz channel of EIL using a higher frequency band (Fig. 3c). In order to provide a discrimination analysis

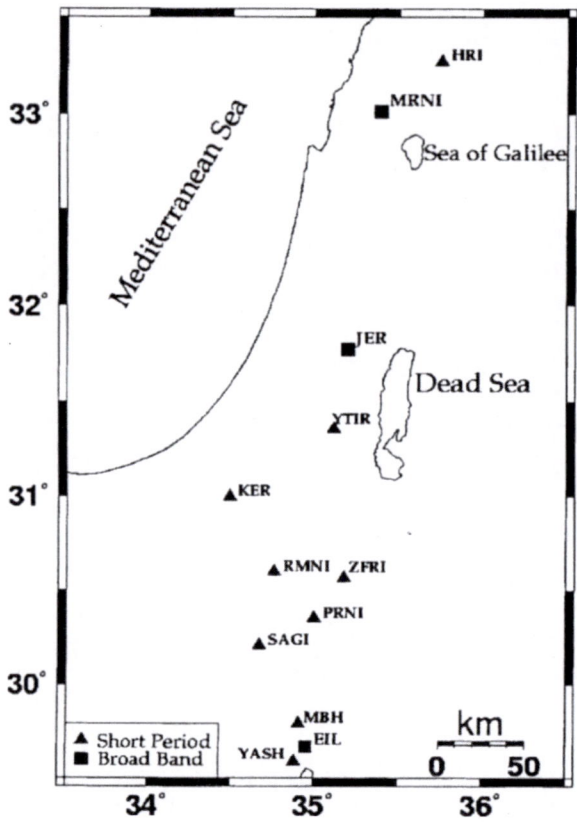

Figure 2
Location of short-period and broadband stations of the ISN used in the analysis.

we added several recordings of earthquakes occurring near the explosion sites in Pakistan (Figs. 1 and 3b). We also analyzed EIL recordings of the chemical calibration explosion of 100 tons at the STS, on August 22, 1998 (shown below on Fig. 9).

Travel Times and Peak Amplitudes

The same set of teleseismic events from southern Asia was used to estimate station corrections based on *P* travel times and NEIC source data. The *P* arrivals to the IMS station EIL showed regular delays with an average of about 4 seconds relative to the IASPEI91 model (see Table 2, Fig. 3c). Analogous delays of about 3.7 seconds on average were found in the analysis of former nuclear explosions at the Semipalatinsk Test site (Table 2). The explosions were recorded at the short-period

Table 1

Source parameters of nuclear tests and earthquakes used in the study (location, origin time and magnitude are taken from PDE)

#	Date	Origin time	Lat. N	Lon. E	Depth km	Mag. m_b	Dist. to EIL, deg.	Event, region
				New events				
1	1997/12/04	10:17:01.33	29.09	64.11	33	5.0	25.4	Earthquake, Pakistan
2	1998/01/05	16:58:35.29	29.01	64.35	18	4.9	25.6	Earthquake, Pakistan
3	1998/05/11	10:13:41.78	27.10	71.80	0	5.2	32.4	Nucl. expl., India
4	1998/05/28	10:16:15.23	28.90	64.79	0	4.8	26.0	Nucl. expl., Pakistan
5	1998/05/28	20:32:46.51	26.58	62.23	47	4.7	24.2	Earthquake, Pakistan
6	1998/05/30	06:54:57.1*	28.50	63.74	0	4.6	25.2	Nucl. expl., Pakistan
7	1998/08/22	05:00:18.90	49.77	77.99	0	3.8 3.7[+]	38.06	Chem. expl. 100 ton, STS
				Old STS nuclear tests				
1	1987/03/12	01:57:17.3	49.93	78.79	0	5.4	38.5	
2	1987/04/17	01:03:04.8	49.89	78.69	0	6.0	38.5	
3	1987/05/06	04:02:05.5	49.78	78.09	0	5.5	38.1	
4	1988/04/03	01:33:05.8	49.92	78.95	0	6.1	38.6	Kazakhstan
5	1988/11/12	03:30:03.7	50.08	78.99	0	5.3	38.7	
6	1988/12/17	04:18:06.9	49.89	78.93	0	5.9	38.6	
7	1989/01/22	03:57:06.5	49.92	78.86	0	6.0	38.6	
8	1989/02/12	04:15:06.7	49.90	78.76	0	5.9	38.5	

* JHD solution by WALLACE (1998). [+] pIDC REB estimate

station **MBH** situated within 14 km North of EIL station (Fig. 2), which was not yet installed. Figure 3c also presents single cases of similar lags 3.5–4 sec for the IMS station **MRNI** and **BB** station **JER**. Similar delays at different stations evidence a path- rather than site-effect.

Based on the corrected arrival time we tried to identify the P arrival of the STS 100-ton chemical explosion using a band pass filtered seismogram at EIL. We estimated a possible P amplitude for this explosion using the dependence of measured P-wave peak amplitudes for a series of old STS nuclear explosions (see Table 2) with magnitudes in the m_b range 5.3–6.1. The upper limit magnitude-yield relationship for explosions in hard rock (KHALTURIN et al., 1998) for the 100-ton explosion gives $m_b = 3.9$, NEIC reports $m_b = 3.8$, and pIDC REB presents $m_b = 3.7$ (CTBTO, 1998b). Rough extrapolation to lower magnitudes of the magnitude-amplitude dependence observed at MBH station shows that for $m_b = 3.7$–3.9 one should expect at EIL station P amplitudes much lower than 0.01 micron/sec,

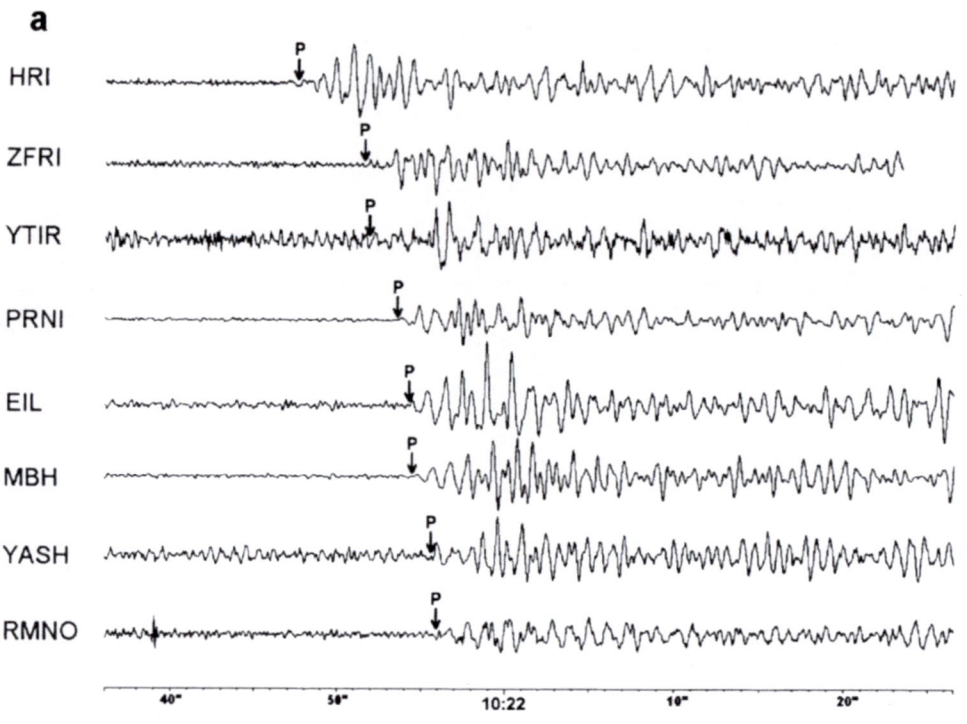

Figure 3

Sample seismograms of different types of seismic events: short-period vertical records of the first Pakistan test (28.5.98) (a), and the same day nearby earthquake in Pakistan (b); filtered broadband vertical records of the three nuclear tests (c). Vertical axis units here (and at all **BB** seismogram plots) are counts (1 count = 1.67 ∗ 10^{-9} m/s).

whereas pre-signal noise level is about 0.01–0.02 micron/sec (in the band 1–3 Hz). A weak phase can be observed on the filtered (2.5–3.5 Hz) vertical component seismogram about 2 sec after the calculated *P*-arrival time, corrected for the 3.7 sec average delay (an arrow on Fig. 9). Nevertheless the phase association with *P* waves from this explosion is not reliable.

Waveform Analysis and Triggering

Station EIL, located inside a 100-m long tunnel, is a quiet station. The broadband and short-period noise velocity spectral density are presented in Figure 4a, along with selected noise velocity values of a quiet station CAD in the Pyrenees (VILA, 1998), located in hard rock tunnel (490-m long, 80-m deep). In the same frequency range the noise level at **MRNI** and **JER** is approximately 10 to 20 times larger, respectively (see filtered seismograms in Figs. 7 and 8). However, for the

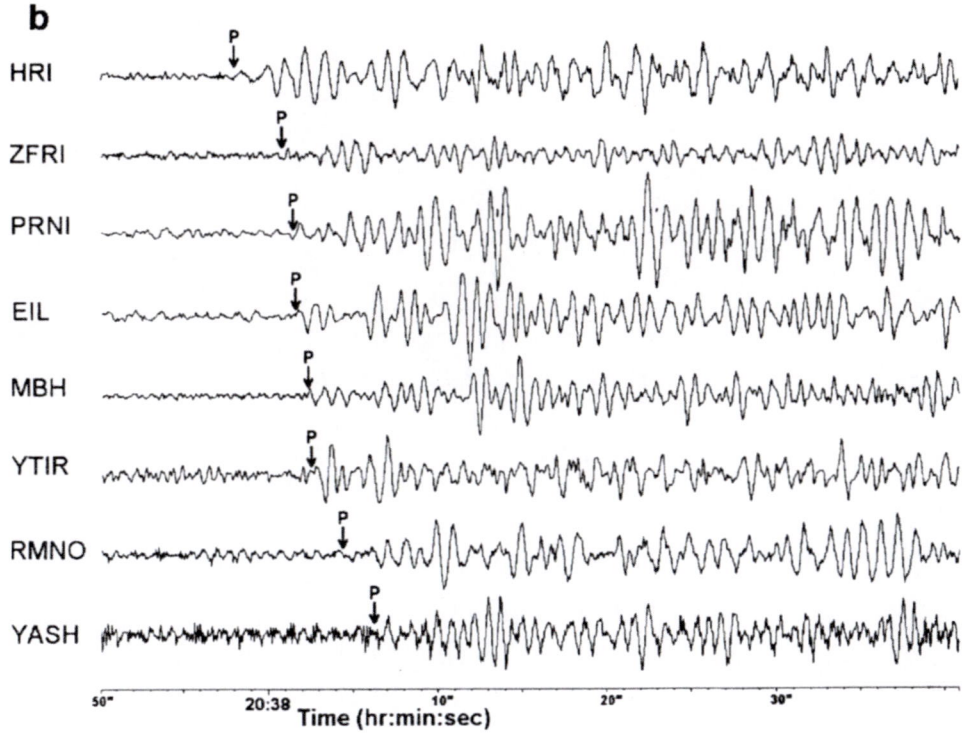

Figure 3b

EIL station the signals of the explosions under consideration appear considerably more attenuated than for more remote GRF, NORES and ARCES (CTBTO, 1998a).

We checked the detectability of the EIL BB station using recordings of the Indian and Pakistan nuclear tests. The 80-Hz channel of the station is supplied with the procedure of MURDOCK and HUTT (1983) for automatic detection (MH), including the following main stages:

1. The input time series is filtered. For simulation of the detector performance we used a specific ARMA (AutoRegressive-Moving Average) filter, applied to the 20-Hz channel of the BB station, instead of the built-in WWSSN-SP standard short-period filter.

2. Relative maxima and minima (peaks and troughs) of the filtered series are found, and successive peaks and troughs are differenced. These differenced values, together with their associated times, are named "*P-T*" time series.

3. Three thresholds (Th1, Th2 and Th3) are calculated. Detection is declared if *P-T* amplitude exceeds Th1 and two other *P-T* values exceed Th2 in a fixed time interval of usually four seconds. Th3 is used to search for onsets of detected events.

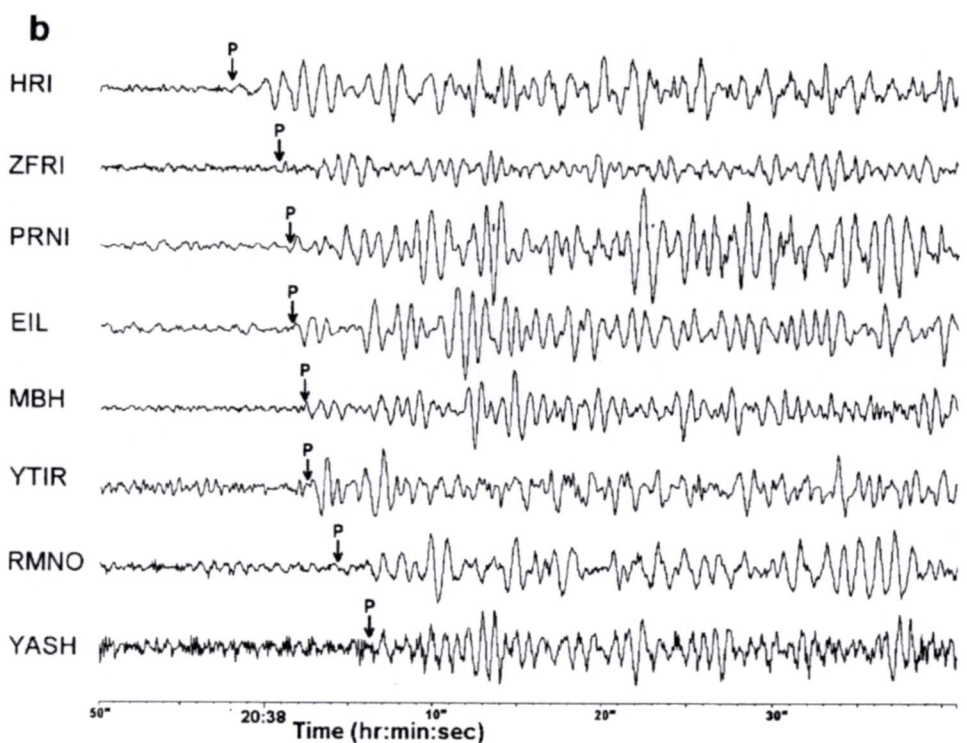

Figure 3c

The values of the thresholds: Th1 $= 4 * s$, and Th2 $= 3 * s$ and Th3 $= 2 * s$ are recommended as typical and provide a high rate of true detections and a low rate of false alarms, where s is one standard deviation of the P-T series for noise.

Figure 5 illustrates the P-T curves extracted from the EIL observed seismograms for the three nuclear explosions. The first two nuclear tests do pass the stated above detection criterion, but the second Pakistan nuclear test (Fig. 5c) does not pass. It would fit with lower Th1 and Th2, but at the expense of a higher risk of false alarms. Therefore, two additionally different detectors were tested using the broadband recordings of the events: the first is the STA/LTA detector (based on JOHNSON, 1979); and the other is adaptive statistically Optimal Detector (OD) by KUSHNIR *et al.* (1990).

To extract waveforms from background noise, we manually bandpass filtered all the seismograms of the broadband stations. Spectral analysis for the vertical channel of EIL (Fig. 4b) shows that the signal of the Indian nuclear test dominates noise in the frequency band 0.5–2 Hz. To provide less distortion of the signal we used the filter 0.5–5 Hz, yielding peak amplitudes $A_m \sim 0.08$ micron/sec (see Fig. 6). We did not manage to extract the signal for MRNI and JER for this explosion.

Table 2

Travel times (IASPEI91) and P-wave peak amplitudes for the selected events

#	Date	P-wave arrival time, sec		$t_m - t_c$, sec	Peak P ampl. mic/sec	Filter band, Hz
		Calculated t_c	measured t_m			
		New events, BB station EIL, vertical component				
1	1997/12/04	10:22:25.6	10:22:30.1	4.5	0.45	0.5–20
2	1998/01/05	17:04:03.5	17:04:07.5	4.0	0.48	0.5–20
3	1998/05/11	10:20:13.5	10:20:17.1	3.6	0.08	0.5–10
4	1998/05/28	10:21:49.8	10:21:54.3	4.6	0.39	0.5–20
5	1998/05/28	20:37:58.9	20:38:01.6	2.7	0.23	0.5–20
6	1998/05/30	07:00:24.0	07:00:28.5	4.5	0.043	2–10
7	1998/08/22	05:07:38.9	?	?	< 0.01	1–3
		Old STS nuclear tests, short period station MBH, vertical component				
1	1987/03/12	02:04:41.4	02:04:44.8	3.4	0.30	
2	1987/04/17	01:10:28.3	01:10:31.3	3.0	1.20	
3	1987/05/06	04:09:25.6	04:09:28.9	3.3	0.44	
4	1988/04/03	01:40:30.7	01:40:33.7	3.0	3.60	0.5–12
5	1988/11/12	03:37:29.2	03:37:33.9	4.7	0.23	
6	1988/12/17	04:25:31.6	04:25:35.5	3.9	0.21	
7	1989/01/22	04:04:30.9	04:04:34.8	3.9	0.77	
8	1989/02/12	04:22:30.6	04:22:35.0	4.4	0.37	

Though of smaller magnitude but somewhat closer, the Pakistan nuclear test on May 28, 1998 was well observed at short-period ISN stations (Fig. 3a), and broad band station EIL (Figs. 3c, 7a), with peak amplitude $A_m \sim 0.4$ micron/sec. A weak signal at JER was observed after a narrow band-pass filtering 0.5–1.5 Hz (Fig. 3c). We could not reveal the signal at MRNI due to a burst of local noise, or possible malfunction of the station.

The case of the second Pakistan nuclear test on May 30, 1998 is particularly complicated due to preceding arrivals from the other two events mentioned above: the strong earthquake in Afghanistan, and the local earthquake. Thus the body waves of the explosion are partially masked by the coda waves of these two events (see Fig. 8). However, the P waves of the nuclear test can be identified on the EIL seismogram in the 2–5 Hz frequency range with amplitude $A_m \sim 0.04$ micron/sec. A strong enough signal with $A_m \sim 0.1$ micron/sec was obtained at MRNI in the narrow frequency range 1.5–3 Hz (Fig. 3c), while at JER this signal was fully masked by noise.

The seismograms, bandpassed filtered in the chosen frequency ranges, were used for testing by the above mentioned STA/LTA detector. The output of this algorithm is set to be negative, and detection is declared once the output is greater than zero. The STA/LTA trigger may fail or be successful, depending on the selected frequency band and the background noise.

The Optimal Detector (OD) algorithm is based on autoregressive representation of seismic noise, preceding the signal, inverse filtering, and producing "whitened"

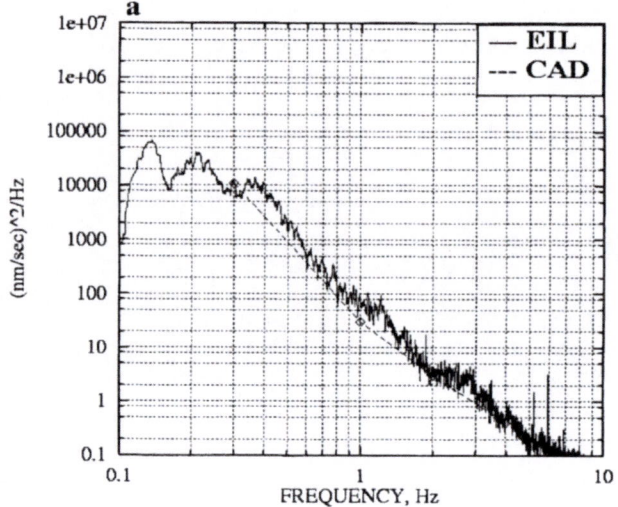

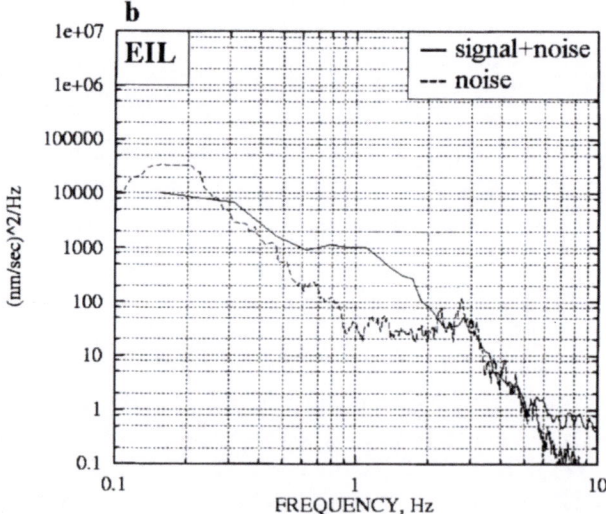

Figure 4

Velocity spectral density of noise at the EIL station compared to: (a) CAD BB station, (b) the Indian test signal + noise on a 2-sec interval.

noise. The OD demonstrates more robust behavior and enhanced sensitivity at relatively wider frequency range than the STA/LTA algorithm. Detection is declared once the autocovariance of the "whitened" noise exceeds a given threshold.

Comparative performance of the OD and STA/LTA, applied to the filtered (0.5–5 Hz) BB records (20 Hz) of the considered explosions, is presented in Figures 6–8. Both detectors were successful at EIL records of the Indian and the first Pakistan

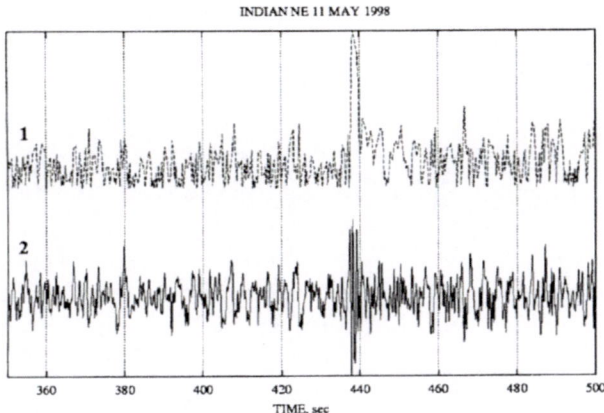

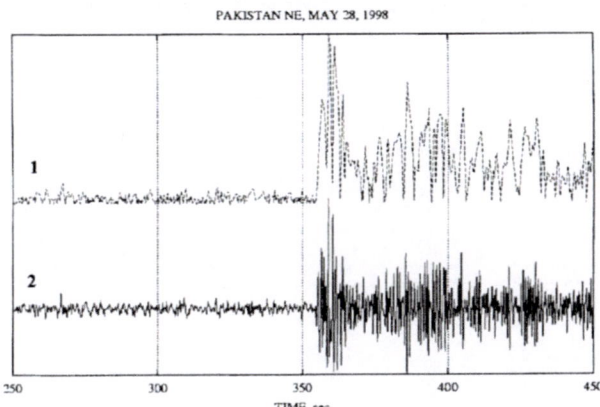

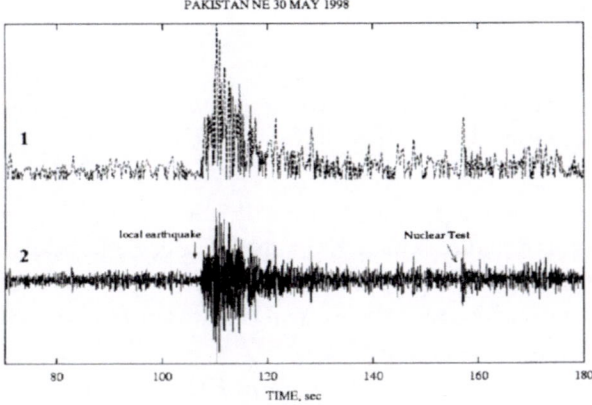

Figure 5
Application of the Murdock-Hutt detector algorithm to the BB EIL seismograms for the three nuclear explosions. *P-T* curves (1) are obtained using output (2) of the WWSSN_SP filter.

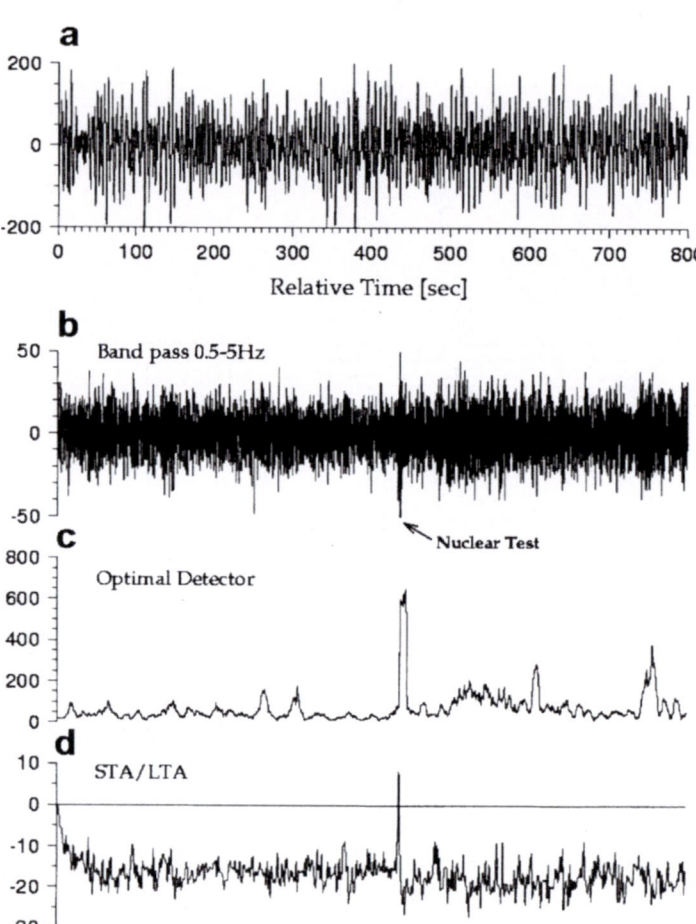

Figure 6

Detection of the Indian test at EIL station: (a) vertical unfiltered 20 Hz channel record; (b) BP filtered data; (c) Optimal Detector output; (d) STA/LTA output.

tests (Figs. 6, 7a). Detection of the second Pakistan NT at EIL was complicated by the nearby local earthquake, which blurs NT triggering, though the OD demonstrates better performance. Records of the Pakistan explosions at JER (Fig. 7b) and MRNI (Fig. 8b) show a poor SNR, resulting in a failure of the STA/LTA, whereas the OD provides reliable detections.

Using the records of the most quiet station EIL, the sensitive OD algorithm and different band-pass filters, we tried in addition to detect the Kazakh chemical explosion. The detector output showed some uprising in the expected signal time interval, however it was still too low for reliable decision (Fig. 9).

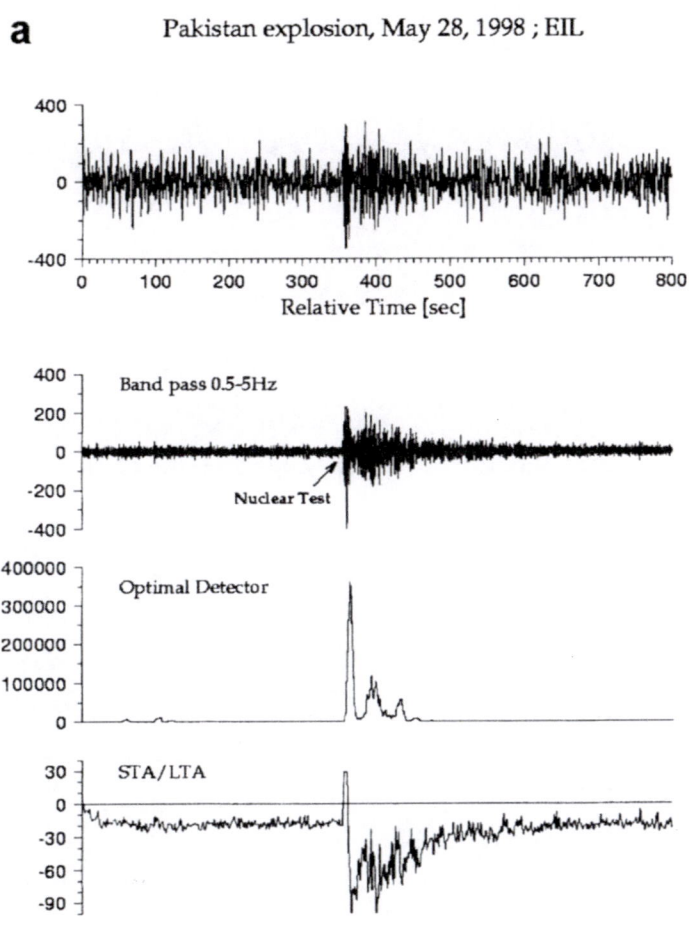

Figure 7
Detection of the first Pakistan test (28.5.98) at EIL (a) and JER (b) BB stations.

Spectral Null Feature

We calculated *P*-wave spectra from seismograms (after removing instrument response) of the test on May 28, 1998; the only one that triggered short-period ISN stations (see Figs. 2 and 3a). All spectra were computed using a ~15 sec time window at vertical recordings, including *P* and *P*-coda waves, and smoothed in the 0.5 Hz triangular window (Fig. 10a). The spectra at all stations showed pronounced coherent spectral minima (nulls) at about 1.7 Hz, which may be interpreted as interference of *P* and *pP* phases, possibly complicated by nonlinear surface effects such as spall (LAY, 1991). KULHANEK (1971) reported similar observations of 1 Hz spectral nulls for Nevada nuclear tests, treated as the interference effect. The

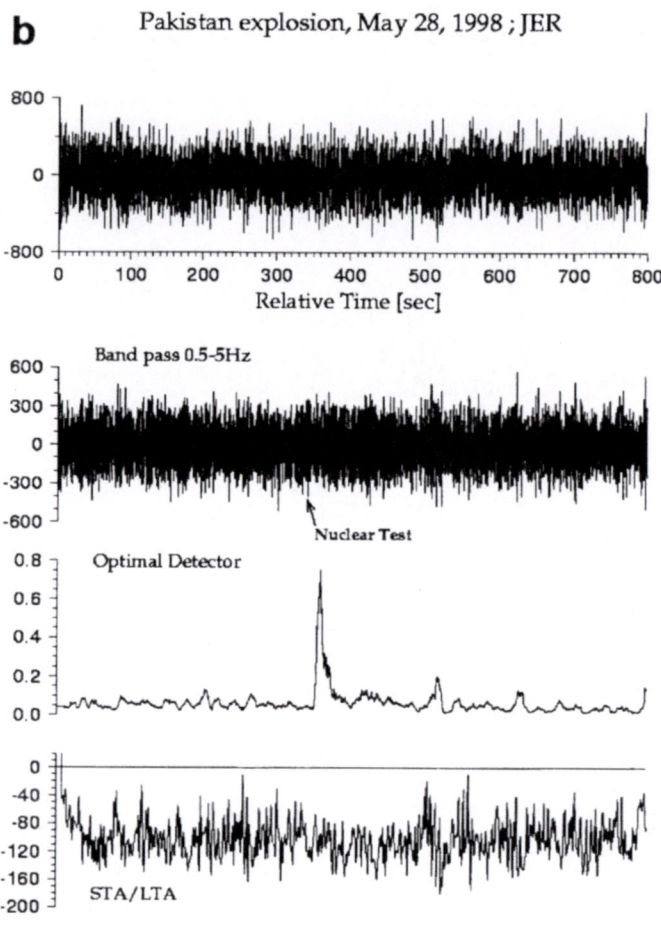

Figure 7b

relatively high null frequency 1.7 Hz possibly corresponds to the reported shallow depth of the test placed in a horizontal shaft of a steep mountain (WALLACE, 1998), compared with deeper Nevada tests. The same 1.7-Hz coherent spectral nulls were also observed at all three components of the BB EIL station for the May 28 explosion, but not for the Indian test.

The same spectral analysis was done for a nearby earthquake in Pakistan with similar magnitude and waveforms (Fig. 3b). The earthquake spectra (Fig. 10b) obtained at the same short-period stations do not show coherent nulls (also as spectra at BB EIL), thus rejecting the site-effect version. WALLACE (1998) notes decidedly more complex waveforms for the May 28 Pakistan test, than for the Indian explosion (see also Fig. 3c), explained possibly by wave scattering from the complex topography in the test region.

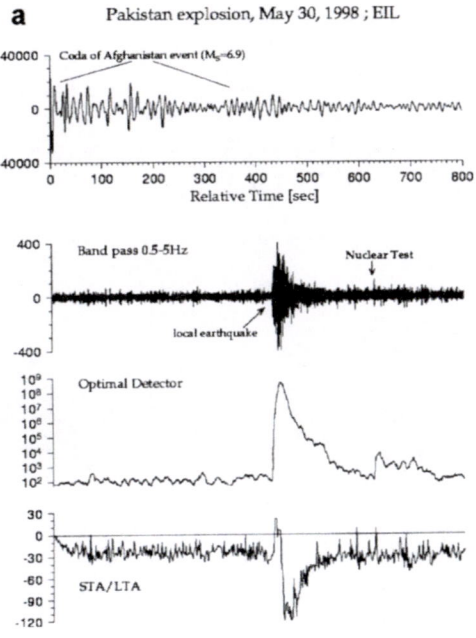

Figure 8
Detection of the second Pakistan test (30.5.98) at EIL (a) and **MRNI** (b) BB stations.

Nevertheless, GITTERMAN et al. (1999), based on ISN observations of Semipalatinsk and Chinese nuclear tests and nearby earthquakes, showed that this spectral
null feature, considered separately, cannot serve as a reliable identification
characteristic of nuclear explosions due to its non-stability over the entire nuclear
test population and its presence in the recordings of some earthquakes.

Spectral Discrimination

TAYLOR and MARSHALL (1991) applied spectral ratio for discrimination between
Soviet (STS) nuclear tests and earthquakes recorded at the United Kingdom short-
period teleseismic arrays. Using 5-sec windows (including *P* waves), the spectral
ratios from an array beam in the 0.5–1 Hz and 2–3 Hz frequency bands provided
the best discrimination performance. It was concluded that the explosions are
characterized by more high-frequency energy (resulting in lower spectral ratios) than
the shallow Central Asia earthquakes.

Similar results were obtained by GITTERMAN et al. (1999) using slightly different
processing parameters, with higher ratios (low-to-high frequency energy) for
teleseismic earthquakes, compared to nuclear explosions. The database included

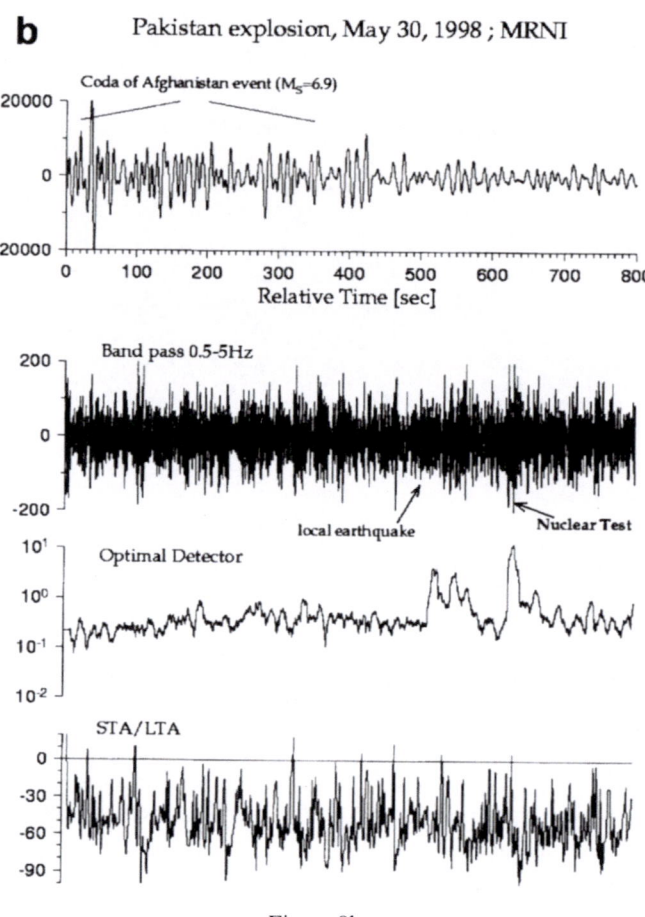

Figure 8b

tests from STS, Novaya Zemlya and Lop Nor (China), recorded by the short-period Israel Seismic Network. A 15-sec window, comprising *P*- and *P*-coda waves, was selected for computation of spectral ratios of energy in low (0.6–1 Hz) and high (1–3 Hz) frequency ranges, providing the best performance. Additionally, this study included the application to the same spectra of a recently proposed spectral semblance discriminant (GITTERMAN *et al.*, 1998). The semblance statistic, commonly used in seismic prospecting for phase correlation of seismic traces in the time domain (e.g., NEIDELL and TANER, 1971), was modified to assess the coherency S_f of smoothed spectral shapes for different stations (channels) in a frequency band [F1, F2]:

$$S_f = \frac{1}{N} \sum_{F_1}^{F_2} \left[\sum_{k=1}^{N} (S_{ki} - \bar{S}_k) \right]^2 \bigg/ \sum_{F_1}^{F_2} \sum_{k=1}^{N} (S_{ki} - \bar{S}_k)^2 \qquad (1)$$

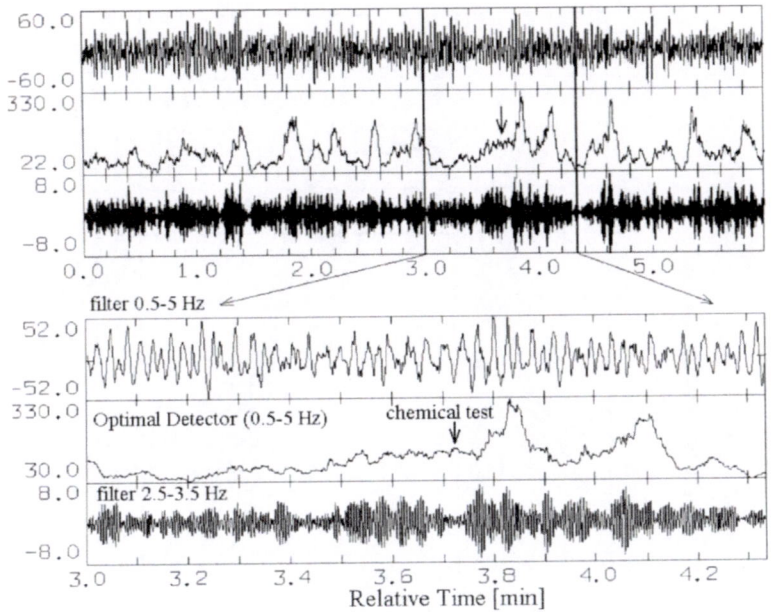

Figure 9
Detection trial of the Kazakh chemical explosion at EIL BB station.

where $S_{ki} = \log_{10} S_k(f_i)$ is $\log_{10}$ (spectral amplitude at the k-th station), $\bar{S}_k$ is the average spectral level and N is the number of used stations (channels). The network-based spectral semblance estimates showed higher coherency for teleseismic earthquakes than for nuclear explosions (GITTERMAN et al., 1999).

For the selected recent events (Table 1), the two discrimination statistics were estimated for a subset of nine short-period vertical ISN stations (shown in Fig. 2), and also for three components (channels) of the EIL broadband station. The ratios were then averaged for a given event over the stations, or over the three BB components. The second Pakistan nuclear test and the STS chemical explosion did not trigger the short-period and broadband stations. Both signals on the 20 Hz continuous recording are weak and comparable with the noise level (see Figs. 3c and 9), therefore the spectral ratio and semblance were not calculated.

Both multi- and single-station estimates correspond well to the successful discrimination results obtained for the former Semipalatinsk and Chinese nuclear tests (GITTERMAN et al., 1999). The new nuclear tests also demonstrate lower values of energy ratio and spectral semblance than nearby earthquakes (Table 3, Fig. 11). The analyzed waves (P and P coda) of teleseismic earthquakes have relatively more low-frequency energy, as predicted from source physics phenomena and more coherent spectral shapes than nuclear explosions. The short-period explosion signals (including scattered waves in the P coda) are more influenced by inhomogeneities in

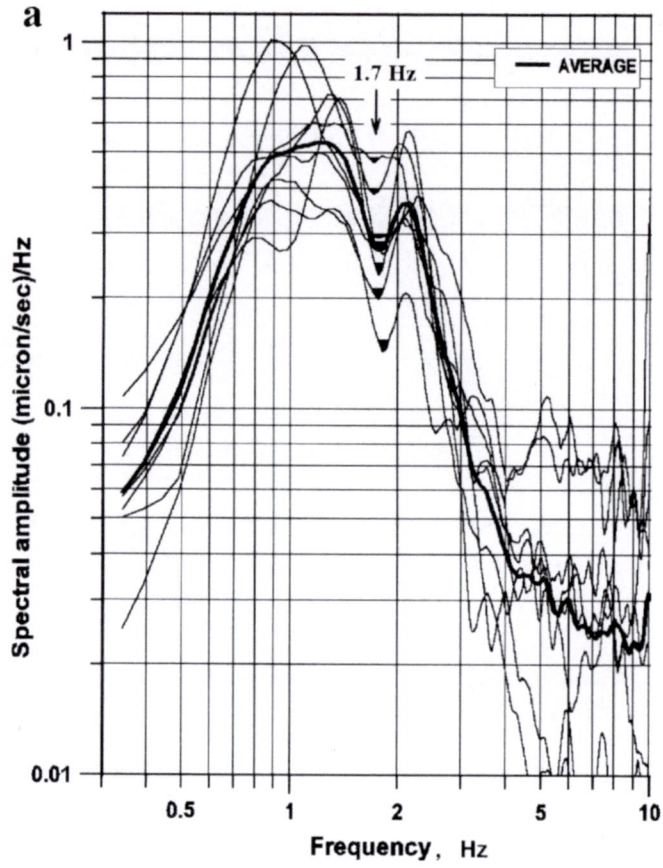

Figure 10

Spectra of the first Pakistan test, observed at short-period stations, showing a spectral null source feature at 1.7 Hz (a), compared with the same day nearby Pakistan earthquake of a similar magnitude (b).

the upper crust beneath the ISN stations, with varying site structure, thus leading to a decrease of spectra coherency and associated semblance values. The results show that both types of discriminant estimates, using the short-period (Fig. 11a) or broadband records (Fig. 11b), being rather different for a given event (especially the energy ratios), nevertheless provide a reliable separation of nuclear explosions and earthquakes.

Discussion and Conclusions

We investigated detectability at the two IMS auxiliary stations EIL and MRNI and cooperating station JER using 20-Hz recordings of the Indian ($m_b = 5.2$) and the

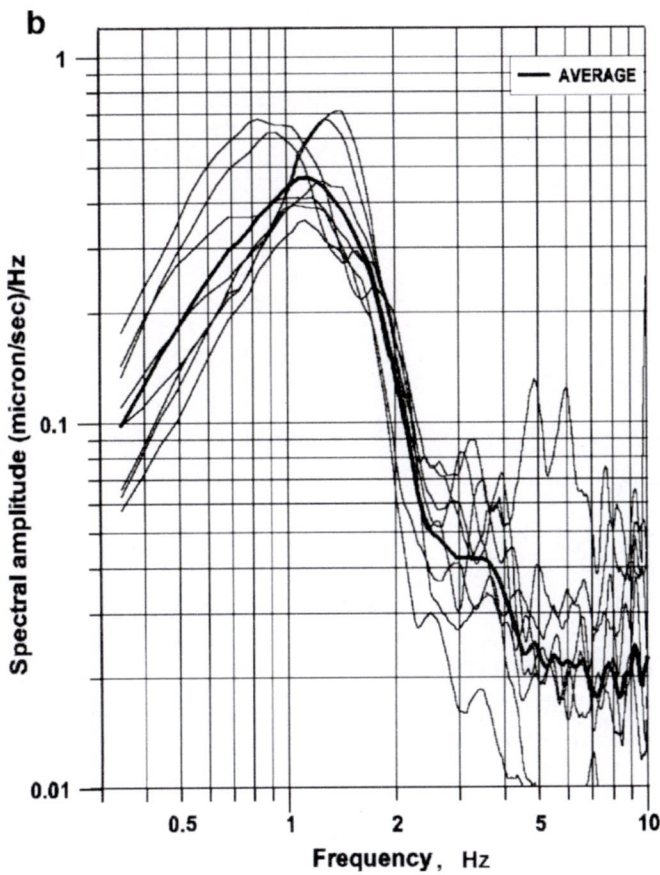

Figure 10b

two Pakistan nuclear tests (m_b = 4.6, 4.8). Additionally the detection of the Kazakh chemical test (m_b = 3.7) was tested. Three detection procedures were compared: STA/LTA, Murdock-Hutt and Adaptive Statistically Optimal Detector. All nuclear tests were detected at EIL, while at MRNI we succeeded only with the first Pakistan NT, and at JER with the second one. A station detectability is determined by the maximum of frequency-dependent SNR, therefore the OD proved to be the most efficient procedure, due to its more accurate adaptation to ambient noise spectra.

Unfortunately, we cannot precisely state detection of the Kazakh explosion at EIL, whereas it was clearly observed at IMS stations NOA (37.7°) and ILAR (60.6°). Therefore, due to unfavorable propagation conditions for the Asian events, the EIL detectability may be approximated by the magnitude threshold m_b = 4.0–4.5, within a 25°–38° distance range. Evidently the magnitude threshold for MRNI and JER should be higher for the same type of events.

Table 3

Average spectral discrimination results for Indian and Pakistan tests based on ISN and BB EIL station records

No.	Date	Energy ratio (0.6–1)/ (1–3 Hz)		Semblance (0.6–2 Hz)		Region, event
		ISN	EIL	ISN	EIL	
1	97/12/04	1.98	0.903	0.840	0.942	Pakistan earthquake
2	98/01/05	5.39	1.890	0.876	0.879	Pakistan earthquake
3	98/05/11	–	0.532	–	0.847	India nuclear test
4	98/05/28	1.248	0.588	0.737	0.849	Pakistan nuclear test
5	98/05/28	1.591	0.666	0.866	0.946	Pakistan earthquake

The case of the second Pakistan test simulates in a manner the scenario of a clandestine test masked by two other events: a strong earthquake in Afghanistan and a local earthquake near EIL. In spite of a poor SNR, the explosion signal was detected by the OD at both IMS stations EIL and MRNI.

The spectra of the first Pakistan NT exhibited a pronounced spectral null at 1.7 Hz for all three components of the broadband EIL station, as well as at the vertical component of the short-period ISN stations. For this ground-truth explosion with a reported shallow source depth (WALLACE, 1998), the interference of P and pP phases seems the most reasonable explanation for this phenomenon. Other possible reasons, the spallation process (LAY, 1991) and wave scattering from the complex topography in the source zone, also relate to surface source effects from very shallow seismic events, i.e. explosions.

However, the spectral null feature, considered separately, cannot serve as a reliable identification characteristic of nuclear explosions, because not all the tests produce the nulls, whereas some earthquakes display this effect. The multi-channel spectral discrimination analysis, based on the spectral ratio of low-to-high frequency energy and semblance of spectral curves, provided a reliable discrimination between the recent NT and several nearby Pakistan earthquakes. The results showed compliance with the former analysis of Soviet and Chinese NT recorded at seismic stations in Israel, where nuclear tests demonstrated lower values of energy ratio and spectral semblance than earthquakes (GITTERMAN et al., 1999).

It should be noted that both spectral discriminants originally were successfully applied to small local quarry blasts, underwater explosions, and earthquakes (GITTERMAN et al., 1998), where the semblance (in the 1–12 Hz band) and the ratio (for 1–3 Hz and 6–8 Hz bands) showed higher values for explosions, contrasting to the teleseismic case. The higher ratio was explained by path effects due to local

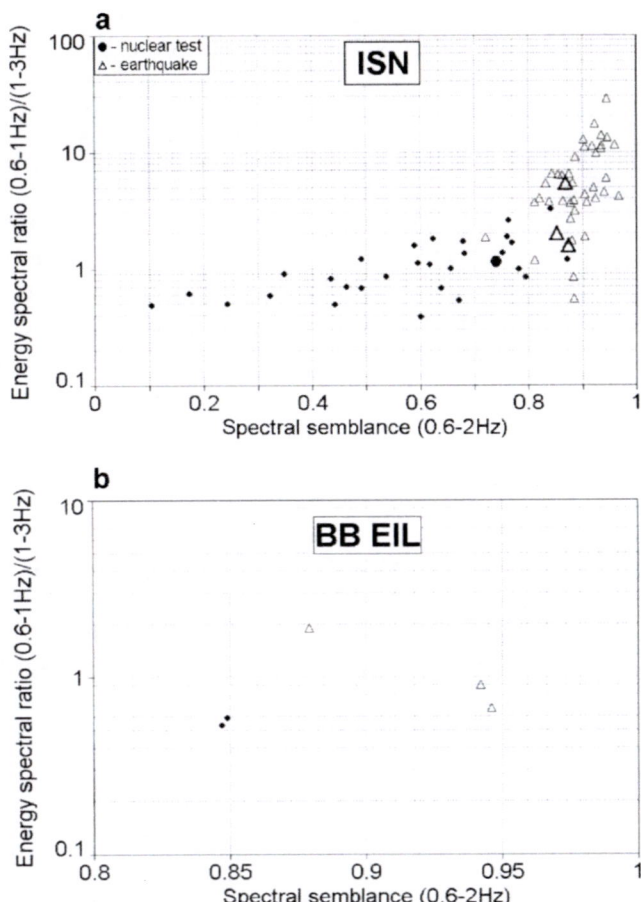

Figure 11

Application of the semblance and energy ratio discriminants to the recent nuclear tests and nearby earthquakes (see Table 3): (a) multi-station ISN estimates (large symbols), compared with the results for old Eurasian events (small symbols) (GITTERMAN *et al.*, 1999); (b) single-station, three-component estimates for the EIL BB station

geological settings (thick soft sediments). The higher semblance was related to specific source features of quarry blasts (ripple-firing) and underwater explosions (bubbling), producing the azimuth-independent spectral modulation, opposites earthquakes having less spectra similarity (at local distances) due to radiation pattern and directivity effects.

In some way the spectral energy ratio is similar to the $M_s:m_b$ discriminant used by the IDC for event screening, however it can be applied (as well as the spectral semblance) to small local events where $M_s:m_b$ technique is not relevant, whereas for teleseismic events the spectral discriminants can provide an additional independent check.

For the recent tests and nearby earthquakes, we observed at the EIL stations consistent delays of *P* arrivals of about 4 sec relative to the IASPEI91 model. Slightly shorter delays of about 3.7 seconds in average were found for old STS explosions recorded at the ISN station MBH situated within a short distance of EIL. Similar residuals for MBH imply that only a few tenths of this effect are site related. *P*-arrival lags 3.5–4 sec were also observed for the IMS station MRNI and BB station JER. Similar delays at different stations evidence a path- rather than site-effect.

The obtained results will contribute to calibration of the IMS auxiliary stations EIL and MRNI regarding Asian seismic events. Possibly, events from Africa, Europe and other areas have different station corrections relative to IASPEI91, and this should be checked when ground-truth events from these areas become available at Israel stations.

Acknowledgements

We are grateful to William R. Walter for constructive comments resulting in a significant improvement of the manuscript. The study was supported by the Defense Threat Reduction Agency of the U.S. Department of Defense under Contract No. DSWA01-97-C-0151, and the Ministry of the National Infrastructures, Israel.

REFERENCES

BARKER, B., CLARK, M., DAVIS, P., FISK, M., HEDLIN, M., ISRAELSSON, H., KHALTURIN, V., KIM, W. Y., MCLAUGHLIN, K., MEADE, C., MURPHY, J., NORTH, R., ORCUTT, J., POWELL, C., RICHARDS, P. G., STEAD, R., STEVENS, J., VERNON, F., and WALLACE, T. (1998), *Monitoring Nuclear Tests*, Science *281*, 1967–1968.

CTBTO (1998a), *Preliminary IDC reviewed event solution for the announced Pakistan underground nuclear test 30-May-1998*, Preparatory Commission for the CTBTO, Provisional Technical Secreteriat, Vienna.

CTBTO (1998b), *Preliminary IDC reviewed event solution for the announced Kazakhstan Calibration Explosion* 22-August-1998, Preparatory Commission for the CTBTO, Provisional Technical Secreteriat, Vienna.

GITTERMAN, Y., PINSKY, V., and SHAPIRA, A. (1998), *Spectral Classification Methods in Monitoring Small Local Events by the Israel Seismic Network*, J. of Seismol. *2*, 237–256.

GITTERMAN, Y., PINSKY, V., and SHAPIRA, A. (1999), *Spectral Discrimination Analysis of Eurasian Nuclear Tests and Earthquakes Recorded by the Israel Seismic Network and the NORESS Array*, Physics Earth and Planet. Inter. *113*, 111–129.

JOHNSON, C. E. (1979), *CEDAR — An Approach to the Computer Automation of Short-period Local Seismic Networks*, Ph.D. Dissertation, California Institute of Technology, Pasadena, CA.

KHALTURIN, V., RAUTIAN, T., and RICHARDS, P. (1998), *The Seismic Signal Strength of Chemical Explosions*, Bull. Seismol. Soc. Am. *88*, 1511–1524.

KULHANEK, O., (1971), *P-wave Amplitude Spectra of Nevada Undergroud Nuclear Explosions*, Pure appl. geophys. *88*, 121–136.

KUSHNIR, A. F., LAPSHIN, V. M., PINSKY, V. I., and FYEN, J. (1990), *Statistically Optimal Event Detection Using Small Array Data*, Bull. Seismol. Soc. Am. *80*, 1934–1947.

LAY, T. (1991), *Teleseismic manifestations of pP: Problems and paradoxes*. In *Explosion Source Phenomenology*, (S.R. Taylor, H.J. Patton, and P.G. Richards, eds.), Geophysical Monograph *65*, 109–125.

MURDOCK, N. J and HUTT, C. R. (1983), *A New Event Detector Designed for the Seismic Research Observatories*, USGS Open-File Report 83-785, October 1983.

NEIDELL, N. S. and TANER, M. T. (1971), *Semblance and Other Coherency Measures for Multichannel data*, Geophysics *36*, 482–497.

TAYLOR, S. R. and MARSHALL, P. D. (1991), *Spectral Discrimination between Soviet Explosions and Earthquakes Using Short-period Array Data*, Geophys. J. Int. *106*, 265–273.

VILA J. (1998), *The Broadband Seismic Station CAD (Tunel del Cadi, Eastern Pyrenees): Site Characteristics and Background Noise*, Bull. Seismol. Soc. Am. *88*, 297–303.

WALLACE, C. T. (1998), *The May 1998 Indian and Pakistan Nuclear Tests*, Seism. Res. Lett. *69*, 386–393.

WALTER, W. R., RODGERS, A. J., MAYEDA, K., MYERS, S. C., PASYANOS M., and DENNY, M. (1998), *Preliminary Regional Seismic Analysis of Nuclear Explosions and Earthquakes in Southwest Asia*, Proc. 20th Symposium on Monitoring a Comprehensive Test-Ban Treaty, Santa Fe, 442–451.

(Received July 7, 1999, revised January 3, 2001, accepted January 19, 2001)

To access this journal online:
http://www.birkhauser.ch

Pure appl. geophys. 159 (2002) 803–830
0033–4553/02/040803–28 $ 1.50 + 0.20/0

▌**Pure and Applied Geophysics**

Discriminating Between Large Mine Collapses and Explosions Using Teleseismic P Waves

DAVID BOWERS[1] and WILLIAM R. WALTER[2]

Abstract — Some of the most suspicious seismic disturbances under the Comprehensive Nuclear-Test-Ban Treaty (CTBT) are likely to be those associated with mining, as they are shallow, and at least some have an explosion-like m_b:M_s signature. Previous research highlighted the potential of broadband teleseismic P waves as a way of identifying large mine tremors. Broadband teleseismic P from two suspected large mine collapses, one in Germany (1302 UT, 13 March 1989, 5.4 m_b) and another in Wyoming (1526 UT, 3 February 1995, 5.3 m_b), show differences in character despite the similarity of the reported ground failure and mine types. We apply a full moment-tensor analysis to the teleseismic P waves and show that the data are inconsistent with either a shallow explosion or an earthquake (double-couple) at depth, but this method is unable to distinguish between a shallow dip-slip source and a closing-crack moment tensor. However, three-component surface-wave seismograms recorded at regional distances fit the shallow closing-crack model, but are inconsistent with a shallow earthquake source, because strong Love waves, expected from a double-couple source, are not observed at a number of stations well distributed in azimuth. Here, we restate the equivalence for shallow sources of the closing-crack model and a gravitational collapse model. We use the latter to model the broadband P waves from these mine tremors and show that, while non-unique, the differences in the observed broadband P waves from the two tremors can be attributed to the area, amount of collapse, depth, and rate of collapse. The collapse model predicts negative first-motion for all P waves in contrast to the positive polarity expected from explosions. Thus, the broadband teleseismic P waves have the potential to discriminate between large collapses and explosions.

Key words: Seismology, Comprehensive Test-Ban Treaty, discrimination, mine, collapse, P waves, modelling.

Introduction

Effective verification of the Comprehensive Nuclear-Test-Ban Treaty (CTBT) requires the identification of possible nuclear explosions. Mine tremors (seismic disturbances located in a mining region) are shallow and the limited data available suggest that many may have an explosion-like m_b:M_s signature (e.g., BENNETT *et al.*, 1994). Thus, at least some mine tremors are likely to appear suspicious under a CTBT (i.e., a possible violation), possibly requiring an on-site inspection. Clearly, it

---
[1] AWE Blacknest, Brimpton, Reading, Berkshire. RG7 4RS. UK, E-mail: bowers@blacknest.gov.uk
[2] Lawrence Livermore National Laboratory, P.O. Box 808, L-205, Livermore, CA 94551-0808. USA.
E-mail: bwalter@llnl.gov

is desirable to minimise the number of unnecessary on-site inspections or "false alarms."

BOWERS and DOUGLAS (1997) analysed short-period (SP) and broadband (BB) teleseismic P seismograms from seismic disturbances in many mining regions, and noted that in general the SP P waves are simple (having a short-duration pulse with little or no coda), with negative (dilatational) first-motion. Further, BOWERS and DOUGLAS (1997) showed that even where the SP first-motion was unclear, the BB P onset generally has a clear negative first-motion. BOWERS and DOUGLAS (1997) concluded that the first motion of BB teleseismic P has the potential to discriminate between shallow mine tremors and explosions (as the BB P first-motion from an explosion is expected to be positive). However, the physical basis for the BB teleseismic P discriminant is still not understood fully.

BOWERS (1997) showed that negative-polarity teleseismic P waves and three-component surface waves from a large mine tremor, on 30 October 1994, in the Welkom district of South Africa, are best explained by a double-couple moment tensor at a depth of 2.3 km. However, it is well established that not all mine tremors are due to shallow double-couple sources (e.g., WONG and McGARR, 1990; McGARR, 1992; TAYLOR, 1994; PECHMANN et al., 1995).

BOWERS and DOUGLAS (1997) showed SP and BB teleseismic P seismograms from two unusually large mine tremors. The first occurred on 13 March 1989, in the Werra River potash mining district, near Völkershausen, Germany (Fig. 1a). The second occurred on 3 February 1995 in the Trona mining district of southwest Wyoming, USA (Fig. 1b). Table 1 compares standard seismic parameters and gives a summary of observations from the two disturbances which are suspected of being large mine collapses (e.g., BORMANN et al., 1992; AHORNER 1993; BENNETT et al., 1994; PECHMANN et al., 1995). Here, we attempt to discriminate the two disturbances from explosions using m_b:M_s, and try to understand the physical processes governing the generation of the teleseismic P waves using a moment-tensor analysis and by forward modelling.

PECHMANN et al. (1995) report negative-polarity local P first-motions from the Wyoming disturbance, and show low-pass filtered (0.025–0.050 Hz) three-component regional surface waves that have weak Love-wave energy at a number of stations, suggesting a non-double couple source. This is supported by our analysis of three-component seismograms recorded at the stations shown in Figure 1b, at a wide range of azimuths, which all show weak Love energy. PECHMANN et al. (1995) conclude that the surface waves are best modelled by a horizontal closing-crack representing a large mine collapse at a depth of about 500 m, and that the first-motion observations are consistent with this model. Figure 2a compares the observed seismograms with synthetic (calculated using the horizontal closing-crack source of PECHMANN et al., 1995) at TUC and ANMO. PECHMANN et al. (1995) argue that a collapse mechanism is appropriate by showing that gravitational potential energy is sufficient to explain the observations, without the need for any seismic energy of tectonic origin.

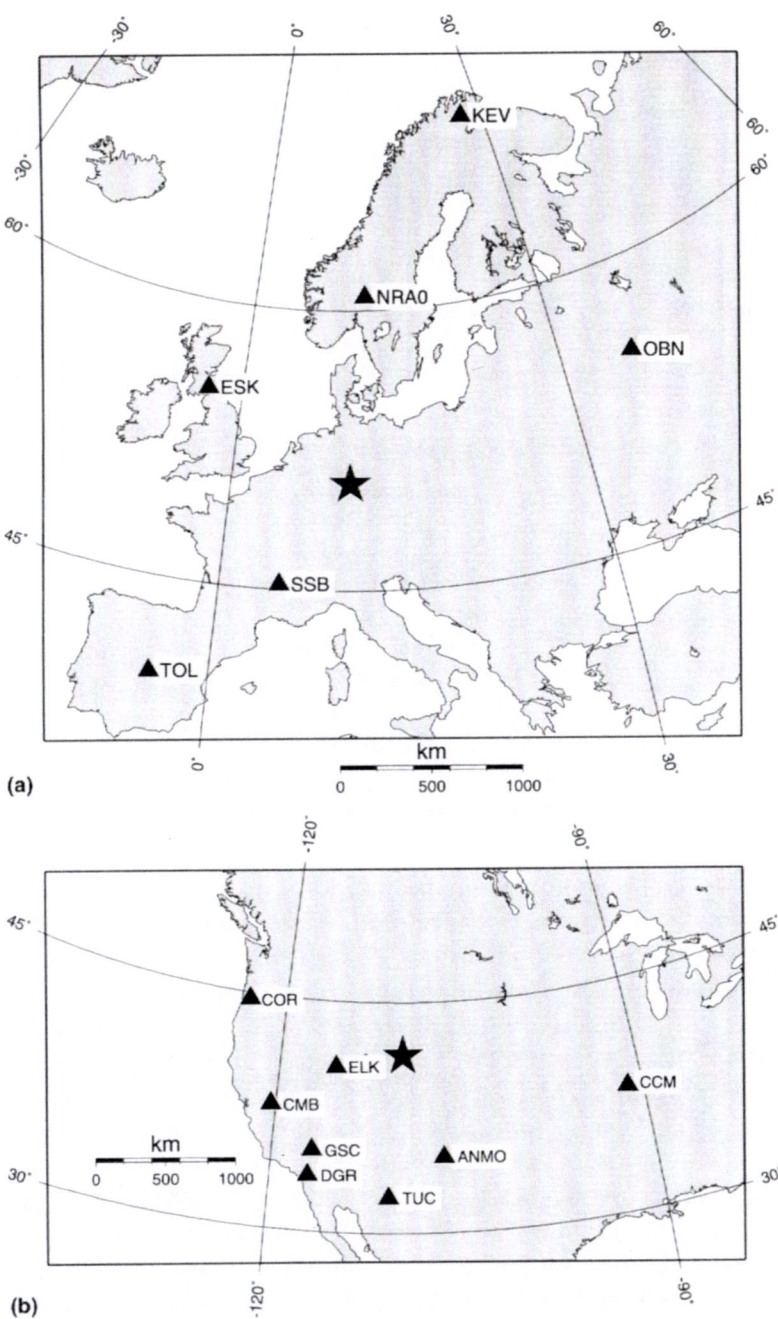

Figure 1
Azimuthal-equidistant plot, centred on the ISC epicentre, showing selected three-component stations recording surface waves from (a) the Völkershausen, and (b) the Wyoming, seismic disturbances.

Table 1

A comparison of observations from the two suspected mine collapses discussed in this article. ISC, International Seismological Center; EDR: Earthquake Data Report; E_p, gravitational potential energy

	Völkershausen, Germany	Solvay Mine, Wyoming
	Seismological Observations	
Date:	13 March 1989	3 February 1995
Origin/Depth (ISC):	13:02:14.8 UT, 1 km	15:26:10.7 UT, 1 km
Location (ISC):	50.72° N, 9.91° E	41.52° N, 109.64° W
Magnitude:	EDR 5.4 m_b, 4.7 M_s	EDR 5.3 m_b, 4.6 M_s
	ISC 5.4 m_b, 5.2 M_s	ISC 5.3 m_b, 4.6 M_s
Local/Regional P:	All Negative (Dilatational)[†ℓ]	All Negative[‡]
Aftershocks:	Few. Max. 1.0 M_L[†]	None ($M_L > 2.5$)[‡]
	Mining Operations and Observations	
Material:	Marine Evaporite Carnallite[†]	Terrestrial Evaporite Trona[‡]
Method:	Room and Pillar[†] ≈ 3200 Pillars[†ℓ]	Room and Pillar[‡] 1000's Pillars
Subsidence:	6.8 km^2 [†ℓ] Maximum 1.0 m[†ℓ] Average 0.7 m[*]	2.0 km^2 [‡] Maximum 0.9 m[‡] Average 0.6 m[‡]
Depth:	850 m[†]	488 m[‡]
Inferred E_p:	1.3 × 10^{14} J[†ℓ] 1.1 × 10^{14} J[*]	1.4 × 10^{13} J[‡]

[†] AHORNER (1993); [*] BENNETT *et al.* (1994); [ℓ] BORMANN *et al.* (1992); [‡] PECHMANN *et al.* (1995).

AHORNER (1993) interpreted regional and local distance P first-motions from the Völkershausen disturbance as evidence of an implosive focal mechanism. AHORNER (1993) proposed a source-time history comprising five sub-events increasing in magnitude, with a total duration of about 2 sec, from analysis of seismograms from two local distance stations (< 9 km). BORMANN *et al.* (1992) also suggest a multiple source, but with six "major onsets," with a total duration of less than 5 sec from analysis of local, regional, and teleseismic P waveforms. BORMANN *et al.* (1992) and BENNETT *et al.* (1994) argue that the available gravitational potential energy exceeds the seismic energy estimated using magnitude-energy relations.

KNOLL (1990) proposes that the Völkershausen disturbance is a "fluid-induced tectonic rockburst," meaning that the collapse process was initiated by slip on a tectonic fault in the overburden, the fault slip being induced by seepage of saline waste water. AHORNER (1993) suggests that the collapse was triggered by a routine blasting operation in the mine (also reported in the Bulletin of the International Seismological Centre; ISC).

We note that regional three-component seismograms from the Völkershausen disturbance, recorded at the stations shown in Figure 1a, also lack significant

Table 2

Stations recording regional distance Rayleigh waves, and M_s^{MB} values calculated using the formula of MARSHALL *and* BASHAM *(1972).* $\overline{M}_s$ *is the mean value*

	Wyoming Surface Waves	Distance Δ	Azimuth	M_s^{MB}
ANMO	Albuquerque, New Mexico, USA	7.0°	158.1°	3.57
BAR	Barrett Dam, Calif., USA	10.5°	214.7°	3.67
CCM	Cathedral Cave, Missouri, USA	14.6°	97.7°	3.63
CMB	Columbia College, California, USA	9.0°	250.7°	3.74
COR	Corvallis, Oregon, USA	10.5°	291.6°	3.60
DGR	Domenigoni Valley Res., Calif., USA	9.8°	219.0	3.79
FFC	Flin Flon, Canada	14.1°	18.4°	3.68
GSC	Goldston, California, USA	8.4°	224.5°	3.66
MLAC	Mammoth Lakes, Calif. USA	8.1°	244.3°	3.69
NEE	Needles, Calif, Calif. USA	7.7°	211.8°	3.69
PAS	Pasadena, Calif. USA	10.0°	225.2°	3.65
PFO	Pinon Flat, Calif. USA	14.4°	203.5°	3.95
RPV	Rancho Palos Verde, Calif. USA	10.4°	224.6°	3.49
TUC	Tucson, Arizona, USA	9.2°	186.0°	3.74
USC	USC, Los Angeles, Calif., USA	10.1°	225.2°	3.60
			$\overline{M}_s$	3.68 ± 0.06 (95%)
	Völkershausen Surface Waves	Distance Δ	Azimuth	M_s^{MB}
ARU	Arti, Russia	28.9°	59.5°	4.63
BHM	Barham, Kent, UK	5.5°	278.5°	4.45
CWF	Charnwood forest, UK	7.3°	290.5°	4.53
EKA	Eskdalemuir, Scotland	9.1°	305.3°	4.53
KEV	Kevo, Finland	20.7°	16.8°	4.66
KIV	Kislovodsk, Russia	23.1°	94.3°	4.21
LLW	Llanuwychllyn, Wales	8.7°	289.5°	4.43
MMY	Middlesmoor, Yorkshire, UK	8.0°	300.2°	4.61
NRA0	Norsar, Norway	10.1°	4.6°	4.52
OBN	Obninsk, Russia	16.6°	64.1°	4.47
SBD	St Breward, Cornwall, UK	9.3°	274.7°	4.62
SCK	South creek, Norfolk, UK	6.1°	294.4°	4.70
SSB	Saint Sauvier, Badol, France	6.5°	215.5°	4.62
TOL	Toledo, Spain	14.6°	227.4°	4.70
			$\overline{M}_s$	4.55 ± 0.08 (95%)

Love-wave energy in the pass band 0.02–0.05 Hz over a range of azimuths that suggests a non-double couple source. Figure 2b compares observed and synthetic seismograms for a horizontal closing-crack moment tensor, in a Poisson solid, at a depth of 850 m (with $M_{zz} = 1.5 \times 10^{17}$ Nm) at ESK and OBN. The fit between observed and synthetic waveforms is reasonable, suggesting that the surface-wave data can be explained (at least to first-order) using a closing-crack moment tensor.

The similarity of the surface-wave modelling results and the parameters in Table 1, from the Wyoming and Völkershausen disturbances, suggests that the two disturbances have a common mechanism. However, in this article we present BB *P* recorded at teleseismic distances that show clear differences. We use the BB data to determine the equivalent moment tensor for each suspected collapse process (and

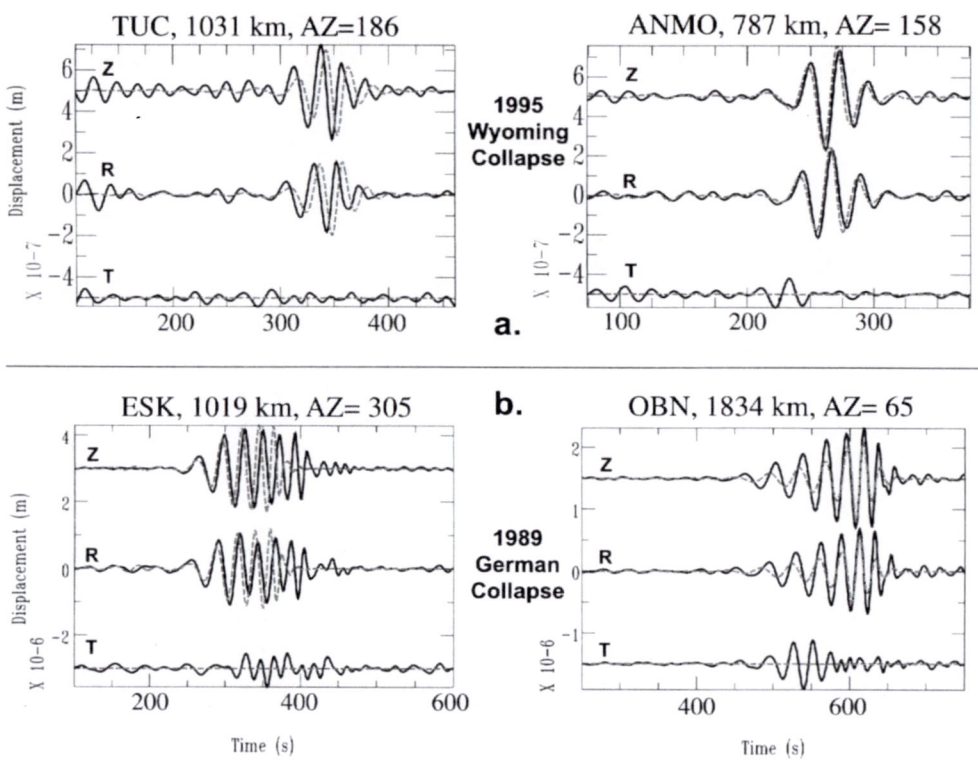

Figure 2

(a) A comparison of the observed (solid line) and synthetic (broken line) three-component (V - Vertical, R - Radial, T - Transverse) surface wave seismograms from the Wyoming disturbance recorded at TUC and ANMO. The source model is identical to that of PECHMANN *et al.* (1995). (b) as (a) above, but for seismograms from the Völkershausen (German) disturbance recorded at ESK and OBN. The source is a horizontal closing-crack moment tensor in a Poisson solid with $M_{zz} = 1.5 \times 10^{17}$ Nm.

thus eliminate formally the possibility that the disturbance was caused by a double-couple (earthquake) moment tensor at depth). Further, we demonstrate, using a modified spall model, that the differences in the observed BB *P* waves can be attributed to the area, amount of subsidence, depth, and failure-rate of the collapse process, driven by the release of gravitational potential energy.

$$m_b:M_s$$

We consider the ISC estimates of M_s from the two disturbances to be unreliable for discrimination purposes, as the mean M_s values are calculated using only seven readings, and show a large scatter around the mean. The ISC m_b values are probably reliable for discrimination purposes as a large number of observations contribute to the estimate (20 stations for Völkershausen; 99 stations for Wyoming).

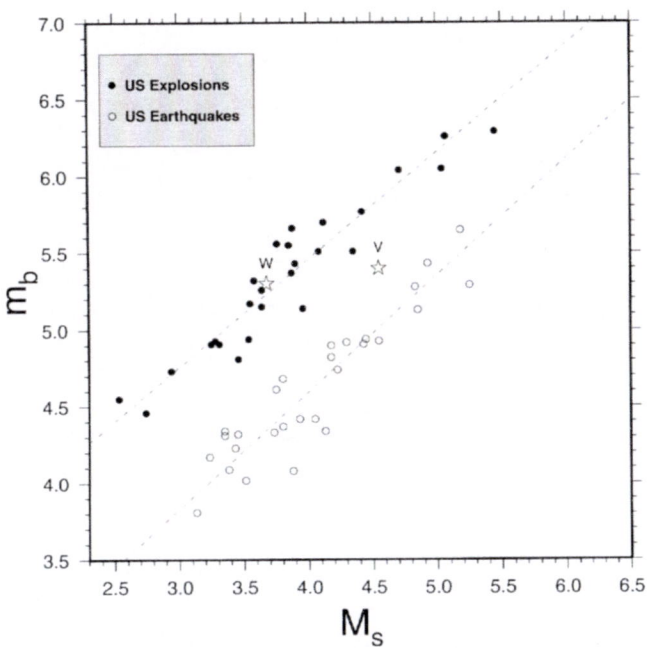

Figure 3
The m_b:M_s diagram modified from Figure 6b of MARSHALL and BASHAM (1972). The disturbances are marked by a star; V = Völkershausen and W = Wyoming.

We collected regional surface-wave seismograms from the Wyoming disturbance from the IRIS archive, and from the IRIS, GEOSCOPE, ORFEUS, and Blacknest archives from the Völkershausen disturbance. We converted the original data to a standard long-period instrument response before measuring M_s. Table 2 shows M_s values calculated from the vertical-component (the Rayleigh wave) using the formula of MARSHALL and BASHAM (1972). Figure 3 shows the position of the Wyoming and Völkershausen disturbances on an m_b:M_s diagram along with a sample of earthquakes and explosions from the United States (US) (MARSHALL and BASHAM, 1972). The Wyoming disturbance plots within the US explosion population and the Völkershausen disturbance plots between the explosion and earthquake populations. Thus, these disturbances are of interest, in the context of verifying the CTBT, as they appear explosion-like using m_b:M_s, and are probably shallow (Table 1).

Analysis of P Recorded at Teleseismic Distances

Analysis of the Wyoming Disturbance

Observations of P

We retrieved from the IRIS archive vertical-component BB and SP P waveforms recorded by stations at distances between 30° and 90° from the ISC epicentre

Figure 4

Azimuthal-equidistant map, centred on the ISC epicentre, showing stations recording digital broadband and short-period teleseismic *P* waveforms from the Wyoming seismic disturbance.

(Fig. 4). We also retrieved SP array waveforms recorded at EKA from the Blacknest archive.

Figure 5 shows the 15 *P* waveforms with the highest signal-to-noise ratio (SNR) recorded from the Wyoming disturbance. The waveforms have all been converted to a common SP instrument response so a direct comparison can be made (the standard UK-array SP response is used). Where first motion is clear, it is always negative (dilatational). Figure 5 shows some "simple" waveforms comprising a few cycles with high amplitude, followed by low-amplitude coda, e.g., the waveform recorded at KIEV suggests a source duration of < 2.5 sec. The SP waveforms in Figure 5 show a correlation between apparently complex signals and low amplitudes, e.g., DPC,

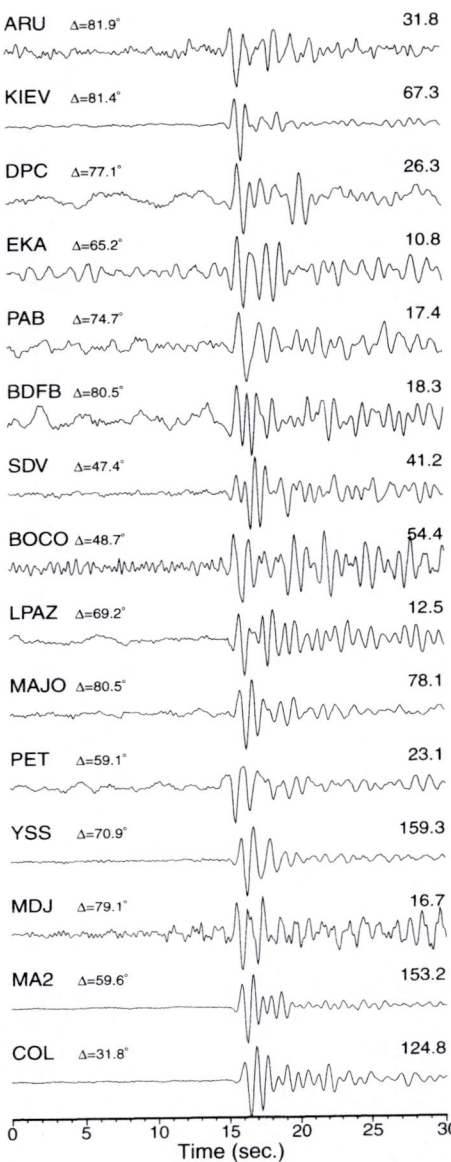

Figure 5

Vertical-component SP *P* waveforms from the Wyoming seismic disturbance recorded by stations at teleseismic distances (ordered by azimuth). The number to the top-right of the seismogram in this, and following figures, is the maximum positive (ground up) displacement in nm (multiplied by the frequency response of the seismometer normalised at 1 Hz). Δ is the distance between the station and the ISC epicentre.

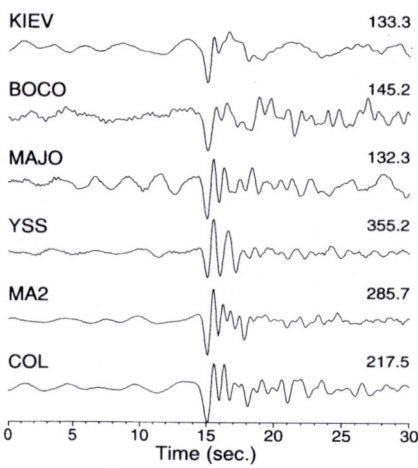

Figure 6

Vertical-component BB *P* waveforms from the Wyoming seismic disturbance recorded by stations at teleseismic distances. The original data have been filtered to simulate BB displacement (0.1–3.5 Hz), to reduce sinusoidal noise and the effects of attenuation.

EKA, and LPAZ. DOUGLAS *et al.* (1971, 1973) reported similar SP *P* observations from underground explosions, that led them to propose that apparently complex SP *P* can be explained if the direct arrival is attenuated relative to later (multi-pathed) arrivals. We assume that the apparent complexity of *P* at other stations in Figure 5 is due to either path or receiver effects, and that the "simple" waveforms are those more likely to contain retrievable information about the seismic source.

We have converted the original data to a "phaseless" BB response (flat to displacement in the pass band 0.1–3.5 Hz), and applied a filter, parameterised by a t^* of 0.20 sec assuming a frequency-independent Q (CARPENTER, 1966), that partially corrects for the effect of attenuation. The data are further filtered using the method described by DOUGLAS (1997) to reduce the effect of low-frequency sinusoidal noise. Figure 6 shows the resulting BB waveforms that are "simple" and have a reasonable SNR.

To further demonstrate the effectiveness of our processing method we retrieved *P* signals, recorded at COL and MAJO, from the IRIS archive for the underground explosion Junction, fired on 26 March 1992 at the Nevada Test Site (37.272°N, 116.361°W, 1630 UT, 5.5 m_b). Figure 7 shows the original velocity broadband data (VBB), the simulated SP and "phaseless" BB (derived as above). Since Junction is a known explosion we expect the first motion of *P* to be positive. While first motion on the SP is clearly positive at MAJO, the first motion on the SP at COL is not at all clear. However, first motion on the BB seismograms are clearly positive. Indeed, the positive-polarity pulse, followed by a negative-polarity pulse on the BB seismograms for COL and MAJO is characteristic of *P* signals from many

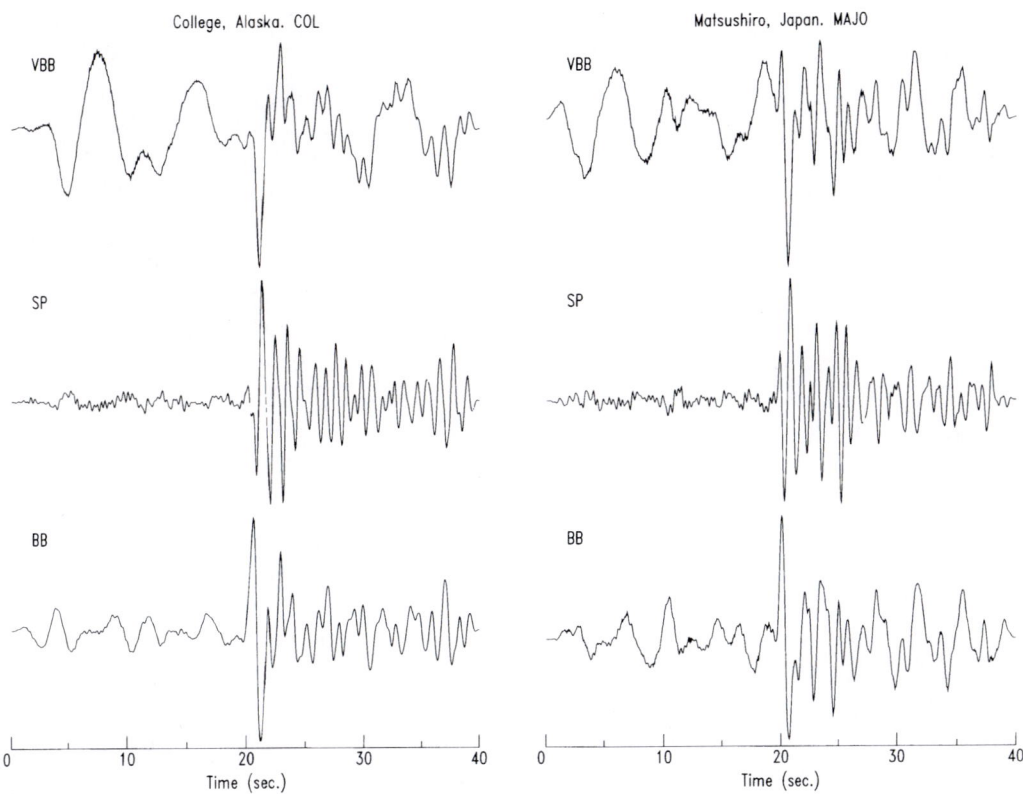

Figure 7

Vertical-component *P* waveforms at COL and MAJO from the Junction underground explosion at the Nevada Test Site. VBB – original data, SP – simulated short-period, BB – simulated broadband displacement.

underground explosions (e.g., DOUGLAS, 1991; DOUGLAS *et al.*, 1987; MCLAUGHLIN *et al.*, 1988).

We interpret the **BB** waveforms from the Wyoming disturbance as comprising a negative-polarity pulse, followed by a constructively interfering positive-polarity pulse (i.e., the opposite of the characteristic explosion *P* waveform). The total duration appears to be < 2.5 sec, and the amplitude spectra show a clear corner frequency at about 1.5 Hz. There appears to be some variation of waveform with azimuth, e.g., COL, MA2, and MAJO show a second positive pulse, but this appears weaker at BOCO and KIEV. The **BB** waveform at YSS appears more complex and longer duration than the other high amplitude waveforms in Figure 6. We suspect that the anomalous nature of the **BB** waveform at YSS may be due to a receiver effect, and/or errors in the specification of the instrument response used to convert the original data to **BB** displacement.

Moment tensor analysis

We use the relative amplitude method (PEARCE 1977, 1980; PEARCE and ROGERS, 1989) to find moment tensors that are consistent with the observed SP and BB P waveforms. The method compares synthetic relative amplitude ratios of pP/P, sP/P, and sP/pP, calculated for a second-order moment tensor, with those observed. Noise, interference, and other effects (such as P-to-R_g conversion near the receiver) contribute to uncertainty in the "true" amplitudes used to calculate the amplitude ratios, so in practice a maximum and minimum amplitude is specified for each phase, within which we are confident the "true" amplitude lies. Each phase is also assigned a polarity, if known. Moment-tensor space is grid-searched and solutions that do not match the specified ranges of amplitude ratios are excluded.

The BB waveforms in Figure 6 comprise a large-amplitude direct P phase, with negative polarity, followed by a large-amplitude pulse with positive polarity. This second pulse could be pP or sP, or the result of interference between pP and sP. Given the inherent uncertainty in the interpretation of this second pulse we restrict our analysis to a search for moment tensors that satisfy negative polarity of P observed on the BB seismograms in Figure 6 and on the SP waveforms at stations DPC, LPAZ, PAB, and SDV.

The six independent variables of the moment tensor can be expressed as two source-type parameters (T [−1.0, 1.0] the deviatoric part, and k [−1.0, 1.0] the volumetric part), three orientation parameters (σ [0°, 360°], δ [0°, 180°] and ψ [0°, 180°]), and the scalar moment M_0 (HUDSON et al., 1989; PEARCE and ROGERS, 1989). Since we use relative amplitudes we cannot determine M_0.

Figure 8 shows the source types (HUDSON et al., 1989) for which at least one orientation fits the observed negative first-motions after grid-searching using steps of 0.2 in T and k, and steps of 10° in σ, δ, and ψ. The take-off angles of P, pP, and sP are calculated using the ISC epicentre, assuming that the source is confined to layer 1 of the structure in Table 3. All source types that require a positive volume change are excluded by the teleseismic P observations. The double-couple and the closing-crack source types are both consistent with the observations. Of the double couples, reverse and steeply dipping orientations are inconsistent with the observations.

We tested the alternative hypothesis that the source is a double couple deeper than 4 km. The time difference between teleseismic P and pP is then ≥ 1.7 sec for sources buried in the structure in Table 3. Thus, the amplitude of the surface reflections (pP and sP) cannot be greater than the amplitude of the coda observed > 1.7 sec after the onset of direct P. Figure 9 shows the amplitude ranges and phase identification for the BB P seismogram recorded at MA2 under the assumption that the source is at 4 km. Table 4 shows the amplitude ranges and polarities used in the moment-tensor search to test our alternative hypothesis.

When double-couple moment-tensor space is grid-searched (with steps of 5° in σ, δ, and ψ) using the relative amplitudes in Table 4 appropriate for a source deeper than 4 km, there are no orientations of the double couple that are consistent with the

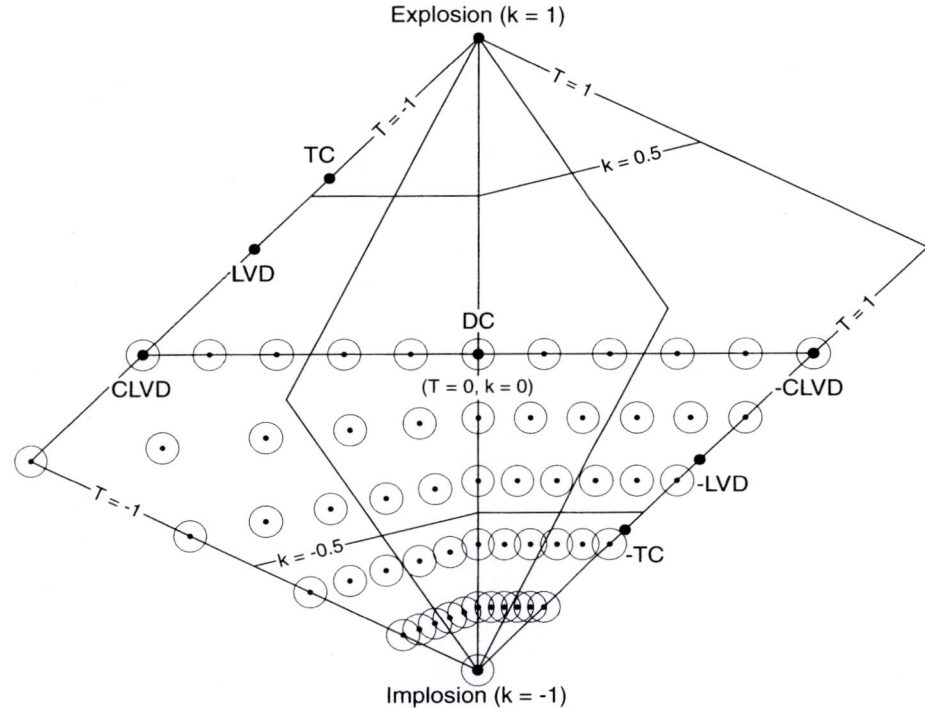

Figure 8

Plot showing source types (dot within a circle) that are consistent with clear negative P first motions on seismograms from the Wyoming seismic disturbance. LVD, Linear Vector Dipole; CLVD, Compensated Linear Vector Dipole; DC, Double Couple; TC, Tension Crack.

Table 3

Structure used to represent the crust in the epicentral region of the Wyoming disturbance (after PRODEHL, *1979)*

Layer No.	P-wave speed (km s^{-1})	S-wave speed (km s^{-1})	Density (kg m^3)	Thickness (km)
1	4.20	2.39	2300	1.0
2	4.60	2.62	2500	1.0
3	4.90	2.79	2600	1.0
4	5.20	2.96	2700	2.0
5	5.70	3.25	2700	2.0
6	6.00	4.42	2700	2.0
7	6.20	3.53	2700	5.0
8	6.60	3.76	2700	8.0
9	6.80	3.87	2700	14.0
10	7.10	4.04	2880	2.0
11	7.40	4.21	2880	2.0
12	7.90	4.37	3330	

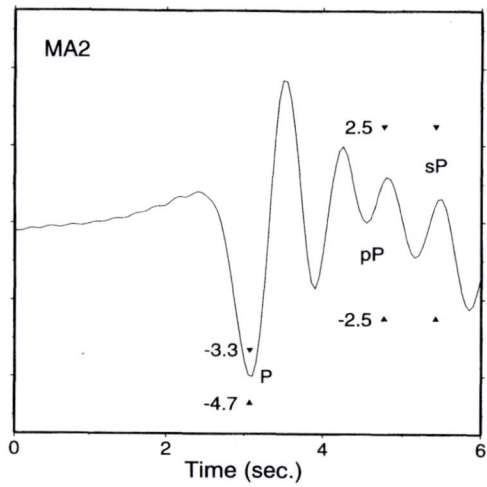

Figure 9

Example showing how the relative amplitude bounds (Table 4) are assigned to the BB waveform from MA2 (see text).

Table 4

The relative amplitude bounds of the P, pP, and sP phases used to test if the data from the Wyoming disturbance are consistent with a double-couple deeper than 4 km. The polarity of each phase is as indicated: – dilatation, U unknown polarity (see Fig. 9). Observations of negative first-motion on the SP waveforms recorded at stations DPC, LPAZ, PAB, and SDV are also used to constrain the moment-tensor grid search

	Δ (°)	Az. (°)	P			pP			sP		
			Pol.	Min.	Max.	Pol.	Min.	Max.	Pol.	Min.	Max.
BOCO	48.7	129.2	–	3.0	5.0	U	0.0	4.0	U	0.0	4.0
COL	31.8	330.1	–	3.7	5.0	U	0.0	2.5	U	0.0	2.5
KIEV	81.4	24.9	–	3.3	6.0	U	0.0	1.5	U	0.0	1.5
MA2	59.6	324.4	–	3.3	4.7	U	0.0	2.5	U	0.0	2.5
MAJO	80.5	310.8	–	2.7	5.3	U	0.0	2.5	U	0.0	2.5
YSS	70.9	316.2	–	2.3	4.3	U	0.0	5.0	U	0.0	5.0

observations. This suggests that the Wyoming disturbance is not due to slip on a fault at a depth greater than 4 km.

Analysis of the Völkershausen Disturbance

Observations of *P*

We retrieved BB and SP waveforms from the IRIS archive recorded by stations at distances between 30° and 90° from the ISC epicentre (Fig. 10). Also, SP array waveforms from GBA (Gauribidanur, India) and YKA (Yellowknife, Canada) were obtained from the Blacknest archive.

Figure 10

Azimuthal-equidistant map, centred on the ISC epicentre, showing stations recording digital broadband and short-period teleseismic P waveforms from the Völkershausen seismic disturbance.

Examination of the data, converted to a common SP response, results in reasonable azimuthal coverage using eight stations recording P with a high SNR (Fig. 11). The SP P waveforms in Figure 11 clearly differ from those observed from the Wyoming disturbance (Fig. 5), having an emergent P onset about 2.5 sec before an impulsive positive-polarity arrival.

We process the original waveforms in the same way as described above to recover an estimate of the "phaseless" BB displacement waveform. The resulting waveforms (Fig. 12) have a broad negative-polarity pulse with slow fall-time (approximately 2.0 sec), followed by a broad positive-polarity pulse with short rise-time (corresponding to the impulsive positive-polarity arrival on the SP waveforms) and a duration of

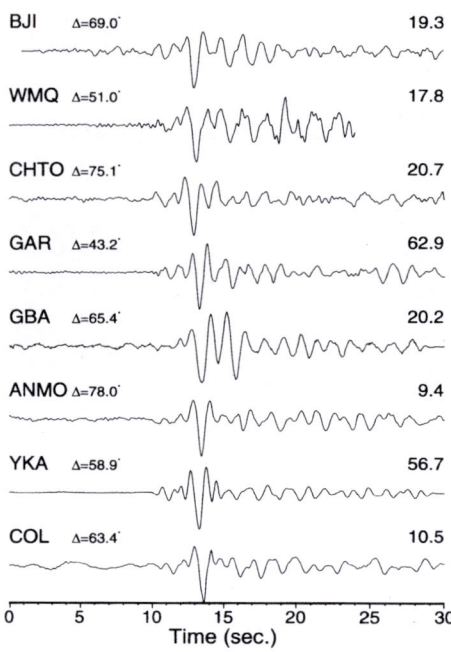

Figure 11

Vertical-component SP *P* waveforms from the Völkershausen disturbance recorded by stations at teleseismic distances (ordered by azimuth).

about 2.5 sec. The amplitude of the coda is low. The stations recording the highest amplitudes (e.g., GAR and YKA) have pulses with high-frequency structure, presumably indicating low attenuation along the direct *P* path. The high frequencies (>1 Hz) riding on the long-duration (2.5 sec period) pulse may be evidence of a complex source function, with the high-frequency arrivals corresponding to the sub-events proposed by BORMANN *et al.* (1992) and AHORNER (1993). The maximum amplitude of the SP and BB responses in Figures 11 and 12 differ by a factor of about 10. This contrasts with the SP and BB observations from the Wyoming disturbance which showed differences of a factor of about two.

Moment tensor analysis

We interpret the BB waveforms in Figure 12 as comprising a negative-polarity large-amplitude direct *P* constructively interfering with large-amplitude *pP*, suggesting a shallow source. If the source is shallow then the time delay between *pP* and *sP* will be shorter than the duration of the source-pulse (assumed here to be about 2.5 sec), thus the amplitude of *sP* is allowed to vary from zero to the maximum amplitude of *pP*. The polarities of *pP* and *sP* are unknown.

Table 5 shows the amplitude ranges and polarities used in the moment-tensor grid search. The take-off angles of *P*, *pP*, and *sP* are calculated using the ISC

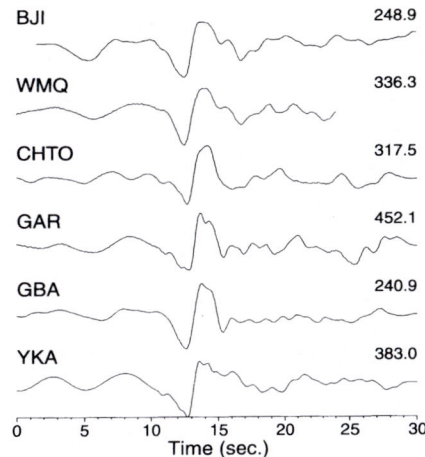

Figure 12

Vertical-component BB *P* waveforms from the Völkershausen disturbance recorded by stations at teleseismic distances. The data are filtered as in Figures 6 and 7.

Table 5

The relative amplitude bounds of the P, pP, and sP phases used in the moment-tensor analysis of the Völkershausen disturbance. The polarity of each phase is as indicated: – dilatation, U unknown polarity

	Δ (°)	Az. (°)	*P*			*pP*			*sP*		
			Pol.	Min.	Max.	Pol.	Min.	Max.	Pol.	Min.	Max.
BJI	69.0	52.1	–	5.0	11.0	U	1.0	11.0	U	0.0	11.0
CHTO	75.1	78.6	–	2.0	8.0	U	2.0	12.0	U	0.0	12.0
GAR	43.2	81.5	–	4.0	10.0	U	5.0	15.0	U	0.0	15.0
GBA	65.4	98.8	–	5.0	7.0	U	4.0	8.0	U	0.0	8.0
WMQ	51.0	65.6	–	8.0	14.0	U	3.0	13.0	U	0.0	13.0
YKA	58.9	333.5	–	7.0	13.0	U	7.0	17.0	U	0.0	17.0

epicentre, and we assume that the source is confined to layer 1 of the structure in Table 6. After our grid-search only moment tensors with zero or negative volume change fit the observations; including both the double-couple and closing-crack source types. Only approximately 45° normal dip-slip double couples fit the observed relative amplitudes.

We tested the alternative hypothesis that the source is a mid-crustal depth earthquake by allowing the amplitudes of the surface reflections to vary between zero and the maximum amplitude of the coda > 5.0 sec after the onset of direct *P*. As for the Wyoming disturbance above, no orientations of the double-couple moment tensor are consistent with the observations if the source depth is greater than about 13.5 km.

Table 6

Structure used to represent the crust in the Völkershausen epicentral region (modified from GAJEWSKI *et al.,*
1987)

Layer No.	P-wave speed (km s^{-1})	S-wave speed (km s^{-1})	Density (kg m^3)	Thickness (km)
1	3.50	2.02	2500	2.0
2	6.00	3.46	2700	13.0
3	6.60	3.81	2800	10.0
4	8.00	4.62	3000	

Modelling the Teleseismic P Waves

We assume that both the Wyoming and Völkershausen seismic disturbances are caused by the failure of 1000's of pillars supporting the mine's roof (Table 1), resulting in significant subsidence of up to 1 m at the Earth's surface. Our assumption that the disturbances were caused by a collapse is supported by the successful modelling of the three-component surface waves using a horizontal closing-crack moment tensor (Fig. 2). Here, we attempt to model the Wyoming and Völkershausen seismic disturbances as multiple pillar-failures driven by gravitational energy release using a single-force representation. DAY and MCLAUGHLIN (1991) showed that, provided the source is shallow, the single-force and moment-tensor (closing crack) representations of a collapse are equivalent.

TAYLOR (1994) modelled the 4.4 m_b (ISC) Gentry Mountain disturbance of 14 May 1981 (Utah, USA) as a single tabular collapse at a depth of 200 m. The collapse model used by TAYLOR (1994) was modified from the single-force spall model of DAY *et al.* (1983) combined with the spall force-time function of STUMP (1985). TAYLOR (1994) adjusted the collapse model parameters (mass of the falling block, the time for the initiation of the collapse and a smoothing operator) until the synthetic seismograms matched the Rayleigh and BB waveforms recorded at regional distances. We use a similar collapse model to generate synthetic teleseismic P seismograms from the two disturbances.

Wyoming Collapse Model

Figure 13a shows the collapse force-time function $f(t)$, with the time between collapse initiation and impact set to 0.35 sec (equivalent to approximately 0.6 m of free-fall, constrained by the average observed subsidence above the mine), with a 0.2 sec smoothing operator. We consider this to be an appropriate force-time function for a single pillar failure during the Wyoming collapse.

Figure 13b shows the vertical component of P displacement $g(t)$, due to an impulsive vertical force at a depth of 500 m in the structure given in Table 3, at a distance of 50°. The seismogram is calculated using the method described by

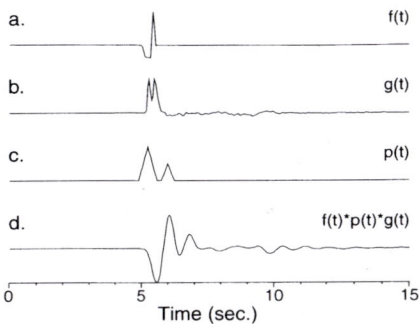

Figure 13

Synthetic seismograms from a collapse model for the Wyoming disturbance. (a) $f(t)$ smoothed (0.2 sec operator) collapse force-time function for a free-fall distance of 0.6 m. (b) $g(t)$ is the **BB** P displacement due to an impulsive vertical force acting at a depth of 500 m in the crustal structure of Table 3. (c) $p(t)$ is the model pillar-failure function. (d) The synthetic P seismogram is calculated by the convolution of $f(t)$, $p(t)$ and $g(t)$.

DOUGLAS *et al.* (1972) using a standard crustal structure for the receiver and an attenuation operator with $t^* = 0.10$ sec.

The spectrum of $f(t)$ contains no zeros and the spectrum of $g(t)$ is approximately flat up to a frequency of about 1.5 Hz. Thus, we can obtain a rough, band-limited, estimate of the pillar-failure history $p(t)$ by deconvolving $f(t)$ from the observed **BB** displacement waveforms. Figure 14 shows the resulting estimates of $p(t)$ filtered with a pass band of 0.2–2.0 Hz. There appears to be some evidence of at least two separate episodes of pillar failure, especially in the $p(t)$'s estimated at COL, MA2, and **MAJO**. We approximate the estimates of $p(t)$ by two triangles, the first having three times the area of the second, and the peak of the second triangle delayed 0.75 sec relative to the peak of the first triangle (Fig. 13c).

We construct our synthetic seismograms by convolution of the appropriate $g(t)$ function, for each station recording P waveforms, with $f(t)$ and $p(t)$, and scale the convolution of $f(t)$ and $p(t)$ so that the resulting waveforms match the observed amplitudes (Fig. 13d). The scaled convolution of $f(t)$ and $p(t)$ represents the model force-time function of the whole collapse process, including the total model mass. The total mass of the model represents the mass required to generate the collapse force-time function which excites seismic waves. In general, the total model mass will be less than the true collapsed mass, unless the collapse process is 100% efficient at generating seismic waves.

Figure 15 compares the observed P seismograms with synthetic, calculated using a total model mass of 7.1×10^{11} kg. The SP response is also shown in Figure 15 so that the high frequency part of the spectrum can be compared. The synthetic seismograms match those observed in both the **BB** and SP pass bands.

BOWERS (1996) originally modelled the Wyoming disturbance as a single collapse. However, for this model the fall-times of the positive pulse (pP) on the synthetic

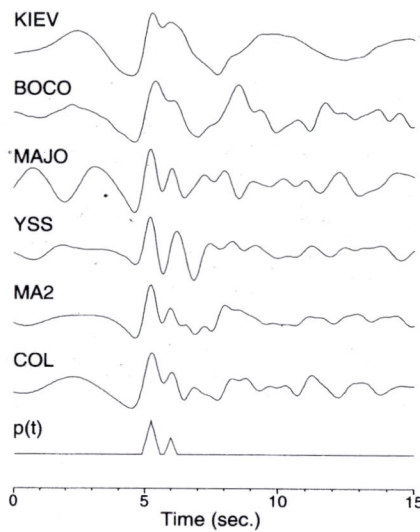

Figure 14

Waveforms resulting from the deconvolution of $f(t)$ from the BB P waveforms in Figure 7. The results from the deconvolution have been filtered with a pass band of 0.2–2.0 Hz. $p(t)$ is the pillar-failure function used to generate the collapse synthetic seismograms for the Wyoming disturbance.

seismograms are a poor fit to those observed. The "double" $p(t)$ function provides a better fit in the time domain, and matches the clear corner frequency of about 1.5 Hz observed in the teleseismic P-amplitude spectra. There may be a physical basis for the "double" $p(t)$ function as the map of subsidence in PECHMANN *et al.* (1995) shows two separate areas of maximum subsidence, one about one-third the area of the other.

DAY and MCLAUGHLIN (1991) showed that the single force and tension crack representations for spall are equivalent, provided the spall depth is shallow. Thus, the vertical single-force model above and the horizontal closing-crack moment tensor, used for the surface-wave forward modelling (PECHMANN *et al.*, 1995), are approximately equivalent. It follows from Equation (16) of DAY and MCLAUGHLIN (1991) that the single-force time history and the moment-rate function $\dot{m}(t)$ are approximately related by:

$$\dot{m}(t) \approx \frac{\alpha_s^2}{h} \int f(t) * p(t)\, dt, \tag{1}$$

where, α_s is the P-wave speed in the source layer, h is the depth of the closing crack, and the asterisk represents convolution. This results in an estimate of $M_0 \,(= \int \dot{m}(t)\, dt$ from the multiple pillar-collapse model of 1.5×10^{16} Nm (using $\alpha_s = 4.2$ km s^{-1} and $h = 500$ m), about one third that estimated by PECHMANN *et al.* (1995) from modelling the surface waves (4.3×10^{16} Nm).

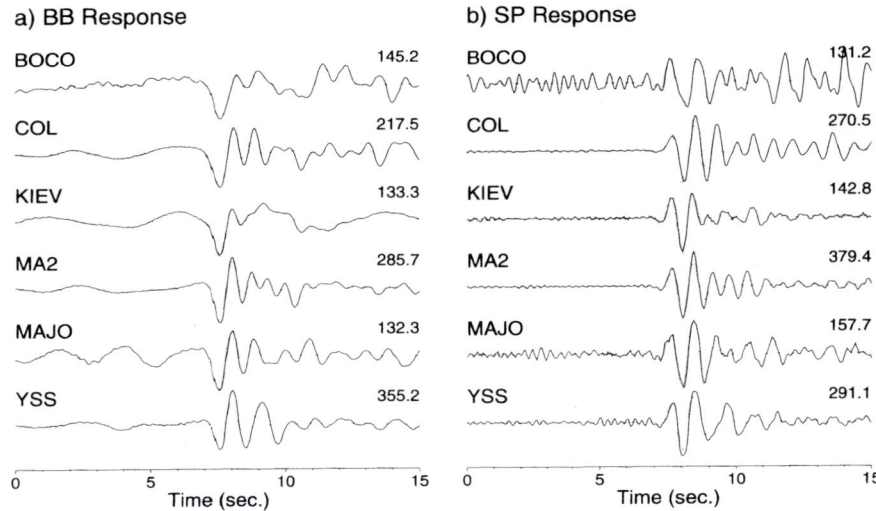

Figure 15
Observed (solid) and synthetic (dotted) BB and SP seismograms for the Wyoming disturbance. The SP seismograms are calculated by converting the BB waveforms (previously corrected for attenuation). The observed and synthetic waveforms are plotted on a common amplitude scale.

Völkershausen Collapse Model

We follow a similar procedure to construct synthetic P seismograms for the Völkershausen disturbance. Figure 16a shows $f(t)$ calculated using 0.40 sec as the time between collapse initiation and impact for a single pillar (this corresponds to approximately 0.8 m of free-fall), with other parameters unchanged from the Wyoming model. The value of 0.8 m for the average subsidence agrees well with a cross section of the subsidence shown in AHORNER (1993).

Figure 16b shows $g(t)$ as above, but appropriate for the Völkershausen epicentral region calculated for an impulsive vertical force at a depth of 850 m in the structure in Table 6. Deconvolution of $f(t)$ from the observed BB P waveforms results in the rough, band-limited estimates of $p(t)$ shown in Figure 17. The estimates of $p(t)$ are remarkably stable considering the range of distances and azimuths of the receivers. We approximate $p(t)$ by a smoothed triangle of 3.5 sec duration (Fig. 16c) and construct synthetic P seismograms by convolution as before.

Figure 18 compares the observed P waveforms with synthetic for a total model mass of 7.2×10^{12} kg. The SP response is shown in Figure 18 so that the high frequency part of the spectrum can be compared. The match between the BB synthetic seismograms and the observed is reasonable. However, while the amplitude of the synthetic SP waveforms appears to be an acceptable fit to the observed, the phase match is poor. The observed SP waveforms contain information about the fine detail of the multiple-collapse process not included in our simple approximation of

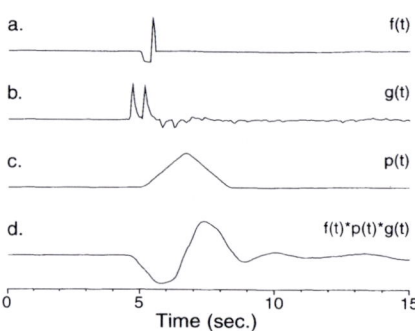

Figure 16
Synthetic seismograms from a collapse model for the Völkershausen disturbance. (a) $f(t)$ smoothed (0.2 sec operator) collapse force-time function for a free-fall distance of 0.8 m. (b) $g(t)$ is the BB P displacement due to an impulsive vertical force acting at a depth of 850 m in the crustal structure of Table 6. (c) $p(t)$ is the model pillar-failure function. (d) The synthetic P-seismogram is calculated by the convolution of $f(t)$, $p(t)$ and $g(t)$.

$p(t)$ by a triangle (apparent arrivals on the synthetic SP seismograms correspond roughly with the turning points of $p(t)$). The equivalent M_0 estimated from the model is 8.3×10^{16} Nm, about half that estimated by our modelling of the surface waves $(1.5 \times 10^{17}$ Nm).

Discussion

We suspect that the discrepancy between estimates of M_0 from the modelling of P waves (0.1–5.0 Hz) and the modelling of the surface waves (0.025–0.050 Hz) for the Wyoming collapse may indicate that the point-source approximation for the high frequency P waves is starting to break down. The teleseismic P waveforms from Wyoming are dominated by energy with frequencies of about 1–2 Hz. The half-wavelength of 2 Hz P waves in the structure used for the Wyoming epicentral region (Table 3) is about 1 km, similar to the dimensions of the observed surface subsidence. If the collapse process has at least two sub-sources (as inferred from the teleseismic P waves) then destructive interference is possible for wavelengths shorter than about 1 km (frequencies >2 Hz), which may result in an underestimate of the total model mass (and hence M_0). We examined the low frequencies of the P waves (0.100–0.125 Hz) to test this hypothesis, unfortunately the waveforms are dominated by noise at these frequencies so a comparison between observed and synthetic waveforms at frequencies lower than about 0.3 Hz is inconclusive. Similar discrepancies between M_0 determined from high- and low-frequency data are reported by EKSTRÖM (1989), who suggests that P-wave data can be insensitive to M_0 as they do not contain sufficient low-frequency energy. Thus, we prefer the estimate of $M_0 = 4.3 \times 10^{16}$ Nm determined by PECHMANN et al. (1995) from their modelling of the surface waves.

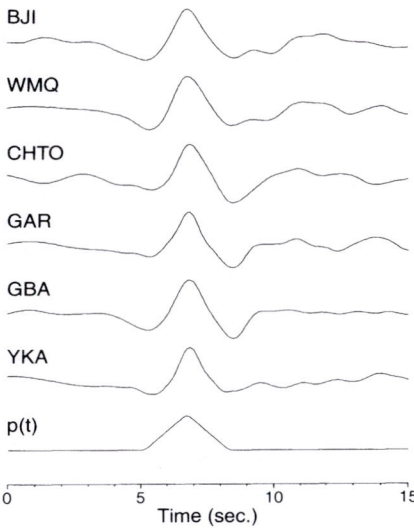

Figure 17
Waveforms resulting from the deconvolution of $f(t)$ from the BB *P* waveforms in Figure 12. The results
from the deconvolution have been filtered with a pass band of 0.2–2.0 Hz. $p(t)$ is the pillar-failure function
used to generate the collapse synthetic seismograms for the Völkershausen disturbance.

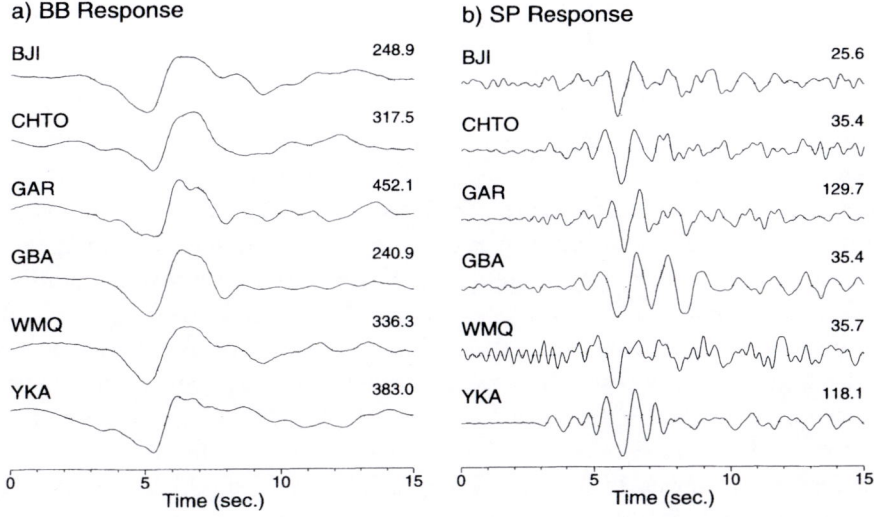

Figure 18
Observed (solid) and synthetic (dotted) BB and SP seismograms for the Völkershausen disturbance. The SP
seismograms are calculated by converting the BB waveforms (previously corrected for attenuation). The
observed and synthetic waveforms are plotted on a common amplitude scale.

The explanation for the discrepancy between relatively high-frequency (0.1–5.0 Hz) and low-frequency (0.02–0.05 Hz) estimates of M_0 for the Völkershausen collapse may also be the breakdown of the point-source approximation. However, the dominant frequency of the teleseismic P-waves from Völkershausen is about 0.25 Hz, with a half-wavelength of about 7 km, greater than the dimensions of the observed surface subsidence. An alternative explanation is that α_s in the model is too low. PRESS (1966) states that carnallite has a P-wave speed of between 4.4 and $6.6 \, km \, s^{-1}$, which gives estimates of M_0 of $1.3 - 3.0 \times 10^{17}$ Nm, similar to the estimate of 1.5×10^{17} Nm from the surface-wave modelling. Experiments show that adding a thin layer of high P-wave speed material at about 850 m to the structure in Table 6 has little effect on the synthetic teleseismic P waveforms. We prefer the estimate of $M_0 = 1.5 \times 10^{17}$ Nm determined by our surface-wave modelling as this is not as sensitive to the source-region structure as M_0 estimates from the P waves.

It is worth noting that both disturbances have earthquake-like signatures if North American regional discriminants are used; such as $M_L:M_0$ (the ratio of local magnitude to scalar moment) of WOODS et al. (1993), and $m_b(P_n):M_0$ (the ratio of the magnitude calculated from P_n to the scalar moment) of PATTON and WALTER (1993, 1994). However, the M_0 estimate used for these discriminants is derived from the surface-wave modelling *assuming* a collapse mechanism. If a double-couple moment tensor is assumed, perhaps a more sensible approach if attempting to identify earthquakes, then the M_0 estimate is reduced, and the two disturbances appear explosion-like using the $M_L:M_0$ and $m_b(P_n):M_0$ criteria.

The seismic energy E_s radiated from the model collapse mechanism is derived from gravitational potential energy E_p. E_p can be estimated if the true collapsed mass (estimated from the average density of the overburden, the mine depth and area of observed subsidence) and the average subsidence are known (e.g., BORMANN et al., 1992; BENNETT et al., 1994; PECHMANN et al., 1995). If E_s is also known then the seismic efficiency $e = E_s/E_p$ of multiple pillar collapses can be estimated.

PECHMANN et al. (1995) estimate the seismic energy E_s from the Wyoming disturbance as 1.1×10^{12} J using the $\log E_s - M_L$ relation determined by KANAMORI et al. (1993). This results in an estimate of e of about 0.08. PECHMANN et al. (1995) report that E_s calculated from integrated velocity-squared seismograms recorded at DUG (Dugway, Utah), using the method of KANAMORI et al. (1993), is 1.4×10^{12} J, similar to that obtained from the $\log E_s - M_L$ relation. However, these estimates should perhaps be treated with caution as E_s is calculated from an empirical equation determined by KANAMORI et al. (1993) using parameters suitable for mid-crustal earthquakes in southern California, and may not be valid for shallow mine collapses in southwestern Wyoming.

Similar uncertainties are encountered with estimates of E_s from the Völkershausen collapse. For example, AHORNER (1993) uses the RICHTER (1958) magnitude-energy relation $\log E_s = 2.9 + 1.9 \, M_L - 0.024 \, M_L^2$, which gives $E_s = 6.1 \times 10^{12}$ J.

BENNETT *et al.* (1994) prefer $\log E_s = 2.4\, m_b - 1.2$, which gives $E_s = 5.8 \times 10^{11}$ J, an order of magnitude less than AHORNER (1993).

For a gravitational collapse model $E_s = mgs$, where m is the total model mass, g is the acceleration due to Earth's gravity, and s is the average free-fall distance in the model. Further, $\int \int f(t) * p(t) = ms$. Combining these relations with the equivalence of the single force and moment-tensor representations for shallow sources implied by Equation (1) leads to the following useful relations:

$$M_0 = ms\alpha_s^2/h, \qquad (2)$$

$$E_s/M_0 = gh/\alpha_s^2 \ . \qquad (3)$$

If we use the estimate of M_0 for the Wyoming collapse from the surface-wave modelling to estimate E_s and e, we get $E_s = 1.2 \times 10^{13}$ J, and $e = 0.86$ (using $\alpha_s = 4.2$ km s^{-1}). The estimates of E_s and e using the Völkershausen collapse surface wave M_0 are dependent on the value of α_s used. If $\alpha_s = 5.0$ km s^{-1}, then $E_s = 5.1 \times 10^{13}$ J and $e = 0.39$. GROVER and JAMES (1972) estimated the seismic efficiency of a mass falling under the influence of gravity (a steel demolition ball dropped onto a concrete foundation) and concluded that $e = 0.44$, perhaps fortuitously close to the values estimated for the Völkershausen collapse, considering the different scale of their experiment (mass of 762 kg falling 9.14 m). We note that while e is potentially high, the estimates of e presented above are poorly constrained due to large uncertainties in our estimates of M_0, α_s and E_p.

Table 7 summarises various parameters determined from the models for the Wyoming and Völkershausen collapses, including parameters from three selected southern Californian earthquakes (MAYEDA and WALTER, 1996) for comparison. It seems that the large Wyoming and Völkershausen collapses have E_s/M_0 ratios similar to high Orowan stress drop earthquakes such as Landers.

Conclusions

M_s measured from Rayleigh waves recorded at regional distances confirms that both the Wyoming and Völkershausen disturbances appear explosion-like using the m_b:M_s criterion. Analysis of the BB teleseismic P waves from the disturbances shows that both are shallow. Thus, the two disturbances appear explosion-like using depth and m_b:M_s (the two most reliable discrimination criteria). However, the first motion of the BB P waves is negative (dilatational), whereas that from an explosion is expected to be positive (compressional). Thus, the BB teleseismic P waves have the potential to discriminate between large collapses and explosions; extending the BB teleseismic P-wave discriminant for mining disturbances proposed by BOWERS and DOUGLAS (1997) to include collapse mechanisms. However, the BB P waves from the two disturbances differ greatly in character. We have modified a non-linear spall

Table 7

Date	Disturbance	M_L	s (m)	h (m)	m (10^{12} Kg)	M_0 (10^{16} Nm)	E_s ($10^{13}J$)	e	E_s/M_0 (10^{-4})
03/13/89	Völkershausen	5.6	0.8	850	6.4	15	5.0	0.39	3.3
02/05/95	Wyoming	5.1	0.6	500	2.0	4.3	1.2	0.86	2.8
02/05/95	Wyoming	5.1	–	–	–	$1.6^\dagger$	$0.7^\dagger$	–	4.1
06/28/91	Sierra Madre	5.8	–	–	–	$25^\dagger$	$2.8^\dagger$	–	1.1
05/04/92	Joshua Tree	4.8	–	–	–	$1.8^\dagger$	$0.2^\dagger$	–	1.0
06/28/92	Landers	6.8	–	–	–	$10700^\dagger$	$2570^\dagger$	–	2.4

† Parameters derived from M_W and $\log E_s$ in Table 1 of MAYEDA and WALTER (1996).

model so as to simulate the BB P waves. We find that the model parameters required correspond closely to the reported area, depth, and amount of collapse. The models suggest differing collapse rates, with at least two short episodes of collapse needed to explain the BB P waves from the Wyoming disturbance (corresponding to two areas of subsidence in the epicentral region), compared with one longer episode of collapse for the Völkershausen disturbance.

Acknowledgements

We thank an anonymous reviewer for their careful and constructive review. DB has benefited from discussions with Alan Douglas, John Hudson and Steve Taylor. T.J. Bennett provided helpful comments on an earlier manuscript. We thank the IRIS, GEOSCOPE, Blacknest and ORFEUS organisations, and contributing stations for providing the data used in this article.

REFERENCES

AHORNER, L. (1993), *Der Gebirgsschlag am 13 März 1989 bei Völkershausen (Thüringen)* (The Rockburst of 13 March 1989 at Völkershausen). In Erdbeben in Deutschland 1989, Hannover (*in German*).

BENNETT, T. J., MARSHALL, M. E., BARKER, B. W., and MURPHY, J. R. (1994), *Characteristics of Rockbursts for use in Seismic Discrimination*, Phillips Laboratory Report No. SSS-FR-93-14382.

BORMANN, P., WYLEGALLA, K., and GROSSER, H. (1992), *The Strong Mining Event of 13 March 1989: Analysis of the Multiple Event*, Proc. XXII ESC General Assembly, Barcelona.

BOWERS, D. and DOUGLAS A. (1997), *Characterisation of Large Mine Tremors Using P Observed at Teleseismic Distances*. In (Gibowicz, S. J. and Lasocki. S. eds.) *Rockbursts and Seismicity in Mines*, Balkema, Rotterdam, pp. 55–60.

BOWERS, D. (1996), *Is the Mechanism of the German Rockburst of March 13, 1989 Similar to that of the Wyoming Rockburst of February 3, 1995?* Supplement to EOS Trans. AGU. 77, F507.

BOWERS, D. (1997), *The October 30, 1994 Seismic Disturbance in South Africa: Earthquake or Large Rockburst?* J. Geophys. Res. *102*, 9843–9857.

CARPENTER, E. W. (1966), *Absorption of Elastic Waves – An Operator for a Constant Q Mechanism*, AWRE Report No. O-43/66. HMSO, London.

DAY, S. M., and MCLAUGHLIN, K. L. (1991), *Seismic Source Representations for Spall*, Bull. Seismol. Soc. Am. *81*, 191–201.

DAY, S. M., RIMER, N., and CHERRY, J. T. (1983), *Surface Waves from Underground Explosions with Spall: Analysis of Elastic and Nonlinear Source Models*, Bull. Seismol. Soc. Am. *73*, 247–264.

DOUGLAS, A., MARSHALL, P. D., and CORBISHLEY, D. J. (1971), *Absorption and the Complexity of P Signals*, Nature, *233*, 50–51.

DOUGLAS, A., HUDSON, J. A., and BLAMEY, C. (1972), *A Quantitative Evaluation of Seismic Signals at Teleseismic Distances — III. Computed P and Rayleigh Wave Seismograms*, Geophys. J. R. astr. Soc. *28*, 385–410.

DOUGLAS, A., MARSHALL, P. D., GIBBS, P. G., YOUNG, J. B., and BLAMEY, C. (1973), *P Signal Complexity Re-examined*, Geophys. J. R. Astr. Soc. *33*, 195–221.

DOUGLAS, A., MARSHALL, P. D., and YOUNG, J. B. (1987), *The P Waves from the Amchitka Island Explosions*, Geophys. J. R. Astr. Soc. *90*, 101–117.

DOUGLAS, A., (1991), *Broadband Estimates of the Seismic Source Functions of Nevada Explosions From Far-field Observations of P Waves*. In Explosion Source Phenomenology, Geophysical Monograph *65*, 127–140.

DOUGLAS, A. (1997), *Bandpass Filtering to Reduce Noise on Seismograms: Is There a Better Way?* Bull. Seismol. Soc. Am. *87*, 770–777.

EKSTRÖM, G. (1989), *A Very Broadband Inversion Method for the Recovery of Earthquake Source Parameters*, Tectonophys. *166*, 73–100.

GAJEWSKI, D., HOLBROOK, W. S., and PRODEHL, C. (1987), *A Three-dimensional Crustal Model of Southwest Germany Derived from Seismic Refraction Data*, Tectonophys. *142*, 49–70.

GROVER, F. H., and JAMES, E. W. (1972), *A Comparison of the Seismic Effects Produced by a Falling Weight and Detonations of Small, Buried, Explosive Charges*, AWRE Report No. O-50/72. HMSO, London.

HUDSON, J. A., PEARCE, R. G., and ROGERS, R. M. (1989), *Source Type Plot for Inversion of the Moment Tensor*, J. Geophys. Res. *94*, 765–774.

KANAMORI, H., Mori, J., HAUKSSON, E., HEATON, T. H., HUTTON, L. K., and JONES, L. M. (1993), *Determination of Earthquake Energy Release and M_L Using TERRASCOPE*, Bull. Seismol. Soc. Am. *83*, 330–346.

KNOLL, P. (1990), *The Fluid-induced Tectonic Rockburst of March 13, 1989, in the 'Werra' Potash Mining District of the GDR (first results)*, Gerl. Beitr. Geophysik *99*, 239–245.

MARSHALL, P. D., and BASHAM, P. W. (1972), *Discrimination Between Earthquakes and Underground Explosions Employing an Improved M_s Scale*, Geophys. J. R. Astr. Soc. *28*, 431–458.

MAYEDA, K., and WALTER, W. R. (1996), *Moment, Energy, Stress Drop, and Source Spectra of Western United States Earthquakes from Regional Coda Envelopes*, J. Geophys. Res. *101*, 11,195–11,208.

MCGARR, A. (1992), *An Implosive Component in the Seismic Moment Tensor of a Mining-induced Tremor*. Geophys. Res. Lett. *19*, 1579–1582.

MCLAUGHLIN, K. L., LEES, A. C., DER, Z. A., and MARSHALL, M. E. (1988), *Teleseismic Spectral and Temporal M_0 and ψ_∞ Estimates for Four French Explosions in the Southern Sahara*, Bull. Seismol. Soc. Am. *78*, 1580–1596.

PATTON, H. J., and WALTER, W. R. (1993), *Regional Moment-magnitude Relations for Earthquakes and Explosions*, Geophys. Res. Lett. *20*, 277–280.

PATTON, H. J., and WALTER, W. R. (1994), *Correction to 'Regional Moment-magnitude Relations for Earthquakes and Explosions'*, Geophys. Res. Lett. *21*, 743.

PEARCE, R. G. (1977), *Fault Plane Solutions Using Relative Amplitudes of P and pP*, Geophys. J. R. Astr. Soc. *50*, 381–394.

PEARCE, R. G. (1980), *Fault Plane Solutions Using Relative Amplitudes of P and Surface Reflections: Further Studies*, Geophys. J. R. Astr. Soc. *60*, 459–487.

PEARCE, R. G., and ROGERS, R. M. (1989), *Determination of Earthquake Moment Tensors from Teleseismic Relative Amplitude Observations*, J. Geophys. Res. *94*, 775–786.

PECHMANN, J. C., WALTER, W. R., NAVA, S. J., and ARABASZ, W. J. (1995), *The February 3, 1995 M_L 5.1 Seismic Event in the Trona Mining District of Southwestern Wyoming*, Seism. Res. Lett. *66*, 25–34.

PRESS, F. (1966), *Seismic velocities*. In *Handbook of Physical Constants* (Clark, S. P., ed.) The Geological Society of America Memoir *97*, 195–218.

PRODEHL, C. (1979), *Crustal Structure of the Western United States*, Geol. Surv. Profess. Paper 1034, 74 pp.

RICHTER, C. F. *Elementary Seismology* (W. J. Freeman and Co., San Francisco, California, (1958)). 768 pp.

STUMP, B. W. (1985), *Constraints on Explosive Sources with Spall from Near-source Waveforms*, Bull. Seismol. Soc. Am. *75*, 361–377.

TAYLOR, S. R. (1994), *False Alarms and Mine Seismicity: An Example from the Gentry Mountain Mining Region, Utah*, Bull. Seismol. Soc. Am. *84*, 350–358.

WONG, I. G., and McGARR, A. (1990), *Implosional Failure in Mining-Induced Seismicity: A Critical Review. In Rockbursts and Seismicity in Mines*, (Fairhurst, C., ed.) Balkema, Rotterdam, pp 45–51.

WOODS, B. B., KEDAR, S., and HELMBERGER, D. V. (1993), $M_L:M_0$ *as a Regional Seismic Discriminant*, Bull. Seismol. Soc. Am. *83*, 1167–1183.

(Received June 29, 1999, revised July 7, 2000, accepted July 20, 2000)

To access this journal online:
http://www.birkhauser.ch

Pure appl. geophys. 159 (2002) 831–863
0033–4553/02/040831–33 $ 1.50 + 0.20/0

| Pure and Applied Geophysics

Identification of Mining Blasts at Mid- to Far-regional Distances Using Low Frequency Seismic Signals

MICHAEL A. H. HEDLIN,[1] BRIAN W. STUMP,[2,3] D. CRAIG PEARSON,[3] and XIAONING YANG[3]

Abstract — This paper reports results from two recent monitoring experiments in Wyoming. Broadband seismic recordings of kiloton class delay-fired cast blasts and instantaneous calibration shots in the Black Thunder coal mine were made at four azimuths at ranges from 1° to 2°. The primary focus of this experiment was to observe and to explain low-frequency signals that can be seen at all azimuths and should routinely propagate above noise to mid-regional distances where most events will be recorded by International Monitoring System (IMS) stations.

The recordings clearly demonstrate that large millisecond delay-fired cast blasts routinely produce seismic signals that have significant spectral modulations below 10 Hz. These modulations are independent of time, the azimuth from the source and the orientation of the sensor. Low-frequency modulations below 5 Hz are seen beyond 9°. The modulations are not due to resonance as they are not produced by the calibration shots. Linear elastic modeling of the blasts that is guided by mine-blast reports fails to reproduce the fine detail of these modulations but clearly indicates that the enhanced "spectral roughness" is due to long interrow delays and source finiteness. The mismatch between the data and the synthetics is likely due to source processes, such as nonlinear interactions between shots, that are poorly understood and to other effects, such as variations of shot time and yield from planned values, that are known to be omnipresent but cannot be described accurately. A variant of the Automated Time-Frequency Discriminant (HEDLIN, 1998b), which uses low-frequency spectral modulations, effectively separates these events from the calibration shots.

The experiment also provided evidence that kiloton class cast blasts consistently yield energetic 2–10 second surface waves. The surface waves are strongly dependent on azimuth but are seen beyond 9°. Physical modeling of these events indicates that the surface waves are due mainly to the extended source duration and to a lesser extent to the slap-down of spalled material. The directionality is largely a path effect. A discriminant that is based on the partitioning of energy between surface and body waves routinely separates these events from the calibration shots.

The Powder River Basin has essentially no natural seismic activity. How these mining events compare to earthquake observations remains to be determined.

Key words: Spectral modulations, attenuation, surface waves, remote source characterization.

1. Introduction

The recent Comprehensive Nuclear-Test-Ban Treaty (CTBT) is unlike any earlier test ban accord, such as the Threshold Test-Ban Treaty (TTBT) which prohibits tests

---
[1] IGPP-University of California, San Diego, La Jolla, CA, 92093-0225, U.S.A.
[2] Southern Methodist University, U.S.A.
[3] Los Alamos National Laboratory, U.S.A.

above 150 kilotons, as it bans nuclear explosions of any yield. The exclusive nature of this treaty and the fact that a 1 kiloton contained explosion in hard rock is well above magnitude 4 motivates interest in the accurate characterization of small (m_b < 4.0) seismic events (MURPHY, 1995). Detonating a nuclear explosion in a cavity can further reduce the magnitude of the seismic waves (U.S. CONGRESS, OTA REPORT, 1988). Interest in small events has increased both the numbers of events that must be considered and the types. It is estimated that, globally, 21,000 events with m_b above 3.5 occur per year (U.S. CONGRESS OTA REPORT; p. 78). Some of these small events will not be associated with natural seismic activity but are due to commercial blasting which occurs globally. The blasting technique favored worldwide is delay firing (LANGEFORS and KIHLSTRÖM, 1978) in which a number of charges are arranged in a spatial grid and detonated in sequence. This technique is favored as it yields efficient fracturing of rock while minimizing damaging seismic and acoustic signals in areas proximal to the mine. Commercial blasting is common but usage varies widely (LEITH, 1994). KHALTURIN *et al.* (1997) surveyed over 30 regions worldwide and found that several hundred industrial blasts each year have a magnitude greater than 3.5. Large blasts, often associated with construction, were relatively common in the Former Soviet Union and in China however the current blasting practice is not yet well known (Bill Leith, personal communication). In Wyoming, very large (>1 kt) surface coal mine blasts are common. The Black Thunder coal mine, one of several mines in the Powder River Basin in NE Wyoming, typically detonates 1 to 2 blasts of this magnitude each month (PEARSON *et al.*, 1995; STUMP, 1995) with a few in the magnitude range of 3.5 to above 4.0.

The Reviewed Event Bulletin (REB), published by the prototype International Data Centre (PIDC), indicates that large mining explosions are commonly detected by IMS seismic stations at all regional distances; some are seen teleseismically. Although there has been very promising progress in identifying mine blasts using correlation techniques (HARRIS, 1991; ISRAELSON, 1991; RIVIERE-BARBIER and GRANT, 1993), significant changes in how mine blasts are detonated at an individual mine are known to occur (MARTIN *et al.*, 1997). This gives us the impetus for exploring other, complimentary, methods. Large mining events are problematic not just because these events will trigger the IMS, but these large, controlled, explosions offer a means to obscure nuclear tests. The "hide-in-quarry blast" evasion scenario (BARKER and DAY, 1990; BARKER *et al.*, 1994; RICHARDS and ZAVALES, 1990; SMITH, 1993), in which a nuclear test is colocated with a mining blast, might be troubling because blasting anomalies, in which a large part of a mining explosion shot grid detonates simultaneously, are not uncommon (MARTIN *et al.*, 1997). A nuclear test colocated with an industrial explosion might be entirely hidden or mistaken for a detonation anomaly.

The CTBT calls for an International Monitoring System (IMS) which will comprise four networks of sensors (seismic, infrasound, radionuclide and hydroacoustic). The seismic network will consist of 50 primary stations and over 100 secondary stations (Fig. 1). This network will place a station within local distance (1°) of 2% of the Earth's landmass. The near-regional (within 5°) coverage will be 34%. The mid-regional coverage (within 15° of the source) is nearly complete at 89% (Fig. 2). If multiple recordings are required for accurate source characterization, the coverage is more limited. Just 22% of the Earth's landmass is within 10° of 3 stations; 74% is within 20°. Some prominent mining regions (e.g., the Kuzbass/Abakan mining region in Russia near the IMS station ZAL) will be monitored at near-regional range (Fig. 2) and a full suite of high- and low-frequency characterization techniques can be brought to bear on suspicious events (STUMP et al., 1999a). In many regions, however, monitoring of man-made and natural seismic activity will rely on recordings made at mid- to far-regional range. Events in these regions will require a more limited set of seismic discriminants: those that operate at low frequencies. Although millisecond delay-fired industrial explosions will produce diagnostic spectral peaks at high frequency, because of interference between shots, the industry standard intershot time delay is 35 msec and the resulting spectral peak is at ~30 Hz (the inverse of the intershot time delay). As the standard sampling rate for IMS stations is 40 samples/sec (sps), this energy will lie beyond the Nyquist frequency of most recordings that will be made under the CTBT.

Quantitative detection capability of the IMS involves many factors, including spatial decay rates, event size, noise levels, instrument type (e.g., array, single station), and phase (e.g., P and S). There have been many studies quantifying these into the detection capability of the IMS (e.g., WUSTER et al., 2000). The analysis in

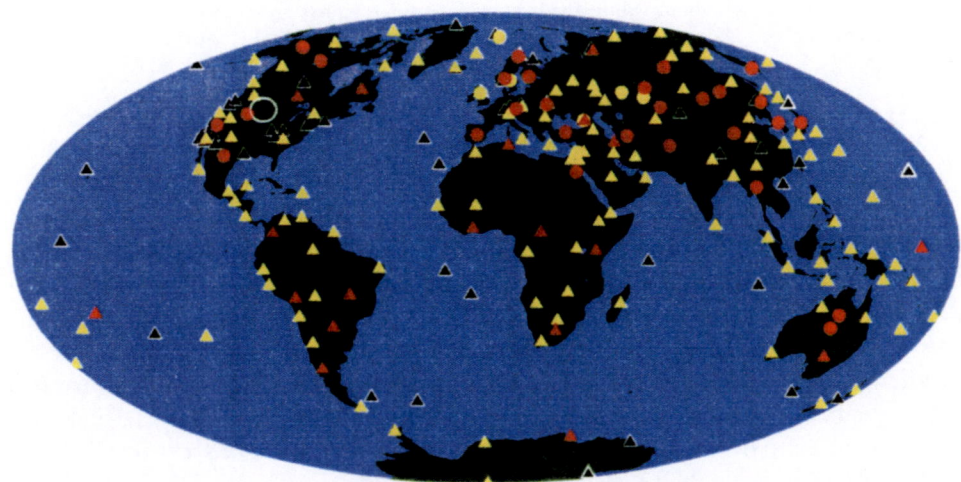

Figure 1

Distribution of current and proposed primary (red) and auxiliary (yellow) IMS seismic stations. Array sites and three-component stations are represented as circles and triangles, respectively. Additional stations (in black) are in the IRIS GSN. A regional network considered in this study was deployed in Wyoming and is represented by the oval.

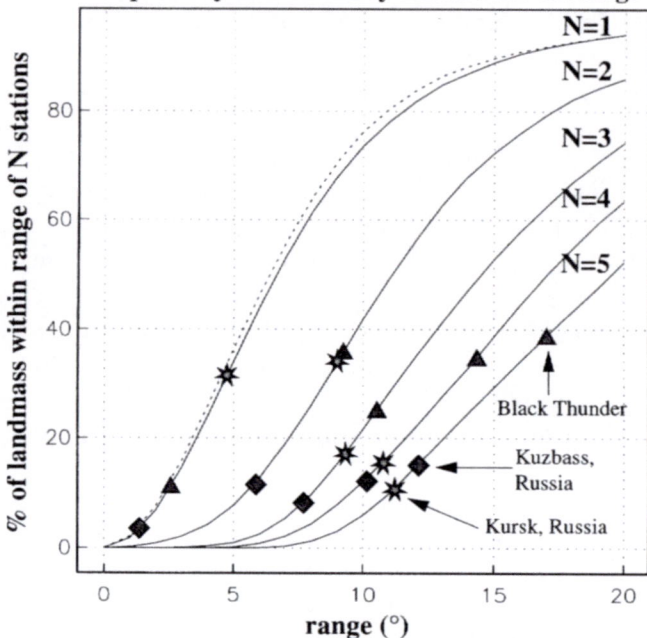

Figure 2

Coverage of the Earth's landmass permitted by the current and proposed IMS primary and secondary seismic networks is indicated by the solid curves. Single-station coverage ($N = 1$) is nearly complete within 10° of the source. If multiple recordings of an event are required for adequate source characterization ($N > 1$), most of the observations will be made at mid- to far-regional or teleseismic range. Coverage of three prominent mining regions is indicated by shaded symbols. Krasnogorsky, the main Kuzbass mine in Russia, at 53.6°N, 87.8°E; Black Thunder in Wyoming at 43.7°N, 105.25°W and Kursk, Russia at 51.8°N, 36.5°E are represented by diamonds, triangles and stars, respectively. These mining regions will be monitored somewhat more closely than the average. For example, three stations are within 8° of the Kuzbass mine. Just 8% of the Earth's landmass has better coverage. The dashed curve is coverage given by the IMS networks (current and proposed) and existing stations in the GSN. Many stations in the GSN that are not already in the IMS are located on oceanic islands and so the improvement is limited.

this case is to emphasize the distance range at which observations of mining explosions are most likely to be made.

It is apparent that large blasts emit seismic energy which can be used for source characterization from local to far-regional distances. Spectral modulations below 10 Hz have been observed by several groups (incl., BAUMGARDT and ZIEGLER, 1988) and are usually attributed to long intershot, or interrow, delays. GITTERMAN and VAN ECK (1993) attributed a low frequency spectral notch in recordings of mine blasts to source finiteness. Unusual time-domain characteristics also have been observed. ANANDAKRISHNAN *et al.* (1997) and STUMP and PEARSON (1997) observed significant surface waves caused by cast explosions in Wyoming. GITTERMAN *et al.* (1997) observed surface waves in recordings of quarry blasts in Israel. ANANDAKRISHNAN *et al.* (1997) modeled the Wyoming blasts and concluded that the surface waves are due to long source duration and significant spall, including material cast into the pit.

This paper presents further evidence for significant seismic signals produced by large mining blasts and not by instantaneous explosions. The paper presents evidence for spectral modulations below 10 Hz produced by documented ("ground truthed") cast blasts in Wyoming. Low-frequency modulations, below 5 Hz, are seen at an IMS station at a range of 9°. The paper identifies observations of significant surface waves also produced just by these events. Source modeling is used to understand these observations and to gauge the sensitivity of these signals to changes in blasting parameters at the source. We assess the usefulness of these signals, and low-frequency spectral modulations, for characterizing mining explosions using the IMS.

2. Regional Monitoring Experiments in Wyoming

To investigate low-frequency seismic signals produced by large mining blasts, researchers from the University of California, San Diego (UCSD), in collaboration with researchers from the Los Alamos National Laboratory (LANL), Southern Methodist University (SMU) and the Air Force Technical Applications Center (AFTAC), conducted two regional monitoring experiments in Wyoming in 1996 and 1997. Five broadband three-component seismic stations (STS-2's; flat response from 0.0083 to 40 Hz) were deployed within near-regional range of the Black Thunder mine (Fig. 3). Four of the sensors were deployed in a ring at 200-km range to study the azimuthal dependence of the seismic signals. The fifth station was deployed (in 1996) at a range of 100 km to the north of the mine along the azimuth to station CUST to allow examination of the range dependence. A three-element, 100-m aperture infrasound array was deployed at station MNTA (Fig. 3). The temporary deployments complemented permanent, nearby, deployments at PDAR, RSSD and more distant stations in the IMS and the IRIS Global Seismographic Network (GSN).

The seismic activity in this region is almost entirely man-made and is concentrated in the Powder River Basin coal mining trend (Fig. 3). Two types of blasting are most common in this trend. The largest blasts are used to cast overburden to the side to expose the coal seam (MARTIN and KING, 1995). Smaller shots are used to fracture the coal seam to facilitate recovery (ATLAS POWDER COMPANY, 1997). Event clusters correlate well with known mine locations (Fig. 3). The events of greatest interest were the large cast blasts and calibration shots, located in the Black Thunder coal mine (PEARSON *et al.*, 1995; STUMP, 1995), as these were closely monitored with video and acoustic equipment by the LANL team and were carefully documented by personnel at the mine. During the two experiments, four large cast blasts were detonated at Black Thunder (Fig. 4; Table 1). These blasts ranged from 2.5 million pounds (Aug. 1, 1996) to 7 million pounds (Aug. 14, 1997) and were used to move overburden. Three of four blasts occurred in the south pit of

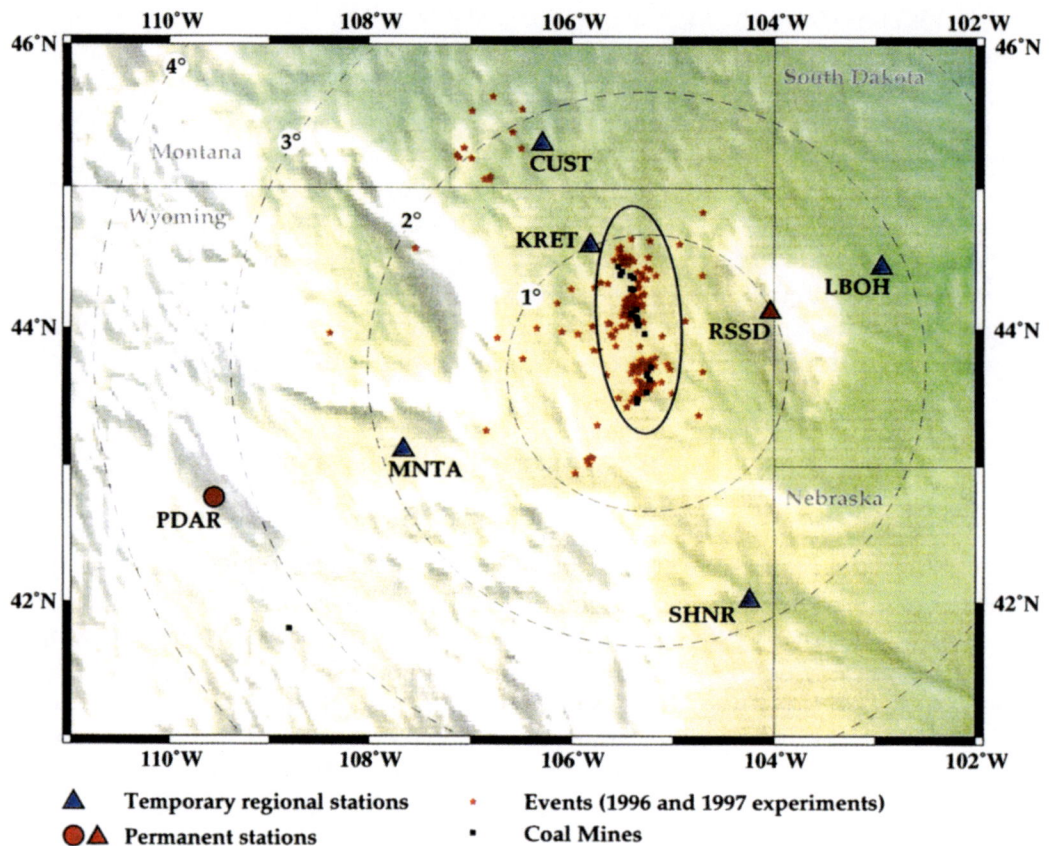

Figure 3

Two regional experiments were conducted in Wyoming by LANL, SMU, AFTAC and UCSD in 1996 and 1997. Four broadband seismic stations (three-component STS2's) were deployed in a ring surrounding the Powder River Basin (PRB) at a range of 200 km from the Black Thunder coal mine. A fifth station (KRET) was deployed at a range of 100 km. Permanent seismic stations are located at PDAR and RSSD. Infrasound sensors were deployed at MNTA. Mining explosions detected by the temporary seismic stations during 49 days in 1996 and 1997 are plotted and correlate well with known mines. The slight northward bias is likely due to the 1-D model. The oval indicates approximate limits of the PRB.

Black Thunder where overburden is cast to the north. Typical cast blasts include several (typically ~7) rows of shots. Intershot time spacing is 35 msec, rows are spaced by 200 to 300 msec. For reasons as yet not fully understood, one cast blast (Aug. 1, 1996) detonated, in part, nearly simultaneously. Six calibration shots (ranging from 5000 to 16,000 pounds) were detonated in the mine in 1997 (Fig. 4; Table 2). The first four calibration shots (yields 5000 to 5500 pounds) consisted of a single cylindrical borehole. The fifth and sixth calibration shots (yields 12,000 and 16,000 pounds) consisted of 3 and 4 boreholes, respectively. The boreholes in the larger shots were spaced 20 to 30 m apart and detonated simultaneously. In the

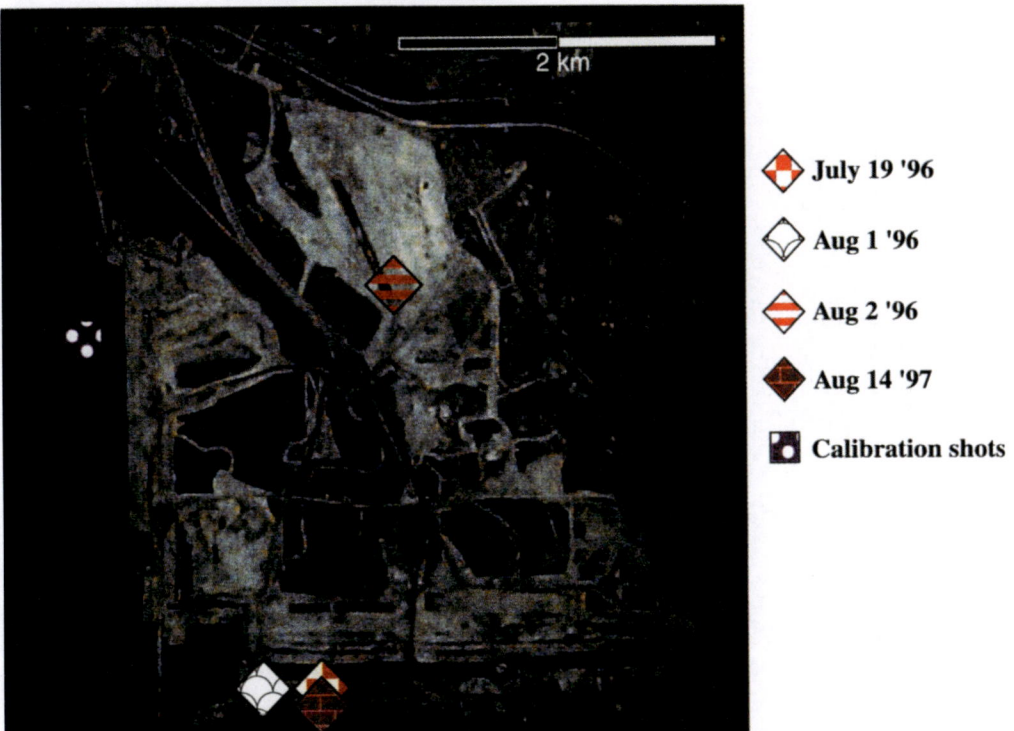

Figure 4

A satellite photo of the Black Thunder coal mine in Wyoming. The symbols give approximate locations of the major events considered in this paper. The 1996 and 1997 experiments produced recordings of four significant cast blasts in the Black Thunder coal mine. The July 19, Aug. 1, 1996 and Aug. 14, 1997 blasts occurred at the west end of the south pit. The west end of the Aug. 1 blast is believed to have detonated simultaneously. In the July 19, Aug. 1, 2 (all 1996) and Aug. 14 (1997) cast blasts 4.5, 2.5, 2.8 and 7.0 million pounds of ANFO were detonated. The two largest calibration tests involved 12,000 and 16,000 pounds of ANFO.

Table 1

Major blasts at Black Thunder

Date	Pit/firing direction	# Rows/# holes	Intershot/interrow delays (m_s)	Total explosive yield (pounds)
July 19, 1996	South/west	7/620 (decked)	35/200–275	4,549,366
Aug. 1, 1996	South/west	7/341	35/200–275	2,460,730
Aug. 2, 1996	NE/south	7/422	35/200–275	2,784,540
Aug. 14, 1997	South/west	7/702	35/200–400	5,958,010

Table 2

Calibration shots at Black Thunder in 1997

Date	Yield (pounds)
Aug. 14, 1997	~5500
Aug. 14, 1997	~5500
Aug. 14, 1997	~5500
Aug. 14, 1997	~6000
Aug. 15, 1997	~12,000
Aug. 15, 1997	~16,000

discussion that follows, we consider the time and frequency domain characteristics of all large Black Thunder events that occurred during the two experiments. However, we will focus on two events; the 4.5 million pound cast blast that occurred on July 19, 1996 and the 16,000 pound calibration shot.

3. Observations of Low-frequency Seismic Signals from Mining Blasts

Time Domain Energy Partitioning

Amplitudes of body waves produced by the 4.5 million pound July 19 cast blast are comparable to those produced by the 16,000 pound calibration shot (Fig. 5). This is expected as the July 19 blast yield is spread out over 620 shots which were arranged in 7 rows and ripple-fired over ~4.8 s. Each shot hole was decked − that is two explosive charges were detonated with a small delay in each hole. Although the two events produced similar body waves, just the cast shot produced significant surface waves (Fig. 5). Unfiltered recordings of the July 19 blast (Fig. 6) show the progression from high-frequency body waves to low-frequency surface waves at four azimuths. The surface-wave amplitudes exhibit a clear dependence on azimuth. Figure 7 indicates that although the low-frequency waveforms are highly dependent on azimuth, these signals do not seem to depend strongly on the fine details of the source if the blasts are in the same pit.

ANANDAKRISHNAN *et al.* (1997) and STUMP and PEARSON (1997) have published PDAR recordings of significant 4 to 12 second surface waves which are produced by cast blasts in Wyoming. ANANDAKRISHNAN *et al.* (1997) used linear source modeling to argue that the surface waves are the result of the long source duration and spall impacts. The 1996 and 1997 regional experiments indicate that these surface waves are routinely produced by the cast explosions and are readily detected at near-regional range.

In Figure 8, peak amplitudes in two pass bands (1 to 10 Hz; centered at P onset and 2 to 10 seconds centered on the surface wave) are estimated from noise-corrected envelopes of recordings made by the stations in the 200-km ring (Fig. 3). The cast blasts and calibration explosions are easily separated despite the strong

CUST vertical component

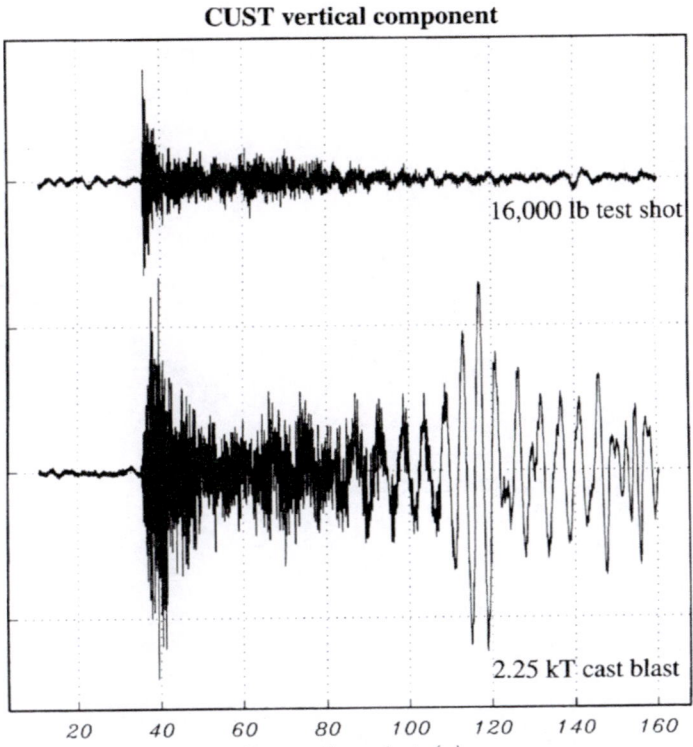

Figure 5

Unfiltered vertical component recordings of a 16,000 pound calibration shot (top) and a 4.5 million pound cast shot made at CUST. Both recordings are plotted at the same scale. The station was located 200 km to the north of the events (Fig. 3) which occurred in the Black Thunder coal mine. The tiny calibration shot rivals the immense cast shot as a source of *P* waves but is an insignificant source of surface waves. The dissimilarity of unfiltered vertical component recordings made at CUST is consistent with previous findings (KIM *et al.*, 1994; ANANDAKRISHNAN *et al.*, 1997) and suggests that a regional variant of the M_s:m_b discriminant could be effective for separating large mine blasts from instantaneous explosions.

dependence of the peak surface-wave amplitudes on azimuth. Although small calibration shots yield *P* waves that are comparable in strength to those produced by the much larger, but not concentrated, cast blasts, just the latter events produce significant surface waves. As indicated in this figure, the calibration shots yielded no surface-wave energy above noise. The lone outlier, among the population of cast blasts, is the Aug. 1, 1996 event. A significant simultaneous detonation, which occurred within this blast, significantly boosted peak *P*-wave amplitudes. The network average surface-wave amplitudes are almost identical, despite the significant differences in how these blasts were detonated. Despite the boosted *P*-wave amplitudes observed in the Aug. 1, 1996 recordings, the amplitude ratio comparing surface waves to body waves still separates this event from the single shots.

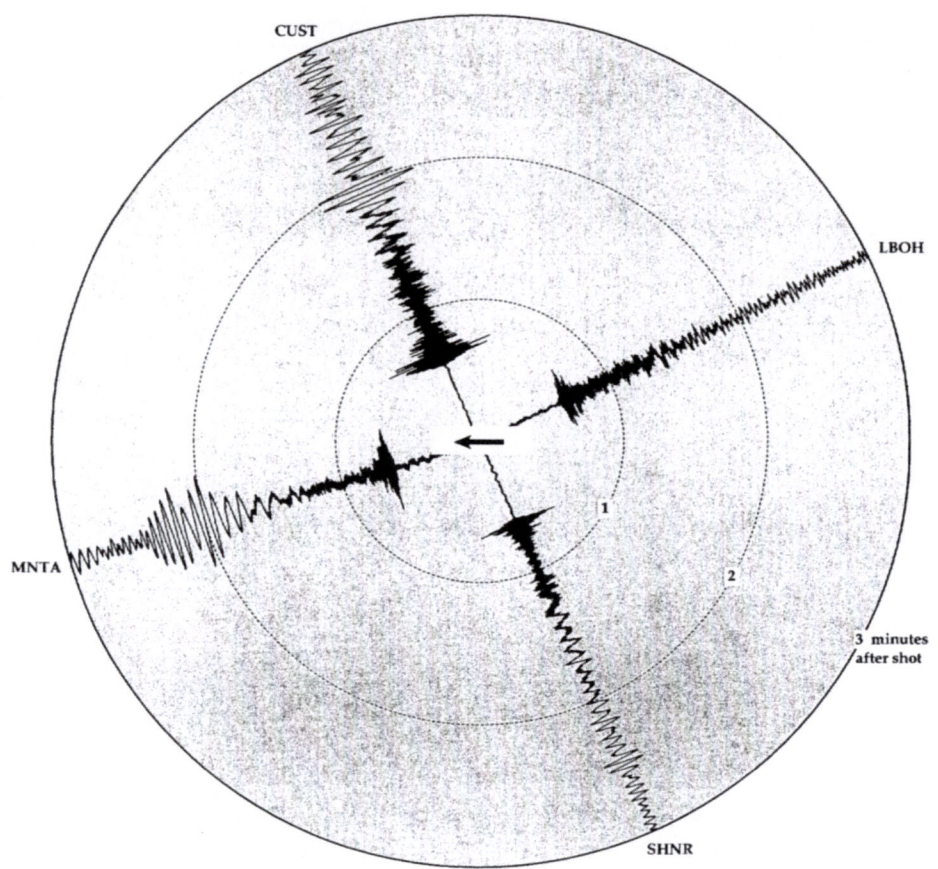

Figure 6
Unfiltered vertical component seismograms from the azimuthal network show the progression from high-frequency body waves to substantial surface waves. The July 19, 1996 blast used 4.5 million pounds of ANFO which detonated as planned. The arrow indicates the direction of shooting. The waveforms are highly dependent on azimuth. Propagation to the east across the Black Hills Pluton resulted in relatively little surface-wave energy.

Frequency Domain Modulations

Previous studies (incl. STUMP and PEARSON, 1997) have shown that the large cast explosions in Wyoming do not yield obvious spectral modulations above 10 Hz. The intershot delays of 35 msec concentrate energy at multiples of 29 Hz (1/0.035 s). Close-in data show a spectral increase that corresponds to the 35 ms delays although the peak is relatively broad, reflective of the different spatial locations of the individual charges and possibly variance in their individual detonation time. Rapid attenuation in this area is an additional cause of the faintness of the high-frequency modulations. These explosions do, however, produce significant modulations below

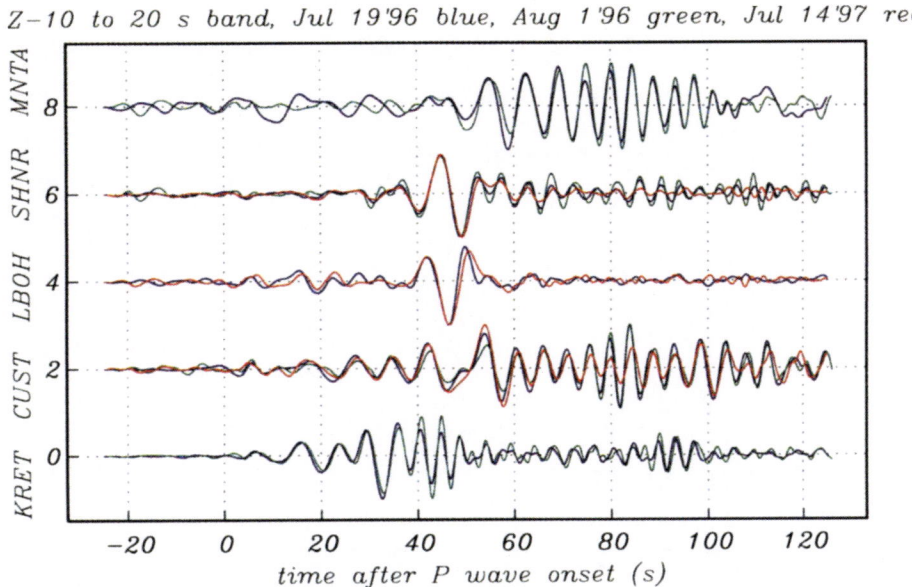

Figure 7

A comparison of band-pass filtered regional seismograms from three different cast blasts at five different stations at varying azimuths from the mine (see Fig. 3). All of these blasts occurred in the south pit of the Black Thunder coal mine. All shots were detonated in the south pit of the Black Thunder coal mine (Fig. 4). Low-frequency seismic signals from the south pit events are robust although highly dependent on azimuth.

10 Hz. Spectral estimates taken from recordings of the July 19, 1996 cast blast made by the 5 station regional network are shown in Figures 9 and 10. These spectra exhibit only a modest dependence on time, the recording direction, and the azimuth from the mine. No organized spectral modulations are seen in the recordings of any of the calibration shots. In this figure we display spectra from the 16,000 pound shot (dashed curves). As will be seen later, the low-frequency modulations below 5 Hz are seen out to 9°.

4. Waveform Synthesis

To give these basic observations a physical basis, we turn to synthetics. The synthesis of extraordinarily complex mining explosions has become relatively easy given the early work of BARKER and DAY (1990), BARKER et al. (1993) and McLAUGHLIN et al. (1994) and recent work by X. Yang who has modified the linear elastic algorithm of ANANDAKRISHNAN et al. (1997) and packaged it into an interactive MATLAB package (MineSeis; YANG, 1998). The algorithm assumes the linear superposition of signals from identical single-shot sources composed of

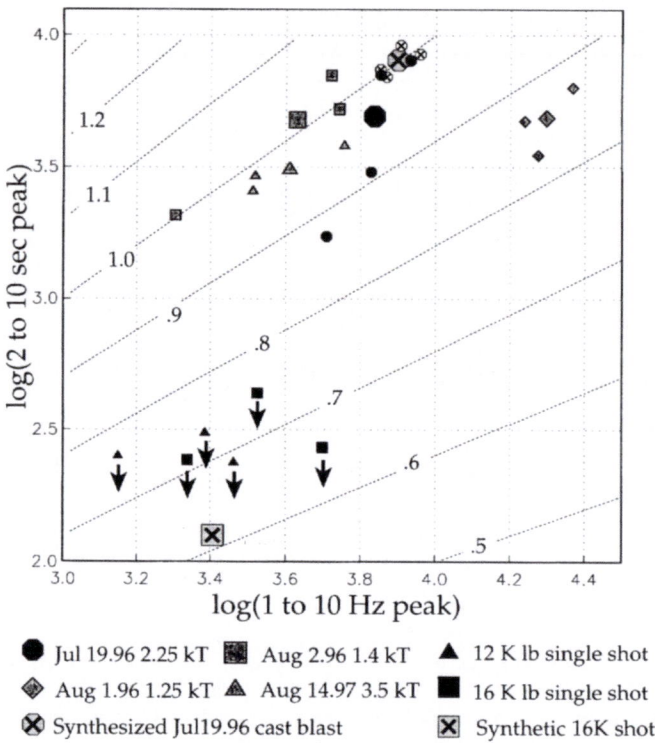

Figure 8

A comparison of 2 to 10 second surface wave and 1 to 10 Hz *P*-wave peak amplitudes using recordings made at a range of 200 km by MNTA, CUST, LBOH & SHNR (Fig. 3). All events occurred in the Black Thunder coal mine. Each trace is filtered, converted to an envelope and adjusted downward by an amount determined by pre-onset noise. Above are displayed the logarithms of the individual station peak amplitudes. The large symbols represent the network average for each event. Each labeled curve indicates a constant ratio of surface wave to *P*-wave amplitude [i.e. log(2 to 10 s peak)/log(1 to 10 Hz peak)]. As expected the calibration shots yield essentially no surface-wave energy above noise. The downward arrows indicate that the maximum *P*-wave amplitude is well constrained but the surface-wave amplitude lies below noise. For this reason, no network averages are displayed for these events. The Aug. 1, 1996 cast shot appears as being somewhat explosion-like due to the sympathetic detonation. The sympathetic detonation greatly boosted *P*-wave amplitudes but left the surface waves untouched. Unadjusted amplitudes are plotted as all stations are at the same range from the mine. We also display peak amplitudes from a synthetic version of the July 19 cast blast and from a synthetic 16,000 pound calibration shot. The energy partitioning from the synthetic events is in agreement with observations.

isotropic and spall components. Both shooting delays and location differences among individual shots are taken into account in calculating delays of the superposition, although the Green's functions are assumed to change slowly so that a common Green's function is used for all the single shots. We used a reflectivity method to calculate the Green's functions. A one-dimensional velocity model was used (PRODEHL, 1979; ANANDAKRISHNAN *et al.*, 1997).

To model the July 19 cast blast we used a blast report issued by the Black Thunder mine. The blast consisted of 7 rows and a total of 620 decked shots with a total yield of 4.5 million pounds (Table 1). In a decked shot, more than one charge is detonated in the same hole. In this event, each hole contained two charges separated by 50 to 200 msec. Although the number of shots in each row varied from 85 to 93, for simplicity we assumed each row had 89 shots and that each decked shot had a total yield of 0.0033 kt. We assumed that all rows were spaced by 9.1 m and that all adjacent shots in the same row were 10.4 m apart. Intershot delays were 35 msec, interrow delays ranged from 200 to 275 msec. SOBEL (1978) estimated that 9.6×10^9 kg of material is spalled by each kt detonated. In our experiment, each decked shot

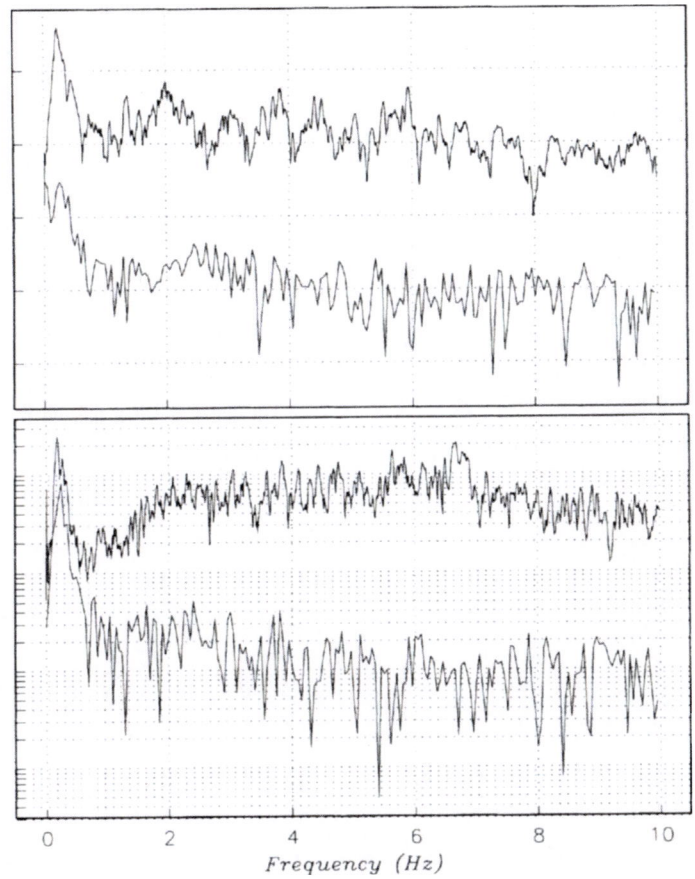

Figure 9
Spectral estimates taken from pre-event noise and signal recorded at CUST. In the upper panel, we display spectra taken from the recording of the July 19, 1996 cast blast (see Fig. 5). In the lower panel we display spectra taken from the CUST recording of the 16,000 pound calibration shot. The noise spectral estimates were untapered. Both P-wave spectra have been convolved with a boxcar function spanning 0.04 Hz to give a clearer view of the spectral modulations in the upper panel.

July 19.96 cast vs 16,000 lb calib. shot

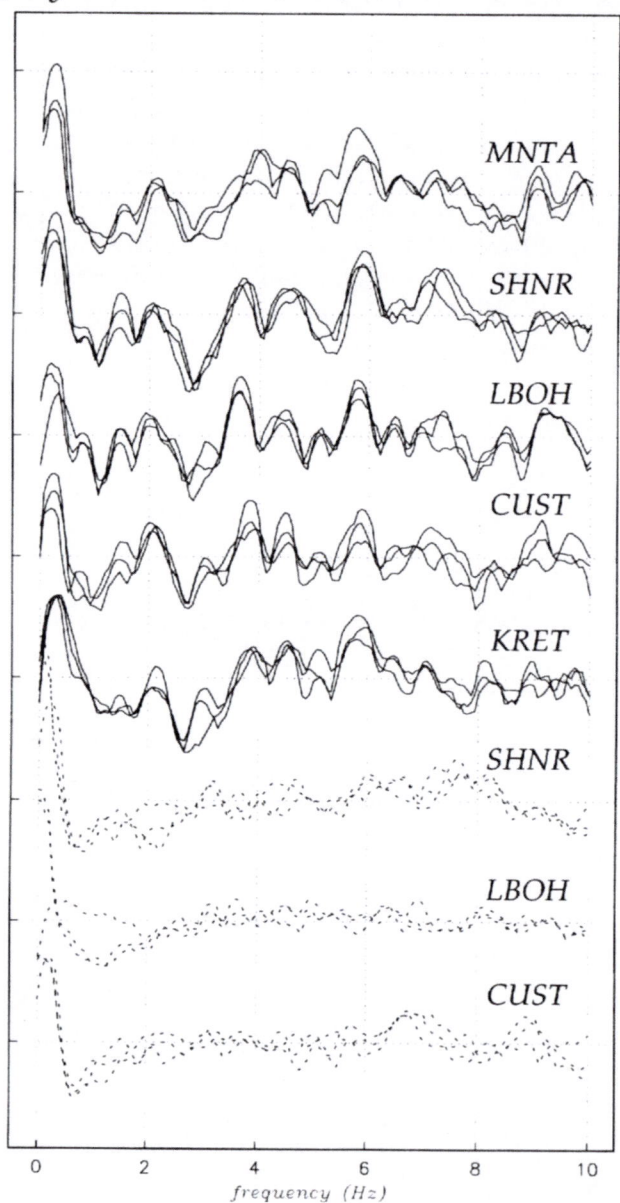

Figure 10

Obvious time-independent modulations exist below 10 Hz in the spectra of recordings of the July 19 cast blast (upper solid lines). At each station three curves, each representing the log of the spectral amplitude from a single component, are plotted. These are similar to the high-frequency modulations observed by HEDLIN *et al.* (1989) although these are likely due to source finiteness and interrow delays. Both these spectra and those observed by HEDLIN *et al.* (1989) are independent of the recording direction. A very slight dependence of the modulations on azimuth can be seen. The 16,000 pound calibration shot (dashed lines) produced no discernable spectral modulations. Each detrended multitaper spectral estimate was taken from 125 seconds of *P* and *S* coda. All components, from each three-component station, are plotted.

had a total yield of 0.0033 kt and thus the Sobel relation gives ~31 kt of spalled material. For the July 19 Black Thunder event we found this figure yielded surface waves that were more energetic than those that were recorded and so we reduced the spall figure to 20 kt. As will be discussed later, this discrepancy can be due to the incomplete conversion of spalled kinetic energy into seismic or to inadequacy of the velocity model. We assumed that the spalled material was cast at an angle 10° above the horizontal and fell 20 m.

Some vertical component synthetics are shown in Figure 11. This figure illustrates the relatively energetic surface waves that can be expected from cast blasts. The 16,000 pound point synthetic source, modeled as an isotropic source, produces a weak surface wave. The figure also illustrates the slow attenuation of the surface wave produced by the synthetic cast blast.

The peak amplitudes of the point source and the July 19, 1996 cast blast synthetics at a range of 200 km have been calculated for the four outer stations in the regional network (Fig. 3) and are displayed in Figure 8 with the observed amplitudes from the recorded events. Despite the assumptions listed above, we found the surface- and body-wave amplitudes produced by the synthetic event were consistent with the recorded waves at MNTA and CUST. The other two stations (LBOH and SHNR) yielded broadly dispersed surface waves which had lower peak amplitudes in the frequency band from 2 to 10 seconds. This pronounced mismatch is not due to unmodeled source effects but results from the propagation of the energy through a crust that differs significantly from the one-dimensional Prodehl model. Any comments at this time on what these differences imply about the crustal velocity model would be imprecise. A surface-wave inversion is currently being conducted by Rongmao Zhou and Brian Stump at Southern Methodist University. Although a significant azimuthal dependence of the surface-wave amplitudes is observed, the surface waves produced by the cast blasts are significantly more energetic at all azimuths than those produced by the single shots. The relatively minor azimuthal dependence that is seen in the synthetics is due to source directivity.

Experiments in Varying Source Parameters

Although we have reproduced the relative peak amplitudes of body and surface waves generated by the July 19 cast blast, several important questions remain unanswered. We have yet to determine which of the many source parameters included in the model play a leading role in generating the energetic surface waves. We need to gauge the sensitivity of the observed signals to changes in source parameters and to assess the utility of these signals for source characterization. ANANDAKRISHNAN *et al.* (1997) considered the effect of changes in several source parameters on regional waveforms. In this paper, we conduct a similar exercise; however, we are most interested in how these changes affect the partitioning of energy between the surface and body waves. We focus on the frequency bands considered in Figure 8. We

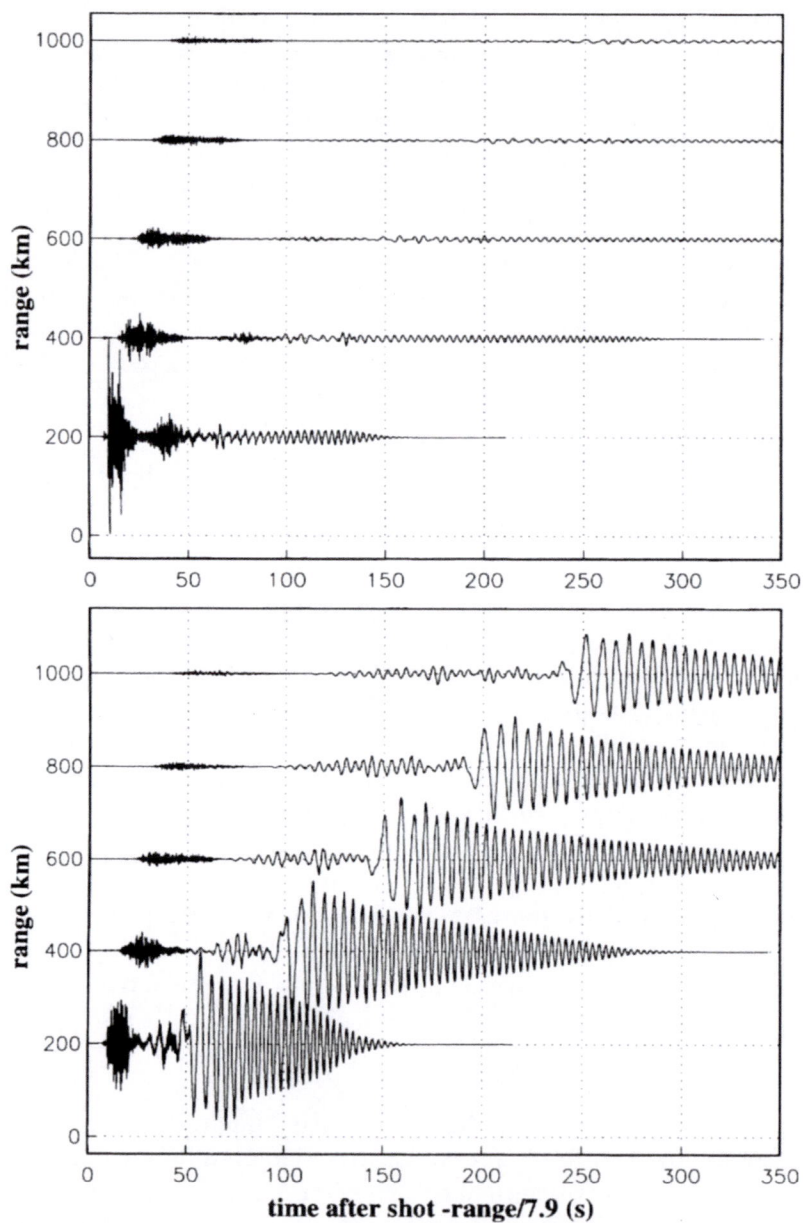

Figure 11

Simulations of two events. A single 16,000 lb shot is displayed on top, the July 19, 1996 cast shot is displayed on the bottom. The single shot is an insignificant source of surface waves. The surface waves excited by the cast blast decay relatively slowly and dominate the waveform of the cast shot at all ranges from 200 to 1000 km. The single shot synthetics are magnified for clarity. The range dependence of all synthetics was reduced by simply scaling all amplitudes by the range.

consider a broad suite of mine blasts. The blasts we synthesize are all the same as the presumed correct one reviewed in the previous section except one source parameter is allowed to vary while the others are held fixed. In turn, we vary the spalled mass, individual shot yield, and the duration of the blast. In a separate experiment we also allowed shot times to deviate from the planned 35 msec delays. All synthetics are calculated for CUST, the station located 200 km to the north of the mine. The results for the other three stations are essentially the same and are not plotted.

An example is shown in Figure 12 which displays synthetics for a suite of cast blasts in which the total yield of each shot hole is varied from 20% of the actual value to 200% (from 1468 to 14,676 pounds). The synthetics suggest that the body-wave amplitudes will scale rapidly with individual shot yield however the surface waves show a weaker dependence. This is consistent with ANANDAKRISHNAN *et al.* (1997) who concluded that the surface waves are mostly due to the different yield scaling in the explosion and spall source models.

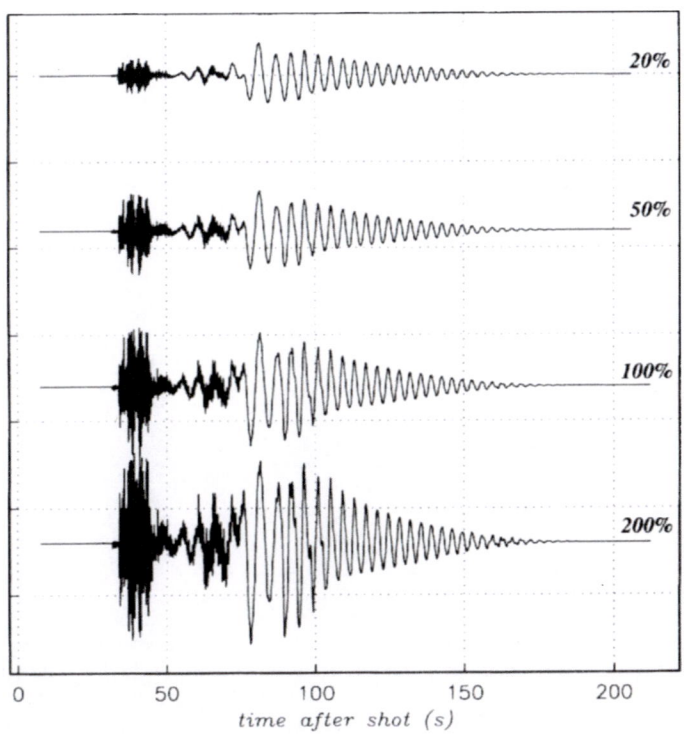

Figure 12

The dependence of vertical component waveforms on shot parameters is easily simulated using the MineSeis Matlab package. These plots show the dependence of the waveform on the yield of the individual shots. The standard cast, with individual shots of 7338 lb, is the third trace from the top. The scaling of the shot yield is indicated by the text to the right of each trace. The body-wave amplitudes depend strongly on shot yield. The surface waves show a weaker dependence.

A more general illustration of source effects on body and surface-wave amplitudes is given in Figure 13. In the upper panel of this figure we consider peak amplitudes in two filtered versions of the synthetic traces. We filter the synthetic traces between 2 and 10 s, and between 1 and 10 Hz and calculate the log of the peak amplitude in each trace. In Figure 13, we display the ratio of the peak amplitude in the the low-frequency trace to that in the high-frequency trace. In the upper panel, we

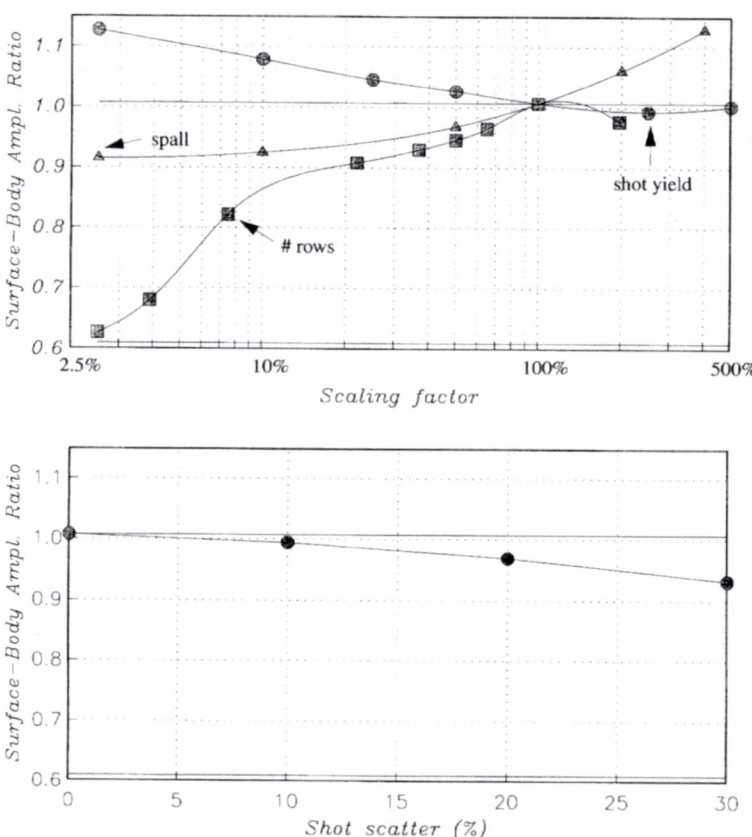

Figure 13

In the upper panel we display the surface-body wave ratios for a suite of blasts in which a single source parameter is varied while all others are held fixed at levels believed to be appropriate for the July 19, 1996 cast blast. The filled circles represent sources in which individual shot yield is varied from 182 to 36,300 lb. The filled triangles represent blasts in which the spalled mass per shot is varied from 0.5 kt to 80 kt. The squares represent blasts grids that range from a single row with three shots to a 10 row blast with 200 shots in each row. Varying blast duration clearly has the most significant impact on energy partitioning. Only the severely restricted grid partitions energy like the synthetic 16,000 pound shot (represented by the horizontal line at a ratio of 0.61). All of these simulations assumed the subshots detonated exactly when planned. The curve in the lower panel represents blasts in which the shot times are distributed normally about the planned times. The shot scatter considered ranges up a variance of 30% of the planned intershot delay of 35 msec.

display the ratios obtained from synthetic events in which a significant source parameter has been varied between 2.5% and 500% of the standard value. As seen in Figure 12, the surface wave to body-wave amplitude ratio decreases with increasing shot yield. This effect is represented by the filled circles in the upper panel of Figure 13. This effect is seen to be relatively minor as varying the yield over more than three orders of magnitude (from 182 to 36,300 pounds) changes the body-surface ratio by ~10%. As seen in Figure 12, this ratio change results from changes in both the surface- and body-wave amplitudes.

Varying the spalled mass has the opposite effect. When the spalled mass is varied from 0.5 kt to 80 kt (from 2.5% to 400% of the standard) the surface-body wave ratio changes from ~0.9 to 1.13 (filled triangles in the upper panel of Fig. 13). The change in this ratio results from the effect spalled mass has on the surface waves as the spalled mass is predominantly a source of low-frequency energy. Changing the spalled mass had little effect on the body-wave amplitudes. Although the production of surface waves is reduced when essentially no material is spalled into the pit, the restricted blast still produces surface waves that are significantly more energetic than those produced by the single shot.

The most significant changes in the waveform result from variations in the total duration of the blast. To vary the blast duration we scaled both the number of rows and the number of shots in each row. The standard blast had 7 rows and 89 shots in each row and lasted a total of ~4.8 s. The largest blast grid we considered had 10 rows, each with 200 shots. The extended blast spanned ~9.5 s – ~200% of the actual duration and had a total yield of 14 million pounds. The reduced blast grids consisted of 5 rows, 50 shots per row; 4 rows, 40 shots per row; 3 rows with 30 shots per row; 2 rows with 20 shots per row and 1 row with 10 shots. To reduce the blast further we decreased the number of shots in the final row to 5 shots and then to 3. The duration of these reduced blasts ranged from 3.1 s (for the blast with 5 rows and 50 shots per row) to 70 msec (for the smallest blast). As we see in Figure 13, extending the blast beyond the standard one that was used on July 19 yielded essentially no change in the surface-body wave amplitude ratio. However reducing the scale of the grid has a significant effect on the partitioning of energy between surface and body waves. Reducing the scale of the blast has a modest effect on body-wave amplitudes but the surface waves are strongly dependent on this source attribute. The ratio drops most rapidly when the blast duration is reduced below 2/10 of the actual duration (down to ~0.5 s from 4.8 s). The production of 2–10 s surface waves is reduced significantly when the blast duration is reduced below 2 seconds. When the blast duration extends much beyond 5 seconds, production of these surface waves is not increased substantially. Of all the blasts considered, just the brief blast resembled the single shot (which is represented by the horizontal line at a ratio of 0.61).

All synthetics considered thus far assumed a perfect temporal and spatial adherence of the actual shot grid to the design grid. Introducing shot time scatter, which is believed to be omnipresent (STUMP and REAMER, 1988; STUMP et al., 1994,

1996), is predicted to have no effect on the surface-wave amplitudes but could increase the body-wave peak amplitudes significantly (Fig. 13; lower panel). Shot scatter increases the likelihood that shots will detonate simultaneously. An extreme example of shot scatter is the simultaneous detonation of a portion of the shot grid. This occurred in the August 1, 1996 blast and, as predicted by the synthetics, the surface waves were not affected (Fig. 8). The network averaged peak surface-wave amplitudes for the three mine blasts shown in this figure are very similar, despite the wide range of explosive yields. It appears that surface waves are just generated by temporally extensive mine blasts without regard for exactly how the blasts are detonated, and whether the blast sequence includes any significant detonation anomalies. The anomalous event was assigned a body wave magnitude of 4.0 (REB Bulletin).

5. Synthesis and Automated Recognition of Spectral Modulations

Synthesis

Seismic signals produced by delay-fired sources have spectral modulations at a wide range of frequencies. High-frequency modulations result directly from intershot delays. These delays are typically 35 msec and the modulations occur at multiples of 1/35 msec ($\sim$30 Hz). Interrow delays are typically longer. The cast blasts we consider in this paper have interrow delays of 200 to 300 msec which give modulations every $\sim$3 to 5 Hz. Source finiteness will also cause subtle modulations spaced at the inverse of the duration of the event (e.g GITTERMAN and VAN ECK, 1993). Spectral modulations can also be acquired during propagation to the receiver (HEDLIN *et al.*, 1989). To illustrate this problem, and to illustrate the modulations that can result from source finiteness, we consider two synthetic sources – a single shot and a simple, 1 row, 25 shot delay-fired source. As shown in Figure 14, a single shot yields spectra with subtle modulations. These modulations are due to resonance in the near-surface layer of the model. This layer has a two-way travel time of $\sim$0.4 s and thus produces modulations spaced at $\sim$2.5 Hz. Much more significant modulations below 10 Hz are produced by the delay-fired event. The simple delay-fired event consisted of a single row of 25 shots. Intershot spacing of 35 msec produces a high-frequency modulation starting at 29 Hz (1/0.035 s). The shot sequence lasts for 0.875 s. The source finiteness causes a spectral modulation at the inverse of the source duration (peaks every 1.14 Hz).

To understand better the modulations produced by the July 19 event, we contrast signals from three different blasts with those produced by an instantaneous shot. Starting at the top of Figure 15, the upper three black traces are modulations between 1 and 10 Hz predicted for 1, 4 and 7 row cast blasts, respectively. These synthetic blasts are modeled after the July 19 decked blast, we have just altered the number of rows. The dashed spectra were taken from a synthetic 16,000 pound

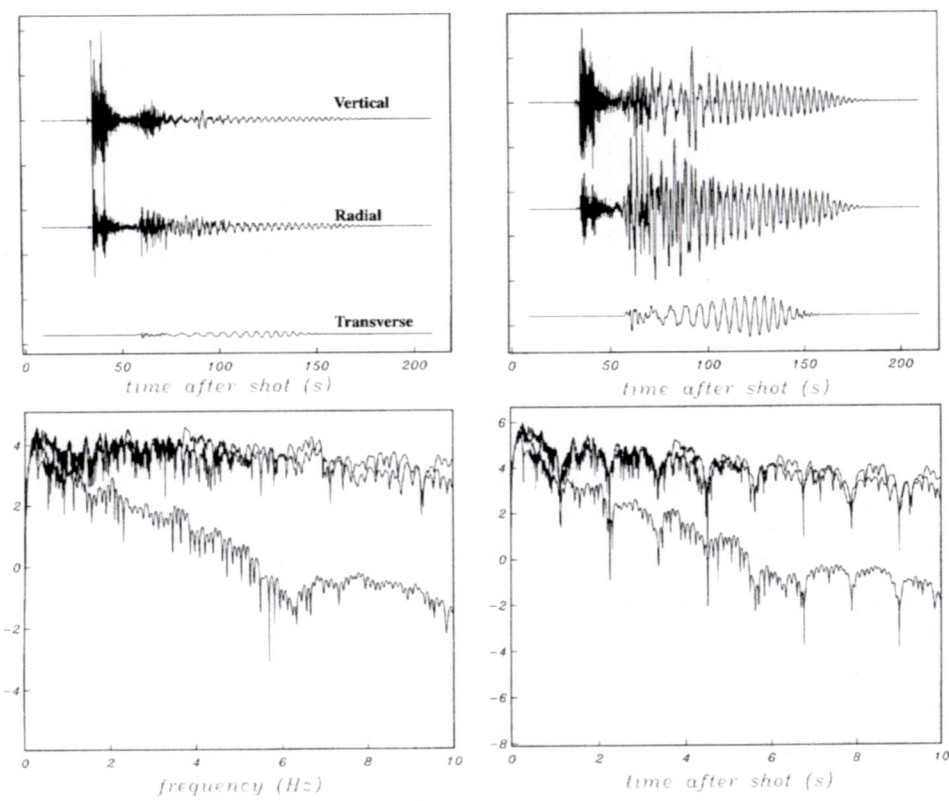

Figure 14

A synthetic 16,000 lb single shot at a range of 200 km is shown on the left. The faint spectral modulations seen in the spectra are due to resonance in near-surface low-velocity strata. A simple cast blast consisting of 1 row of 25 shots spaced at 35 msec is shown on the right. This simulation is for a station at a range of 200 km. The prominant modulations seen in the untapered spectral estimates are due to source finiteness. This shot lasted 0.875 s and produced modulations at the inverse of the duration (1.14 Hz). The first peak due directly to the intershot delays lies at 29 Hz.

instantaneous shot which was located at the same point. In the single row delay-fired event we see broad modulations spaced at $\sim$2.5 Hz and fine-scale modulations spaced at $\sim$0.3 Hz. The fine scale modulations are due to the finiteness of the source (which spanned $\sim$3 s). The broader modulations are seen in the spectra from both sources and are due to resonance. Strong modulations spaced at $\sim$0.75 Hz appear in the spectrum of the 7 row blast. These are not continuous across the band from 0 to 10 Hz but appear to be strongest at $\sim$3 Hz. These modulations are not present in the spectra of the single-row blast and are due to the combined effect of the interrow and interdeck delays (which range from 0 to 300 msec) and source finiteness. Although the 7 row event had a total duration of $\sim$5 s, the rate at which explosives were detonated was strongly dependent on time. The first shot in the final row detonated

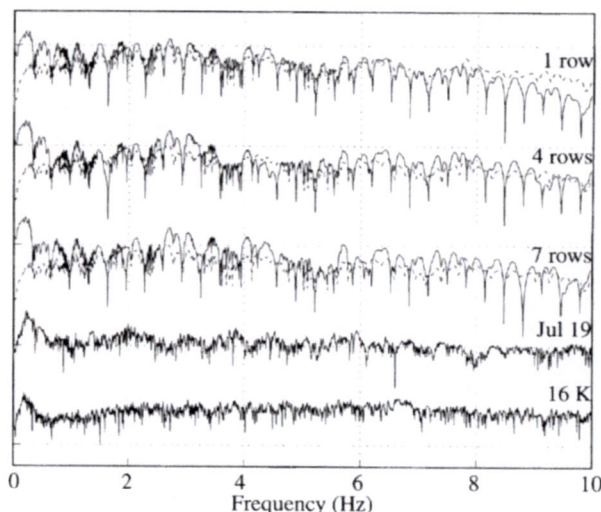

Figure 15

A comparison of spectra from synthetic and real events. The upper three black traces are vertical component spectra from synthetic cast blasts that are based on the July, 1996 south pit blast. The third spectrum is calculated from the full 7 row shot pattern. The middle and upper traces were calculated by using the front 4 and 1 rows, respectively. The grey curve plotted with each spectrum is from a synthetic 16,000 pound calibration shot. All synthetics were calculated for a receiver located 200 km to the north of the mine at the CUST station. The fourth spectrum is from the CUST vertical component of the July 19 event. The lowest trace is from the recorded 16,000 pound calibration shot. Broad modulations seen in all synthetic spectra result from resonance in the one-dimensional model. The fine-scale modulations are a source effect. Although the synthetics do not reproduce the fine detail of the low-frequency spectra recorded by the regional network, they do reproduce the general periodicity and variance of the modulations. In this figure, all spectral estimates are untapered.

~1.9 s after the shot sequence began. The final shot in the first row detonated ~3.2 s into the sequence. For 1.3 s in the middle of the shot grid, all 7 rows were being detonated and the explosive yield per time delay was at a peak (~25,000 pounds per 8 msec delay period). The spectral modulations produced by this trapezoidal source time function are not spaced at the inverse of the total shot duration (~ every 0.2 Hz) but rather are more broadly spaced at ~0.75 Hz (the inverse of the 1.3 s period during which explosive yield is at a peak). These fine scale modulations are not continuous across the band from 0 to 10 Hz because of destructive interference with long interrow and interdeck delays.

The fourth spectrum displayed in Figure 15 was taken from the vertical component recording of the July 19 event made at CUST. At the bottom of this figure we display a spectrum taken from the CUST recording of the 16,000 pound calibration shot. Strong peaks are observed in the cast blast spectrum at 2, 4, 6 and 10 Hz (see also Fig. 10). Smaller modulations are seen across the band from 1 to 10 Hz at a spacing of ~0.75 Hz. These modulations are not seen in the spectrum of the calibration shot and are clearly a source effect. These modulations require source delays of >1 s to 500 msec. The synthetic test displayed in the upper part of this figure suggests that these modulations are due to the combined effects of source finiteness and the complex interference between the 7 rows and 2 decks at this source; however, the fine spectral details are not reproduced.

We lack the necessary ground-truth information to reach definitive conclusions regarding probable causes of the mismatch between the observed and synthesized modulations; however, unmodeled source effects such as shot time and yield scatter seem most likely. The synthetic events we considered in this experiment consisted of shots that had exactly the same yield and detonated exactly when they were supposed to. Any deviations from uniform spacing of identical shots will change the manner in which signals from the different source processes (finiteness, interrow, interdeck and intershot delays) interfere with one another and will cause the recorded modulations to differ substantially from those predicted by synthetics. The synthetics suggest that most modulations observed in the recorded cast blast are due to long source delays and to source finiteness; however, this observation remains tentative as the fine details of the spectral modulations are not matched.

Automated Recognition

HEDLIN (1997, 1998a, b) used a variant of cepstral analysis to quantify time-independent spectral modulations at high frequencies (up to 40 Hz). The standard cepstrum is the Fourier transform of the log of a single spectrum. Hedlin used the two-dimensional Fourier transform of sonograms to isolate spectral energy that is periodic in frequency and independent of time. The same processing technique can be used for the low-frequency modulations observed here. Figure 16 shows discrimination parameters output by the Automated Time Frequency Discriminant (ATFD) for individual stations. The three parameters displayed are the autocorrelation, which measures the independence of spectral modulations with time; the cross correlation, which measures the independence from recording component; and cepstral extreme, which indicates the strength of time-independent spectral modulations. Although there were too few ground-truthed events to put the output to a statistical test, we tentatively conclude that the network averaged parameters separate single explosions from cast explosions.

6. IMS Recordings of the July 19, 1996 Cast Blast

The regional experiment provided evidence that large mining events will routinely yield significant low-frequency seismic signals. This network was deployed within 2° of the Powder River Basin so this test is rather unrealistic. Under a foreseeable monitoring scenario, where the bulk of the data comes from IMS stations, these events will be detected from mid-regional range. Detection statistics from the PIDC

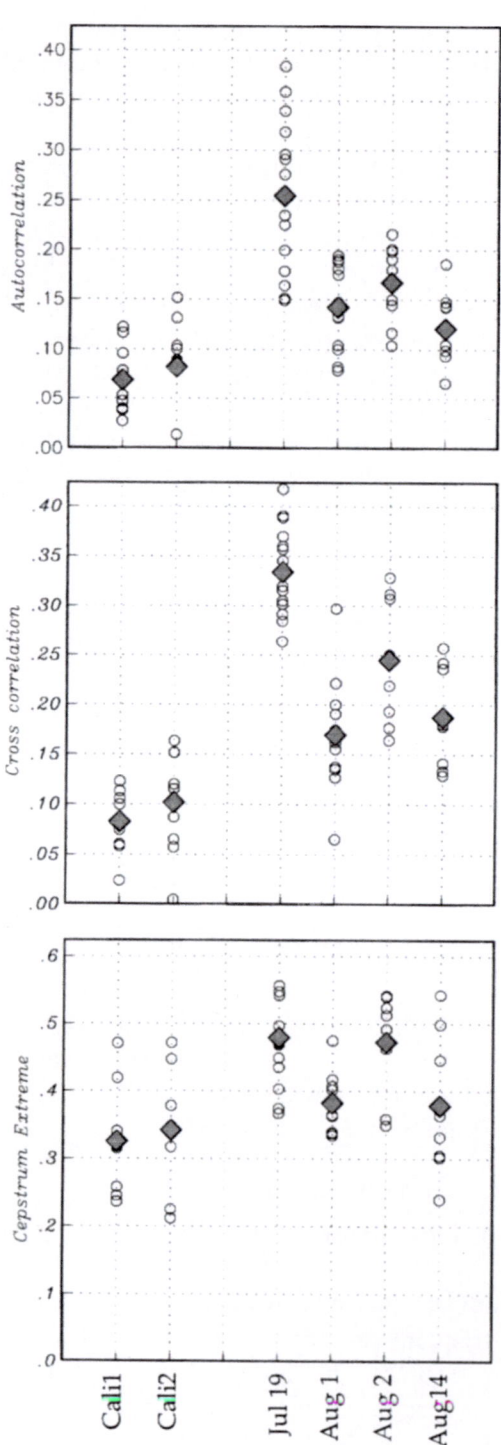

show that many of the PRB explosions are seen at far-regional to teleseismic range but can these attenuated signals be used for source characterization?

IMS recordings of the Wyoming events can give some indication of the range from which these signals might be used for source characterization. As shown in Figure 17, a low-frequency surface wave packet from the July 19, 1996 Black Thunder blast are seen out to ULM at a range of 9.1°. Figure 18 shows that the low-frequency spectral modulations also survive to this range; however, those above 6 Hz have fallen below noise.

7. Discussion and Conclusions

Mining Explosions and the CTBT

Under a CTBT, mining explosions will be problematic. As these events occur worldwide, many produce significant seismic signals which can be detected at mid- to far-regional and to teleseismic distances. Mining blasts are exceedingly complicated events. The complexity in the cast blasts considered in this paper stems primarily from the interaction of delay-fired explosives, rock fracture, and spall. The challenge mining events pose for the monitoring community is heightened because, through time, mine operators will occasionally experiment with new shot patterns. Furthermore, mine blasts will typically not detonate exactly as planned. Most deviations from the planned shot grid, such as shot scatter (e.g., STUMP and REAMER, 1988), are not highly significant. Other anomalies, such as the sympathetic detonations discussed in the introduction, are relatively infrequent but can be significant. A constraint is also placed on characterization methods by the IMS stations which sample the signal at 40 sps. Most diagnostic signals due to intershot delays are beyond the recording Nyquist frequency and are likely to be attenuated. For these reasons, we believe that it is important that a suite of approaches are developed to characterize these events. Under the CTBT, monitoring of mining blasts at any range will have to rely heavily on low-frequency signals.

◄

Figure 16

Discrimination parameters calculated by the Automated Time-Frequency Discriminant (ATFD, are described in detail in HEDLIN (1998). Each panel shows the results of applying a single operator to the time-frequency expansions of the data. The autocorrelation operator (top panel) assesses the dependance of the spectral modulation pattern on time. The cross correlation (middle panel) is a measure of the independence of the modulation pattern from the recording component. The cepstral extreme (bottom panel) is a measure of the amount of energy in the coda that is periodic in frequency and independant of time. Each three-component recording gives rise to nine parameters – three from each operator. Two calibration shots (left) and four cast blasts (right) were considered. The large symbols represent unweighted network averages.

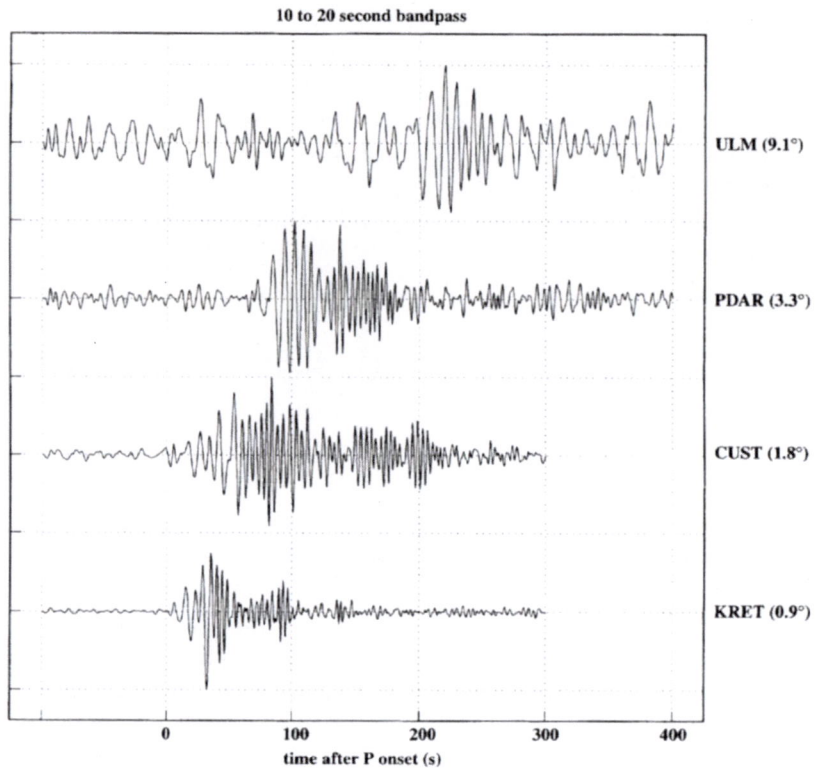

10 to 20 second bandpass

ULM (9.1°)

PDAR (3.3°)

CUST (1.8°)

KRET (0.9°)

0   100   200   300   400

time after P onset (s)

Figure 17

IMS/WYnet recordings of the Jul. 19, 1996 cast blast. The low-frequency bandpassed recordings exhibit an energetic dispersed surface wavetrain out to the IMS station ULM at 9.1°.

The Relative Merits of the Discriminants

Methods based on spectral modulations are potentially useful. For this approach to have value, it is necessary to separate modulations acquired at the source from those acquired during propagation and from those present in the background noise. Techniques that are dependent on high frequency modulations are typically limited to near-regional ranges and settings where the propagation Q is high (e.g., stable cratons). For this reason, this paper has focused on the causes of modulations below 10 Hz. It is apparent that complex mining blasts produce spectral modulations at a broad range of frequencies. The Wyoming field experiments have provided evidence that significant mining explosions will produce spectral modulations below 10 Hz that are not single-explosion like. Low-frequency modulations below 5 Hz will be detected beyond near-regional range. Just how common these signals are beyond near-regional range, and whether they are noticeable at lower blast yields is implied by our analysis of synthetics and by the recordings at ULM but will require further study of recorded events. The path from Wyoming to ULM probably has a relatively

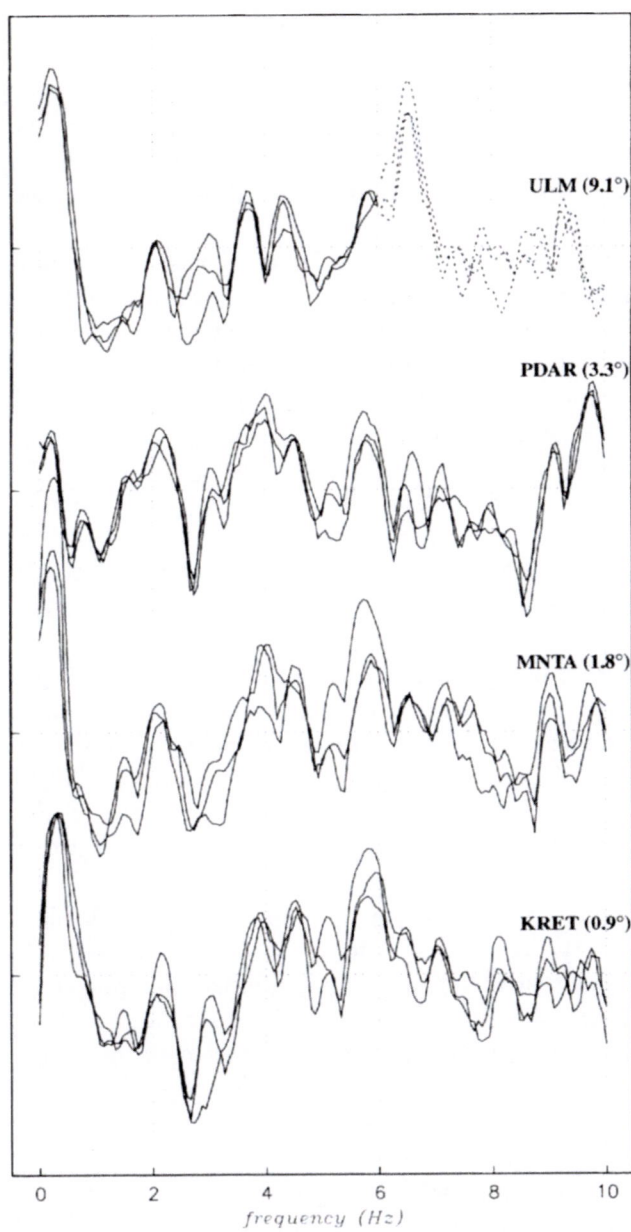

Figure 18
Multitaper spectral estimates taken from IMS and WYnet recordings the July 19 cast blast. The dashed portion of the ULM spectrum is noise. This robust low-frequency spectral roughness is not produced by small single explosions (Fig. 10).

high propagation Q. As a result, this observation cannot be taken as representative most regions but is the kind of scenario one might expect in a stable continental region. One region in which large mining blasts are not uncommon and propagation Q is relatively high is the Kuzbass/Abakan mining region in Kazakhstan. In regions where propagation Q is low, such as Wyoming, the high-frequency modulations due to intershot delays are quickly lost, although those due to longer duration source processes remain.

Long-period surface waves from large mining blasts in Wyoming are readily seen at mid-regional distances. The expenditure of a lot of energy spread out in time results in body waves that are relatively inenergetic and surface waves that are large. Our experiment indicates that the energy partitioning of mining blasts is not explosion-like and might be useful at mid-regional distances. It is clear that in geologically complex regions like Wyoming, surface waves are highly dependent on the path (Figs. 6–8). Even in this region, however, our study indicates that peak amplitudes along paths where surface waves are rapidly dispersed, are high.

Our analysis of a small number of events indicates that the observables (spectral roughness, energy partitioning) can be reduced to simple discrimination parameters. This result indicates that the ATFD approach developed by HEDLIN (1997, 1998a, b) can be simply scaled to lower frequencies where necessary. We have not yet attempted to define similar discrimination parameters for the body-surface amplitude ratios. In large part, this is due to a shortage of events included in the analysis and a shortage of recordings at ranges other than 200 km. A number of researchers (incl. ANANDAKRISHNAN *et al.*, 1997; STUMP and PEARSON, 1997) have pointed out anomalous surface waves. If this observation appears to be routine from studies of events from other regions the next logical step is to take source-receiver range into account and define a $M_b{:}m_s$ relationship for mining blasts.

The potential for misidentifying a mining blast by using any one of these individual tools is rather high. It will, no doubt, be necessary to regionalize these techniques by examining, in detail, blasting practices and propagation in each area of interest. These tools will, perhaps, some day be used alongside other tools, such as cross correlation, that have been proven to be powerful by a number of researchers.

Simulations

The state-of-the-art in simulating mine blasts is not able to match wiggle for wiggle the seismic waves from these very complex events in the time or in the frequency domain. However, general features of these events (energy partitioning between body and surface waves, enhanced spectral roughness below 10 Hz) have been reproduced. We have used the synthetics as an interpretation tool to reproduce the general character of these events to better understand which source processes are important, but an exact replication might never be realized.

There are several likely causes of the misfit. The code makes a number of assumptions about these sources that limit how well we can fit the data. Nonlinear interactions between shots are not taken into account and are likely important (MINSTER and DAY, 1986). The model assumes 100% of the spall kinetic energy converts into seismic. It is known that much of this energy is used to compact the spalled mass by collapsing voids and fracturing the rock, more energy is irretrievably lost to friction; however, these losses are not well understood at this time and are not taken into account in this code. As a remedy, this loss of energy can be modeled to some degree by assigning a smaller amount of spalled mass. The spalled mass is directly proportional to the seismic energy the spall generates. All shots in the blast are assumed to be identical. However it is well known that explosive yield is highly variable. STUMP et al. (1999b, c) analyzed single-fired shots ranging in explosive weight from 5500 to 50,000 pounds. They found amplitudes from several 5000 pound shots varied by up to 2 orders of magnitude. The calibration shots considered by STUMP et al. (1999b, c) were, ~30 m apart at the same vertical depth and so it is unlikely that changes in the physical properties of the medium are the cause of this variable performance. Velocity of detonation measurements in the holes were not made although video footage indicates a lot less borehole response in terms of motion and permanent displacement around the borehole for the event producing the smaller motions. These arguments support degraded explosive performance.

Other processes associated with each shot (e.g., spall tonnage and throw direction) are also assumed to be identical. Decorrelation between shots is also unavoidable (STUMP et al., 1999b, c) and will degrade the constructive interference (BAUMGARDT, 1995). The analysis of single shots conducted by STUMP et al. (1999b, c) indicates that for the shot separations commonly used in these mining blasts (~10 m) we should expect good correlation below ~4–5 Hz. Decorrelation is not taken into account in the current code as all sources are assigned a common Green's function. Shot timing is also an issue. We have concluded that low-frequency modulations result from long source delays and source finiteness; however, the fine details of these modulations are dependent on the interplay, or interference, between these processes. Marked changes in the modulation patterns can be caused by making seemingly minor adjustments in the shot pattern. Shot scatter, or the mismatch between intended and actual shot detonation times, is ever present (STUMP and REAMER, 1988) and is not taken into account in our simulations.

Despite these limitations, synthetics reproduce the general character of the spectral modulations observed and indicate that, in the absence of strong crustal resonance which will also yield spectral modulations, this trait can be useful for separating delay-fired blasts from instantaneous explosions. The synthetics underscore the need for taking into account seismic resonance. In the one-dimensional model, the low-velocity surface layer is continuous between the source and receiver. In practice, seismic resonance will be observed if the layer is discontinuous and present only at the source or at the receiver.

The energy partitioning between body and surface waves is more easily reproduced. The synthetic test clearly indicates that source duration has the most significant effect on surface wave amplitudes. Spall is a second-order effect. This successful simulation gives a physical basis to the surface waves observed. The source parameter tests are a first step in using synthetics to gauge the sensitivity of surface waves to changes at the source.

Outstanding Issues

This empirically based study has identified a number of characteristics of seismic waves from large-scale cast blasting that can be used for identifying the source. Within the context of the CTBT, the extent to which this observational experience can be used to assess discrimination techniques in other regions where propagation path effects are different and blasting practices may vary needs to be assessed. In some of these other areas there may be little or no access to information concerning blasting practices and so it is hoped that these detailed studies may provide the foundation for the interpretation of the observations after consideration of propagation path effects.

A number of outstanding issues associated with the generation of regional waveforms from mining explosions remain independent of specific information on propagation paths. Mechanisms for the generation of regional surface waves have been demonstrated but need further investigation and constraint. The impact of a range of blasting practices from long-time duration cast blasting in coal mines to short duration rock fragmentation blasts in surface coal mines needs to be explored. The modeling and data analysis have assumed that design blast parameters are those that are implemented. For purposes of assessing the reliability of discriminants, the quantification of anomalies in blasting and their implications for regional seismograms needs additional exploration. Finally this study has focused on observations from mining and single-fired explosions but has not compared these observations to those from earthquakes along similar propagation paths. Studies which include earthquake and mining explosion sources along comparable propagation paths will allow the assessment of the proposed discriminats for separating earthquake and mining explosion populations.

This study has focused on seismic observations from mining explosions. There is increasing evidence that infrasonic observations may help in the identification of surface mining explosions (SORRELLS *et al.*, 1997). It is possible that the colocation of seismic and infrasonic instrumentation surrounding mining regions may significantly add to the identification capability of mining explosions.

Mining events are controlled and thus might provide an opportunity for clandestine tests as part of an advanced nuclear weapons program. There is a strong incentive for learning how to use the IMS to distinguish anomalies from clandestine nuclear tests and avoid On-Site Inspections.

Unannounced mining events that include significant simultaneous explosive energy releases are problematic for several reasons. Such blasts are unwanted by the mining community as these events have both reduced rock-fracturing efficiency and increased seismic efficiency. Some of these blasts will trigger the IMS and cause some to question whether the anomalous energy release was chemical in nature. Effective characterization techniques, which rely on remote IMS observations, will reduce the need for on-site inspections. Our preliminary analysis suggests that such events can be identified using spectral modulations and the relative strength of surface- and body-wave seismic energy. A fuller analysis of errant blasts will be the subject of a forthcoming paper.

Acknowledgments

The combination of regional and close-in measurements for solving problems related to mining explosions could not have been made without the close collaboration with Vindell Hsu at the Air Force Technical Applications Center (AFTAC). Bob Martin, David Gross, Al Blakeman and Terry Walsh at the Black Thunder mine provided essential support. Portable deployments were made possible by CL Edwards, Diane Baker and Roy Boyd (LANL) and Adam Edelman, Aaron Geddins and John Unwin. The authors would like to thank Doug Baumgardt and an anonymous reviewer for helpful comments. Funding and equipment provided by LANL (under contracts 1973USML6-8F and F5310-0017-8F) and the efforts of Frank Vernon and the IGPP north lab permitted us to deploy the azimuthal network. Data processing and analysis by MAHH was funded by DTRA under contracts DTRA01-97-C-0153, and DTRA01-00-C-0115 and LANL under contract F5310-0017-8F.

REFERENCES

ANANDAKRISHNAN, S., TAYLOR, S. R., and STUMP, B. W. (1997), *Quantification and Characterization of Regional Seismic Signals from Cast Blasting in Mines: A Linear Elastic Model*, Geophys. J. Internat. *131*, 45–60.

ATLAS POWDER COMPANY (1997), *Explosives and Rock Blasting*, Dallas, TX.

BARKER, T. G. and DAY, S. M. (1990), *A Simple Physical Model for Spall from Nuclear Explosions Based upon Two-dimensional Nonlinear Numerical Simulations*, PL Report SSS-TR-93-13859.

BARKER, T. G., MCLAUGHLIN, K. L., STEVENS, J. L., and DAY, S. M. (1993), *Numerical Models of Quarry Blast Sources: The Effects of the Bench*, PL Semiannual Report, SSS-TR-93-13915.

BARKER, T. G., MCLAUGHLIN, K. L., STEVENS, J. L., and DAY, S. M. (1994), *Numerical Simulation of Quarry Blasts Part 2: Implications for Discrimination*, Seismol. Res. Lett. *65*, 71.

BAUMGARDT, D. R. (1995), *Case Studies of Seismic Discrimination Problems and Regional Discriminant Transportability*, Phillips Lab Report PL-TR-95-2106.

BAUMGARDT, D. R. and ZIEGLER, K. A. (1988), *Spectral Evidence for Source Multiplicity in Explosions: Application to Regional Discrimination of Earthquakes and Explosions*, Bull. Seismol. Soc. Am. *78*, 1773–1795.

GITTERMAN, Y. and VAN ECK, T. (1993), *Spectra of Quarry Blasts and Microearthquakes Recorded at Local Distances in Israel*, Bull. Seismol. Soc. Am. *83*, 1799–1812.

GITTERMAN, Y., PINSKY, V., and SHAPIRA A. (1997), *Application of Spectral Semblance and Ratio Discriminants to Regional and Teleseismic Events Recorded by ISN and NORESS*, AFTAC/DOE/DSWA Seismol. Res. Symp. *19*, 369–378.

HARRIS, D. B. (1991), *A Waveform Correlation Method for Identifying Quarry Explosions*, Bull. Seismol. Soc. Am. *80*, 2177–2418.

HEDLIN, M. A. H. (1997), *A Global Test of a Time-frequency Small Event Discriminant*, AFTAC/DOE/DSWA Seism. Res. Symp. *19*, 390–399.

HEDLIN, M. A. H. (1998a), *Identification of Mining Blasts at all Regional Distances using Low-frequency Seismic Signals*, 20th Ann. Seis. Res. Symp. *20*, 335–344.

HEDLIN, M. A. H. (1998b), *A Global Test of a Time-frequency Small-event Discriminant*, Bull. Seismol. Soc. Am. *88*, 973–988.

HEDLIN, M. A. H., MINSTER, J.-B., and ORCUTT, J. A. (1989), *The Time-frequency Characteristics of Quarry Blasts and Calibration Explosions Recorded in Kazakhstan, U.S.S.R.*, Geophys. J. Int. *99*, 109–121.

ISRAELSSON, H. (1991), *Correlation of Waveforms from Closely Spaced Regional Events*, Bull. Seismol. Soc. Am. *80*, 6, 2177–2193.

KHALTURIN, V. I., RAUTIAN, T. G., RICHARDS, P. G., and KIM, W. Y. (1997), *Evaluation of Chemical Explosions and Methods of Discrimination for Practical Seismic Monitoring of a CTBT*, Phillips Lab. Final Report.

KIM, W. Y., SIMPSON, D. W., and RICHARDS, P. G. (1994), *High-frequency Spectra of Regional Phases from Earthquakes and Chemical Explosions*, Bull. Seismol. Soc. Am. *84*, 1365–1386.

LANGEFORS, U. and KIHLSTRÖM, B. *The Modern Technique of Rock Blasting* (Halsted Press, Wiley, New York. 1978).

LEITH, W. (1994), *Large Chemical Explosions in the Former Soviet Union and Blasting Estimates for Countries of Nuclear Proliferation Concern*, Arms Control and Nonproliferation Technologies, 1st quarter 1994, 25.

MARTIN, R., GROSS, D., PEARSON, C., STUMP, B., and ANDERSON, D. (1997), *Black Thunder Coal Mine and Los Alamos Experimental Study of Seismic Energy Generated by Large-scale Mine Blasting*, Proc. Twenty-third Ann. Conf. on Explosives and Blasting Technique, Las Vegas, Nevada, Feb. 2–5, 1997, International Society of Explosive Engineers, Cleveland, Ohio, *pp. 1–10*.

MARTIN, R. L., and KING, M. G. (1995), *The Efficiency of Cast Blasting in Wide Pits*, Proc. Twenty-first Conf. on Explosives and Blasting Technique, Nashville, Tennessee, Feb. 5–9, 1995, International Society of Explosive Engineers, Cleveland, Ohio, pp. 176–186.

MCLAUGHLIN, K. L., BARKER, T. G., STEVENS, J. L., and DAY, S. M. (1994), *Numerical Simulation of Quarry Blast Sources*, PL Final Report SSS-FR-94-14418.

MINSTER, J.-B. and DAY, S. (1986), *Decay of Wavefields near an Explosive Source due to High-strain, Nonlinear Attenuation*, J. Geophys. Res. *91*, 2113–2122.

MURPHY, J. R. (1995), *Types of Seismic Events and their Source Description, Monitoring a Comprehensive Test Ban Treaty*, NATO ASI Series E: Applied Sciences, Vol. 303, (E. Husebye and A. Dainty ed.) 225–245.

PEARSON, D. C., STUMP, B. W., BAKER, D. F., and EDWARDS, C. L. (1995), *The LANL/LLNL/AFTAC Black Thunder Mine Regional Mining Blast Experiment*, 17th Annual PL/AFOSR/AFTAC/DOE Seism. Res. Symp., 562–571.

PRODEHL, C. (1979), *Crustal Structure of the Western United States*, US. Geological Survey Professional Paper, *1034*.

RICHARDS, P. G. and ZAVALES, J. (1990), *Seismic Discrimination of Nuclear Explosions*, Annu. Rev. Earth Planet. Sci. *18*, 257–286.

RIVIERE-BARBIER, F. and GRANT, L. T. (1993). *Identification and Location of Closely Spaced Mining Events*, Bull. Seismol. Soc. Am. *83*, 1527–1546.

SMITH, A. T. (1993), *Discrimination of Explosions from Simultaneous Mining Blasts*, Bull. Seismol. Soc. Am. *83*, 160–179.

Sobel, P. A. (1978), *The Effect of Spall on m_b and M_s*, Teledyne Geotech Rept., SDAC-TR-77-12, Dallas, TX.

Sorrells, G. G., Herrin, E. T., and Bonner, J. L. (1997), *Construction of Regional Ground-truth Databases Using Seismic and Infrasound Data*, Seism. Res. Lett. *68*, 743–752.

Stump, B. W. (1995), *Practical Observations of U.S. Mining Practices and Implications for CTBT Monitoring*, Phillips Lab Report, PL-TR-95-2108.

Stump, B. W. and Reamer, S. K. (1988), *Temporal and Spatial Source Effects from Near-surface Explosions*, 10th Annual AFGS/DARPA Seism. Res. Symp., 95–113.

Stump, G. W., Riviere-Barbier, F., Chernoby, I., and Koch, K. (1994), *Monitoring a Test-ban Treaty Presents Scientific Challenges*, EOS, Trans. of the Am. Geophys. *Union 75*, 265.

Stump, B. W., Anderson, D. P., and Pearson, D. C. (1996), *Physical Constraints on Mining Explosions: Synergy of Seismic and Video Data with 3-D Models*, Seismol. Res. Lett. *67*, 9–24.

Stump, B. W. and Pearson, D. C. (1997), *Comparison of Single-fired and Delay-fired Explosions at Regional and Local Distances*, 19th Ann. Seism. Res. Symp., 668–677.

Stump, B. W., Hedlin, M. A. H., Pearson, D. C., and Hsu, V. (1999a), *Characterization of Mining Explosions at Regional Distances*, Rev. Geophys., in Review.

Stump, B. W., Pearson, D. C., and Hsu, V. (1999b), *Empirical Scaling Relations for Contained Single-fired Chemical Explosions and Delay-fired Mining Explosions at Regional Distances*, 1999 SSA Meeting.

Stump, B. W., Pearson, D. C., and Hsu, V. (1999c), *Empirical Scaling Relations for Contained Single-fired Chemical Explosions and Delay-fired Mining Explosions at Regional Distances*, 21st Ann Seism. Res. Symp., 764–772.

Stump, B. W. and Reamer, S. K. (1988), *Temporal and Spatial Source Effects from Near-surface Explosions*, 10th Ann. AFGS/DARPA Seism. Res. Symp. *95*, 113.

U.S. Congress, Office Of Technology Assessment (1988), *Seismic Verification of Nuclear Testing Treaties*, OTA-ISC-361 (Washington, DC: U.S. Government Printing Office, May, 1988).

Wuster, J., Riviere, F., Crusem, R., Plantet, J. -L., Massinon, B., and Caristan, Y. (2000), *GSETT-3: Evaluation of the Detection and Location Capabilities of an Experimental Global Seismic Monitoring System*, Bull. Seism. Soc. Am. *90*, 166–186.

Yang, X. (1998), *MineSeis – A Matlab GUI Program to Calculate Synthetic Seismograms from a Linear, Multi-Shot Blast Source Model*, 20th Ann. Seism. Res. Symp. 755–764.

(Received June 15, 1999, revised June 4, 2000, accepted June 15, 2000)

 To access this journal online:
http://www.birkhauser.ch

Pure appl. geophys. 159 (2002) 865–888
0033–4553/02/040865–24 $ 1.50 + 0.20/0

Pure and Applied Geophysics

Experimental Seismic Event-screening Criteria at the Prototype International Data Center

MARK D. FISK,[1] DAVID JEPSEN,[2] and JOHN R. MURPHY[3]

Abstract — Experimental seismic event-screening capabilities are described, based on the difference of body-and surface-wave magnitudes (denoted as $M_s{:}m_b$) and event depth. These capabilities have been implemented and tested at the prototype International Data Center (PIDC), based on recommendations by the IDC Technical Experts on Event Screening in June 1998. Screening scores are presented that indicate numerically the degree to which an event meets, or does not meet, the $M_s{:}m_b$ and depth screening criteria. Seismic events are also categorized as onshore, offshore, or mixed, based on their 90% location error ellipses and an onshore/offshore grid with five-minute resolution, although this analysis is not used at this time to screen out events.

Results are presented of applications to almost 42,000 events with $m_b \geq 3.5$ in the PIDC Standard Event Bulletin (SEB) and to 121 underground nuclear explosions (UNE's) at the U.S. Nevada Test Site (NTS), the Semipalatinsk and Novaya Zemlya test sites in the Former Soviet Union, the Lop Nor test site in China, and the Indian, Pakistan, and French Polynesian test sites. The screening criteria appear to be quite conservative. None of the known UNE's are screened out, while about 41 percent of the presumed earthquakes in the SEB with $m_b \geq 3.5$ are screened out. UNE's at the Lop Nor, Indian, and Pakistan test sites on 8 June 1996, 11 May 1998, and 28 May 1998, respectively, have among the lowest $M_s{:}m_b$ scores of all events in the SEB.

To assess the validity of the depth screening results, comparisons are presented of SEB depth solutions to those in other bulletins that are presumed to be reliable and independent. Using over 1600 events, the comparisons indicate that the SEB depth confidence intervals are consistent with or shallower than over 99.8 percent of the corresponding depth estimates in the other bulletins. Concluding remarks are provided regarding the performance of the experimental event-screening criteria, and plans for future improvements, based on recent recommendations by the IDC Technical Experts on Event Screening in May 1999.

Key words: Seismic event screening, focal depth, body- and surface-wave magnitudes, International Data Center.

Introduction

The Protocol to the Comprehensive Nuclear-Test-Ban Treaty (CTBT) specifies the functions of the International Data Centre (IDC) with regard to event screening.

[1] Mission Research Corporation, 8560 Cinderbed Road, Suite 700, Newington, VA, 22122, U.S.A.
E-mail: fisk@neutron.mrcwdc.com
[2] Australian Geological Survey Organisation, GPO Box 378, Canberra, ACT 2601, Australia.
[3] Maxwell Technologies, Inc., 11800 Sunrise Valley Drive, Suite 1212, Reston, VA 20191, U.S.A.

Among them, it indicates that the IDC shall apply standard event screening criteria to each event, "… with the objective of characterizing, highlighting in the standard event bulletin, and thereby screening out, events considered to be consistent with natural phenomena or non-nuclear, man-made phenomena." It further states that, "The standard event bulletin shall indicate numerically for each event the degree to which that event meets or does not meet the event screening criteria." The procedures and criteria for event screening are to be progressively developed and elaborated in the Operational Manual for the IDC. The procedures and criteria developed initially by the Preparatory Commission are to be approved by the initial Conference of States Parties.

An initial plan for the development of event screening criteria (WGB/TL7/Rev.2) was approved by the Preparatory Commission in 1997. Three key aspects of the plan are: (1) To build upon the capabilities already implemented at the prototype IDC (PIDC) in Arlington, Virginia, USA; (2) that the main mechanism for achieving the approach will be Working Group B expert meetings; and (3) that the results are to be reviewed by Working Group B of the Preparatory Commission at the end of each development phase.

The basic concept of event screening is to screen out those events that may be considered as natural or non-nuclear, man-made phenomena with high confidence, without screening out any explosions that may correspond to potential violations of the CTBT. Events that are not screened out can indicate either that they have explosion-like characteristics or simply that the uncertainties associated with the screening criteria are too large for the event to be screened out with sufficient confidence. No effort is made, within the context of event screening at the IDC, to further distinguish or identify such events that are not screened out.

In this paper we describe the seismic event-screening capabilities being developed and tested at the PIDC on an experimental basis. To guide the development and testing at the PIDC, the IDC Technical Experts on Event Screening have convened, to date, on three occasions and provided summary reports to Working Group B. The first meeting was held in Beijing, China on 4–7 November, 1997. The second meeting was held in Vienna, Austria on 5–9 June, 1998. The experimental procedures and criteria described in this paper are based on the recommendations produced by the second meeting (see CTBT/WGB-6/TL-2/10). Recently, a third meeting was held in Vienna, Austria on 21–25 May, 1999. Recommendations from this meeting are currently being implemented at the PIDC. We plan to report on results of this work in the future.

The PIDC currently produces several automated and reviewed bulletins on a daily, experimental basis. The bulletins based on events reviewed by human analysts include the Reviewed Event Bulletin (REB), the Standard Event Bulletin (SEB), and the Standard Screened Event Bulletin (SSEB). All three of these bulletins are similar in content and format, except that the SEB contains additional event characterization parameters and screening results that are computed in a post-analysis mode (i.e., after the REB is produced) and the SSEB does not include events from the SEB that are screened out. On-line versions of these bulletins may be viewed at the PIDC web site (*http://www.pidc.org*).

We present the experimental seismic screening criteria, based on the difference of body (m_b) and surface (M_s) wave magnitudes (denoted as M_s:m_b) and hypocentral depth estimates. We then present the results of applying these criteria to almost 42,000 events in the SEB and to 121 underground nuclear explosions. We also make comparisons to other global and local bulletins to assess the validity of the screening results based on event depth. Last, we provide some concluding remarks regarding the performance of the screening criteria and plans for future improvements, based on recent recommendations by the IDC Technical Experts in May 1999.

Experimental Event-screening Criteria

In June 1998, the IDC Technical Experts recommended event-screening procedures and criteria, based on M_s:m_b and event depth, for testing at the PIDC, and that the criteria should be applied to seismic events in the SEB of m_b 3.5 or above (CTBT/ WGB-6/TL-2/10). In this section we describe the recommended event-screening criteria, along with some background information on the basis and preliminary calibration of these criteria. The screening criteria utilize the depth and magnitude estimates that are reported in both the REB and the SEB.

M_s:m_b Screening Criterion

M_s:m_b has proved to provide the most robust teleseismic event-screening procedure for shallow seismic events (e.g., MARSHALL and BASHAM, 1972; STEVENS and DAY, 1985). It is based on the well-documented observation that m_b–M_s is typically less than one for crustal earthquakes and greater than one for underground explosions. These generalizations are based on classical m_b and M_s values such as those published by the U.S. National Earthquake Information Center (NEIC) and other agencies. Unfortunately, such prior experience is not directly transportable to the PIDC as it has been demonstrated that the m_b and M_s values estimated at the PIDC are systematically different from those published by the NEIC for the same events, as described below.

At the onset of GSETT-3 (Group of Scientific Experts Third Technical Test) in 1995, the PIDC initiated a program whereby, for the first time in the history of seismology, seismic magnitude measures, m_b and M_s, were estimated in a uniform and fully automatic fashion, using digital data from a global network of stations, employing essentially uniform instrumentation. Not surprisingly, this lead to some differences with respect to the traditional NEIC and ISC magnitude measures. However, the differences for teleseismic m_b measurements proved to be unexpectedly large. This is illustrated in Figure 1, which shows a comparison of PIDC and NEIC

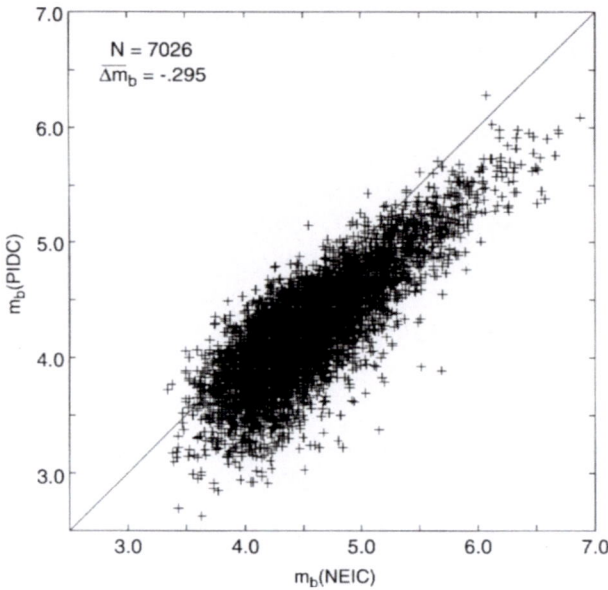

Figure 1

Comparison of PIDC and NEIC m_b values for common events during 1995 and 1996 indicates a significant offset between the two sets of measurements, with the PIDC values being lower on average by approximately 0.3 magnitude units.

m_b values for common events during 1995 and 1996. While the scatter is large, it can be seen that the PIDC values are lower on average by about 0.3 magnitude units. In fact, the offset is even larger than it appears here in that the NEIC began including PIDC data in their m_b estimates as GSETT-3 progressed. This resulted in a systematic decrease in the NEIC m_b values by about 0.1 magnitude units. Thus, the true average offset between the PIDC and NEIC values is actually 0.4 units. Although there is still no consensus on all the reasons for this offset, a number of contributing factors have been identified. These include the facts that the PIDC and NEIC employ different corrections for epicentral distance and focal depth (i.e., Veith-Clawson versus Gutenberg-Richter), different data processing and reduction procedures, and different networks of stations with different response characteristics. In any case, it is clear that there is a systematic offset between PIDC and NEIC m_b values, such that around $m_b = 4$,

$$m_b(\text{NEIC}) \approx m_b(\text{PIDC}) + 0.4 \ . \tag{1}$$

Figure 2 illustrates a similar, but less dramatic offset between PIDC and NEIC M_s values for a sample of common events during 1997. The PIDC M_s values are lower, on average, than the corresponding NEIC M_s values by about 0.1 magnitude units, i.e.,

$$M_s(\text{NEIC}) \approx M_s(\text{PIDC}) + 0.1 \ . \tag{2}$$

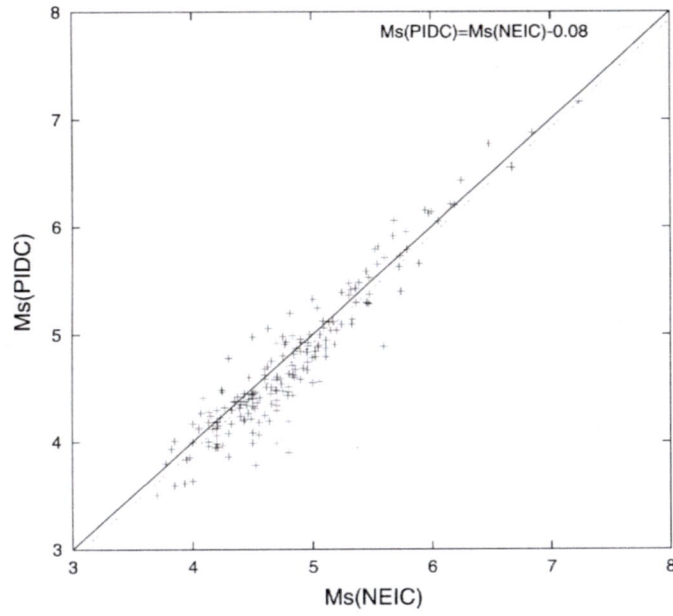

Figure 2

Comparison of PIDC and NEIC M_s values for a sample of common earthquakes in 1997 indicates an offset between the two sets of measurements, with the PIDC values being lower on average by approximately 0.1 magnitude units.

While PIDC and NEIC m_b and M_s values can be directly compared for a very large common set of earthquakes that have occurred since the beginning of GSETT-3 in January 1995, direct comparisons for underground nuclear explosions are limited to the 1996/06/08 explosion at the Lop Nor test site, the 1998/05/11 explosion at the Indian test site, and the 1998/05/28 explosion at the Pakistan test site. Thus, the following approach was used to calibrate the provisional M_s:m_b screening criterion being tested currently at the PIDC.

MURPHY (1997) evaluated time dependencies of average annual differences between NEIC m_b and M_s estimates and corresponding logarithms of Harvard moment estimates from 1977 to 1997. Consideration was limited to 5259 shallow earthquakes (depth <50 km) in the magnitude range, $4.5 \leq m_b \leq 5.5$, for which both Harvard moment solutions and NEIC m_b and M_s values have been published. The average differences are plotted by year in Figure 3, where it can be seen that they are remarkably consistent between 1982 and 1994, showing variations of less than ± 0.05 magnitude units about the mean values represented by the dotted lines. In fact, the results are so stable that it is easy to identify the 0.1 magnitude unit decrease in NEIC m_b values associated with the incorporation of PIDC data into their m_b determinations, starting in 1995. Therefore, it can be concluded that, except for this consistent

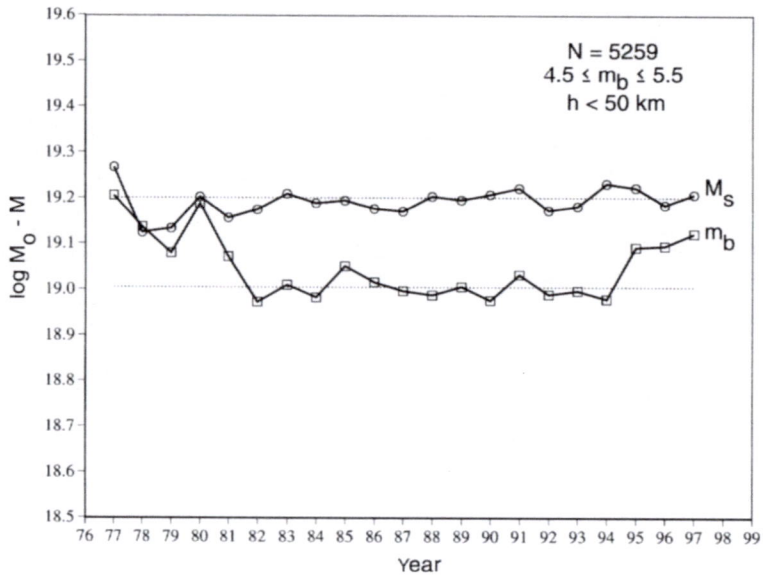

Figure 3

Time-dependent average differences between the logarithms of the Harvard event moment estimates (M_0) and the corresponding NEIC m_b and M_s estimates using 5259 shallow earthquakes (depth <50 km) between 1977 and 1997.

recent shift in m_b, NEIC m_b and M_s determinations have been consistent between 1981 and the present.

Based on the results summarized in Figure 3, a database was assembled that consists of NEIC m_b and M_s estimates for 66 underground nuclear explosions detonated since 1981 at the U.S. Nevada Test Site (NTS), the Former Soviet Union test sites at Semipalatinsk and Novaya Zemlya, and the Chinese Lop Nor test site. The M_s–m_b values for these explosions were then compared with NEIC M_s–m_b values for a subset of 2260 earthquakes (used in Fig. 3) that occurred between 1991 and 1997, a time interval for which the observed variability in NEIC M_s–m_b values is most consistent with the current value. These NEIC results are summarized in Figure 4 where histograms of the observed M_s–m_b values for the two-source types are compared. (Note that 0.1 magnitude units have been added to the NEIC m_b values for the events in 1995–1997 to compensate for the offset documented in Fig. 3). It is evident from this comparison that for events having NEIC m_b values near the sample means of 5.5, an M_s–m_b threshold of about −1.0 adequately separates the two populations.

Extension of the M_s:m_b threshold at m_b 5.5 for application to the entire range of m_b values reported in the SEB is complicated by the fact that observed M_s–m_b values are known to be a function of m_b for earthquakes and explosions. In particular, it has been conclusively demonstrated that m_b and M_s scale differently with explosion yield. To a first approximation,

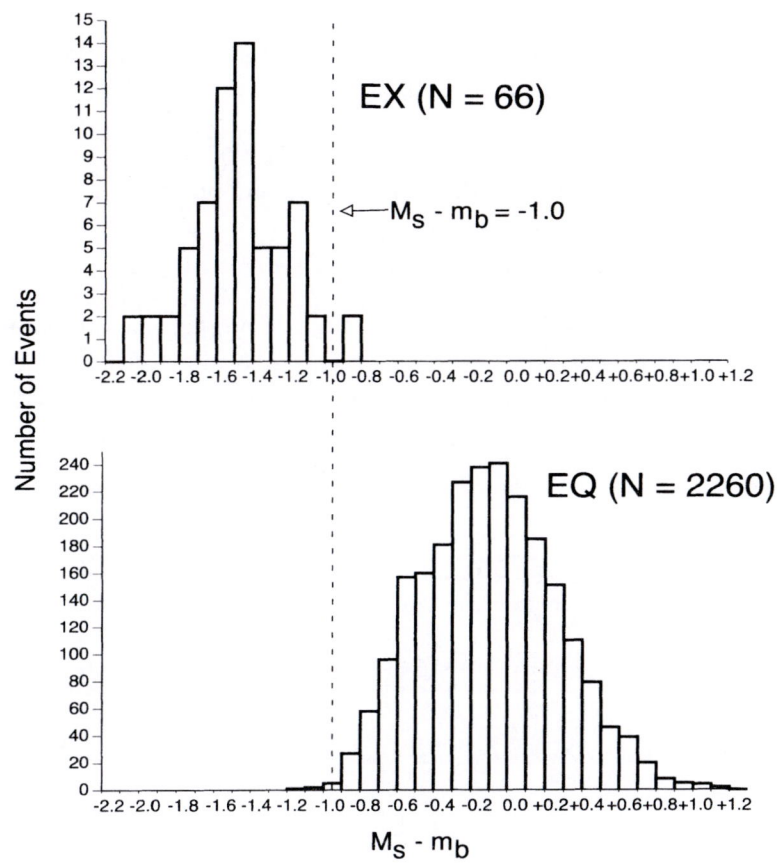

Figure 4

Histograms of NEIC M_s–m_b values for 66 underground nuclear explosions since 1981 and for 2260 shallow earthquakes that occurred between 1991 and 1997.

$$m_b = K + 0.8 \log \ W \ , \tag{3}$$

where W is the explosion yield, and $K \approx 4.0$ for tectonically active regions such as the NTS (MURPHY, 1981), and $K \approx 4.4$ for stable continental interior regions such as the Semipalatinsk test site (MURPHY, 1993). Alternatively, the nominal relation between M_s and yield for underground explosions is roughly independent of source region and is given by (MARSHALL *et al.*, 1979)

$$M_s = 2.0 + 1.0 \log \ W \ . \tag{4}$$

It follows from Eqs. (3) and (4) that the expected mean relation between M_s and m_b for underground explosions has the form:

$$1.25 \, m_b - M_s = 1.25 \, K - 2.00 \ . \tag{5}$$

Equation (5) is depicted in Figure 5, for values of $K = 4.0$ and $K = 4.4$, on a plot of M_s versus m_b data for NTS and Semipalatinsk explosions. Note that these are not NEIC values, but were determined in a carefully controlled study in which all the data were processed using procedures consistent on average with NEIC practice (STEVENS, 1986). It can be seen that Eq. (5), for $K = 4.0$ and $K = 4.4$, provides good fits to the average M_s:m_b values of these two explosion populations. Also shown is an estimated upper bound for the explosions, given by the relation:

$$1.25\,m_b(\text{NEIC}) - M_s(\text{NEIC}) = 2.60 \ . \tag{6}$$

In the absence of any evidence to the contrary, our preliminary assumption is that the upper bound of Eq. (6) corresponds to the global upper bound. Substituting Eqs. (1) and (2) into Eq. (6) leads to a nominal decision line, based on PIDC measures of m_b and M_s, of the form:

$$1.25\,m_b(\text{PIDC}) - M_s(\text{PIDC}) = 2.20 \ . \tag{7}$$

MURPHY (1997) applied this criterion to PIDC M_s and m_b data, compiled by STEVENS and McLAUGHLIN (1997), for 1226 shallow earthquakes during 1997. It can be seen in Figure 6 that Eq. (7) provides a reasonable lower bound to the data for

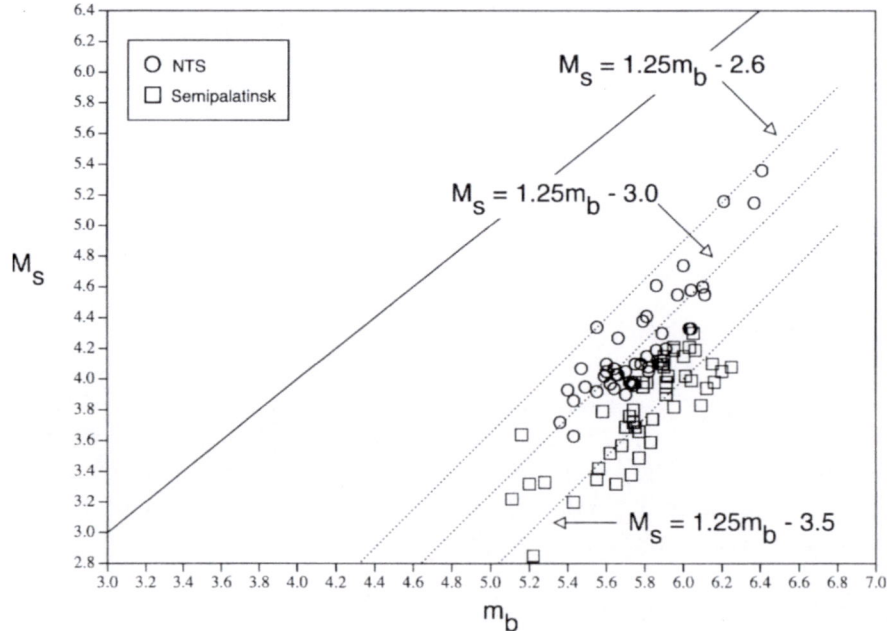

Figure 5

M_s versus m_b values, computed using procedures consistent with NEIC practice, for NTS and Semipalatinsk underground nuclear explosions, including the M_s:m_b scaling relation of Eq. (5), for $K = 4.0$ and $K = 4.4$, and an estimated upper bound line for the explosions of $M_s = 1.25\,m_b - 2.60$.

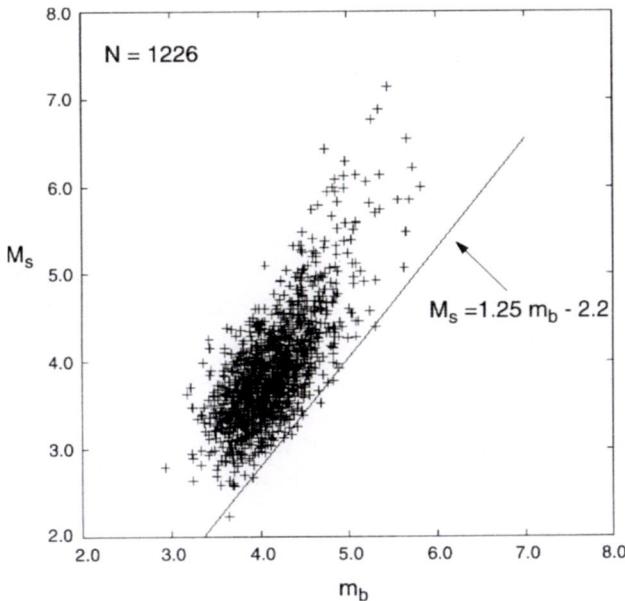

Figure 6
Comparison of the proposed interim PIDC M_s:m_b event-screening line of Eq. (7) with observed PIDC M_s:m_b data for 1226 shallow earthquakes during 1997.

earthquakes with m_b(PIDC) values as low as 3.5. Less than one percent of these earthquakes have values of 1.25 m_b–M_s less than 2.20.

Given this background, the provisional M_s:m_b screening criterion, including treatment of m_b and M_s uncertainties, is defined as follows. Let $\overline{m}_b$ and $\overline{M}_s$ denote network-average values of m_b and M_s from N_b and N_s stations, respectively. An event is screened out if:

$$1.25\overline{m}_b - \overline{M}_s + 2\sigma_M < 2.20 \ , \tag{8}$$

where

$$\sigma_M^2 = 1.25^2 \frac{\sigma_b^2}{N_b} + \frac{\sigma_s^2}{N_s} \ , \tag{9}$$

σ_b and σ_s are the standard deviations of single-station m_b and M_s measurements, respectively. The errors are assumed to be statistically independent. Determination of these uncertainties are still under development. At present, values of $\sigma_b = \sigma_s = 0.3$ are used. The M_s:m_b screening criterion in Eq. (8) is equivalent to requiring that the one-sided 97.5% confidence interval is entirely less than 2.20. It has been applied to all events in the SEB with $m_b \geq 3.5$ and an M_s estimate, as well as to historical underground nuclear explosions. Results are presented below.

Hypocentral Depth Screening Criteria

Events that are confidently deeper than a depth threshold for which it is feasible, with existing technology, to test an underground nuclear explosion, are very likely to be of natural seismic origin. An experimental screening criterion, based on the depth confidence interval, is being tested at the PIDC to screen out such events.

Hypocenter estimation is performed at the PIDC as an iterative least-squares inversion of arrival time, azimuth, and/or slowness measurements, as originally introduced by JORDAN and SVERDRUP (1981) for arrival time data, and extended by BRATT and BACHE (1988) to include azimuth and slowness data. Event location and depth estimates, confidence bounds, residuals, and data importances (e.g., MINSTER *et al.*, 1974) are computed using these measurements from stations at teleseismic, regional, and local distances. A convergence test is applied to the data residuals, weighted by *a priori* uncertainties. The iterative procedure is performed until a maximum number of iterations is exceeded (the default is 20), or convergence or divergence occurs prior to that. The depth is fixed at zero during the first two iterations. If the depth becomes negative, it is fixed at zero during the next iteration. If this occurs for several iterations, the depth is fixed at zero for all subsequent iterations. Similarly, events with depths greater than 650 km are fixed at a value of 650 km during the next iteration. The event depth and other hypocentral parameters can also be constrained based on other documented criteria, e.g., if the depth uncertainty is greater than twice the estimated depth. (For details, see GSE/CRP/ 243.)

The IASPEI91 travel-time curves (KENNETT, 1991) are used at the PIDC for seismic event location. As these one-dimensional curves are global averages, they are not universally applicable to regional data, since heterogeneities in the crust and upper mantle can significantly affect the travel times of regional phases. Source-specific station corrections (SSSC), to treat regional travel-time variations, have been incorporated into the processing capability (e.g., for Fennoscandia and North America, see YANG *et al.*, 1999, 2000), along with slowness-azimuth station corrections (SASC) for nearly all existing IMS seismic stations (BONDAR *et al.*, 1999).

The event error ellipsoid and normalized confidence regions are computed using the approach of JORDAN and SVERDRUP (1981), where the error analysis is represented as either an entirely *a priori* or *a posteriori* distribution, based on the *F*-distribution. BRATT and BACHE (1988) showed that the distribution can be represented as any combination of these two end members through control of two input parameters, the number of degrees of freedom (*num_dof*) and the *a priori* scale variance factor (*est_std_err*). For default values of *est_std_err* set to 1.0 and *num_dof* set, effectively, to infinity, the limiting *F*-distribution becomes a chi-squared. The confidence interval of focal depth is determined from the marginal distribution with respect to epicenter and origin time. The resulting depth error estimates have not proved to be entirely adequate for use in event screening; hence, a temporary empirical correction factor has been incorporated, pending the development of a more reliable uncertainty model.

Given this background, the provisional depth screening criterion is defined as follows. Let $\hat{D}$ and s_{zz} denote the depth estimate and its estimated variance that are reported in the SEB. An event is screened out if:

$$\hat{D} - 2\sigma_D > 10\,\text{km} \ , \tag{10}$$

where

$$2\sigma_D = 2\sqrt{s_{zz}} + k \ . \tag{11}$$

This criterion is equivalent to requiring that the one-sided 97.5% confidence interval for depth is entirely deeper than 10 km. It is applied to all SEB events with $m_b \geq 3.5$ and unconstrained depth estimates. The term k in Eq. (11) is included in the overall depth uncertainty as an interim measure to compensate for the fact that model errors are not adequately accounted for by the s_{zz} estimates currently being produced by the PIDC. It is anticipated that the value of k will be gradually decreased to zero over time as more reliable estimates of depth uncertainty become available. In the meantime, the IDC Technical Experts have recommended that k have different values, depending on whether depth phase observations are used in estimating the event depth.

For events with validated depth phase observations, the value of k is set to zero. The following validation criteria are currently applied to depth phase observations: (1) at least three depth phases must detected for an event; and (2) the moveout of pP–P travel times, for stations in the distance range of 25 to 100 degrees, must be at least 1.5 seconds. (Additional validation criteria for depth phases are being investigated.)

For all other depth solutions, the IDC Technical Experts originally recommended (in June 1998) a value of $k = 40$ km, based on preliminary comparisons by FISK and BONDAR (1998) of SEB depth confidence intervals to depth estimates published in other bulletins for events during 1995 to 1997. Since then, a new *a priori* error model, used in the hypocentral inversion and error analysis, has been implemented into the PIDC system. The new error model includes separate treatment of measurement and model errors that depend on the phase type, signal-to-noise for defining phases, epicentral distance, and the travel-time model, including larger uncertainties for regional phases (for details, see BONDAR, 1998; BONDAR *et al.*, 1999). Based on this error model and more recent comparisons by JEPSEN and FISK (1999), presented below, the IDC Technical Experts recommended in May 1999 that $k = 20$ km be used for free-depth solutions.

In the following comparisons, we assess whether the SEB depth confidence intervals contain the corresponding depth solutions reported in other bulletins that are thought to be reliable and independent. Note that the SEB depth solutions prior to 1998 were reprocessed off-line using the hypocenter location algorithm, *EvLoc*, in its current operational configuration at the PIDC (including the new error model) and identical procedures as current PIDC practice.

First, we compare depth estimates for 54 events in the Northern Hindu Kush region that were reported in both the SEB and by the Chinese National Data Center (NDC), of which 32 events have unconstrained PIDC depth estimates. According to HE DONGMEI (1999), the Chinese NDC solutions are typically based on over 100 stations, with the closest station at about 5 degrees and an azimuthal gap of about 60 degrees; and some include depth phases (pP and sP). Thus, the Chinese NDC solutions are believed to be more accurate than those in the SEB. Figure 7 shows a plot of the PIDC depth estimates versus those reported by the Chinese NDC. The error bars represent the PIDC uncertainties (i.e., plus and minus $2\sigma_D$), with k set to zero for this comparison. Although the PIDC depth estimates are generally deeper than the corresponding estimates by the Chinese NDC, all of the PIDC depth confidence intervals are large enough to be consistent with (or are shallower than) the depth estimates reported by the Chinese NDC. Note also that the upper bounds of all of the PIDC depth confidence intervals are shallower than 50 km, consistent with historical seismicity records for the Northern Hindu Kush region.

Second, we perform a similar comparison using a common set of 418 events in or near Japan during 1997 to 1998 that were reported in both the SEB and the Bulletin of the JMA. Figure 8 shows the locations of these events. The JMA network is a very

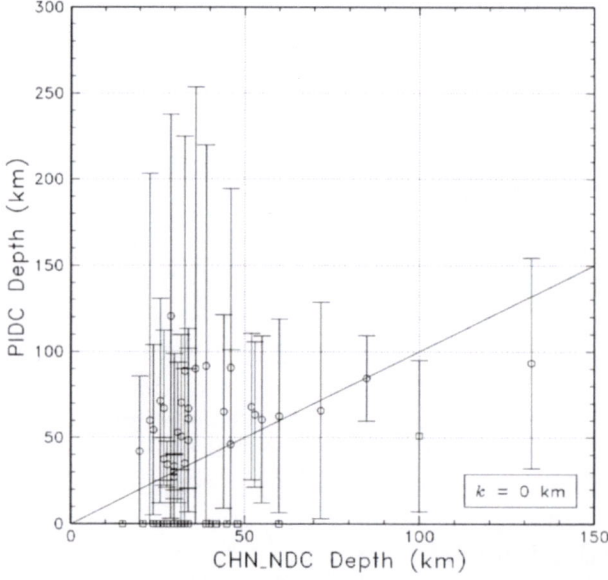

Figure 7
Comparison of PIDC and Chinese NDC depth estimates for 54 seismic events in the Northern Hindu Kush region. All of the PIDC depth confidence intervals are consistent with or shallower than the Chinese NDC depth estimates.

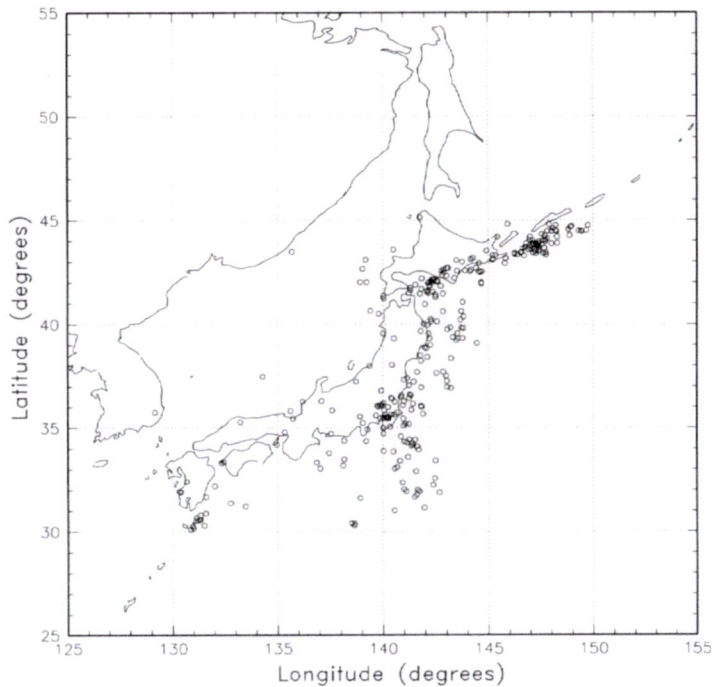

Figure 8

Locations of 418 seismic events above m_b 3.5, in or near Japan during 1997 and 1998, that were reported in both the SEB and the Bulletin of the JMA.

dense set of over 180 seismic stations. JMA depth estimates for events located within the JMA network are considered to be accurate to within 10 percent, typically much smaller than the depth uncertainties in the SEB, based on the sparse global IMS network. JMA depth estimates for events located outside the JMA network are thought to be larger than 10 percent, but specific uncertainties are not provided by the JMA. (See, for example, CTBT/PC/V/WGB/JP/6, submitted by the Japanese delegation to Working Group B.) In the following, we make the conservative assumption that differences in the comparisons for common events are mostly due to errors in the SEB.

Figure 9 shows a plot of PIDC versus JMA depth estimates (provided by the Japanese NDC). Events are restricted to $m_b \geq 3.5$, according to the SEB. Error bars are shown for events with PIDC depth confidence intervals that do not contain the JMA depth estimate. Using $k = 20$ km, all of the PIDC depth confidence intervals are consistent with or shallower than the depth estimates reported by the JMA. Also, none of the events that are shallower than 10 km, according to the JMA, are screened out based on the PIDC depth confidence intervals.

Third, we perform another comparison using 55 common events during 1997 and 1998 in California and Nevada that were reported in the SEB and by the Southern

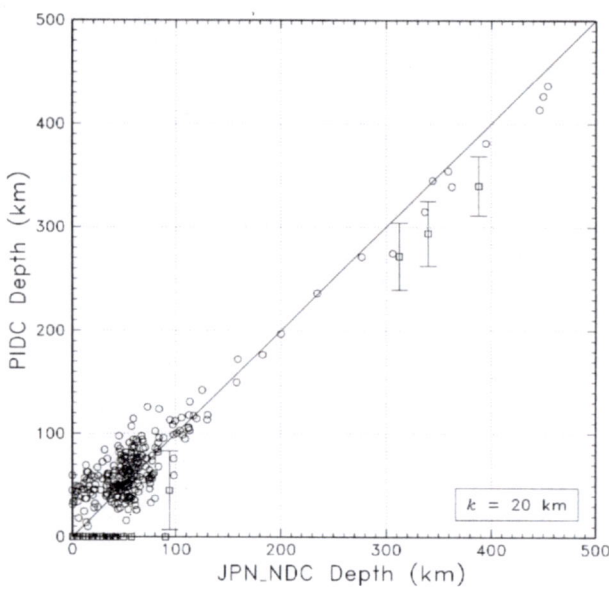

Figure 9

Comparison of PIDC and JMA depth estimates for the events shown in Figure 8. Error bars ($\pm 2\sigma_D$) are shown for events with PIDC depth confidence intervals that do not contain the JMA depth estimate. All of the PIDC depth confidence intervals are consistent with or shallower than the JMA depth estimates.

California Earthquake Center (SCEC), USGS, California Institute of Technology (CIT), University of Nevada, Reno (UNR), or University of California, Berkley (UCB). All but six of the 55 events are constrained to the surface in the SEB. For the remaining six events, all of the SEB depth confidence intervals are consistent with the depth estimates reported by the other local bulletins.

The previous comparisons are somewhat limited in that the data sets are restricted to specific geographical regions and some are relatively small. In the following comparison, we examine over one thousand events worldwide that were reported in both the SEB and the PDE (Preliminary Determination of Epicenters, produced by the U.S. NEIC). Of potential concern is the fact that the PDE solutions incorporate SEB data received from the PIDC. Thus, some of the SEB and PDE depth solutions are not independent, which could lead to agreement that is overly optimistic. Also, depth uncertainties are not reported in the PDE. Thus, it is not clear how to treat the PDE depth errors and potential correlation for various events. To mitigate these problems, we consider only PDE depth estimates that include depth phases. If depth phases have been identified reliably, such solutions are expected to be quite accurate. We then compare these PDE depth-phase solutions to SEB free-depth solutions (i.e., excluding any depth phases). Since depth-phase solutions are highly constrained by the depth phase data, while free-depth solutions are a result of minimizing the residuals, primarily of P arrivals, these two types of solutions are expected to be statistically independent. A set of 1086 common events during 1997, with PDE depth-phase solutions and SEB free-depth solutions, are available for this comparison.

Figure 10 shows the SEB free-depth estimates versus PDE depth-phase solutions. Events are restricted to $m_b \geq 3.5$. Using $k = 20$ km, only three of the SEB depth confidence intervals are significantly deeper than the corresponding PDE depth estimates, all of which are deeper than 160 km. Thus, 99.7 percent of the SEB confidence intervals are consistent with or shallower than the PDE depth-phase solutions, and none of the events that are shallower than 10 km, according to the PDE, are screened out as being deep, based on the SEB free-depth solutions.

Overall, 1613 SEB free-depth solutions were compared to the PDE and other local bulletins. Limiting the events to $m_b \geq 3.5$ and using a value of $k = 20$ km, only three events have depth confidence intervals that are significantly deeper than the corresponding depth estimates in the other bulletins. Thus, based on these comparisons, the 97.5 percent one-sided confidence intervals, used for event screening at the PIDC, are consistent with or shallower than over 99.8 percent of the depth estimates in the other bulletins. Also, no events were screened out that have depth estimates of less than 10 km in the other bulletins. Based on these results, the IDC Technical Experts recommended in May 1999 that $k = 20$ km be used for free-depth solutions.

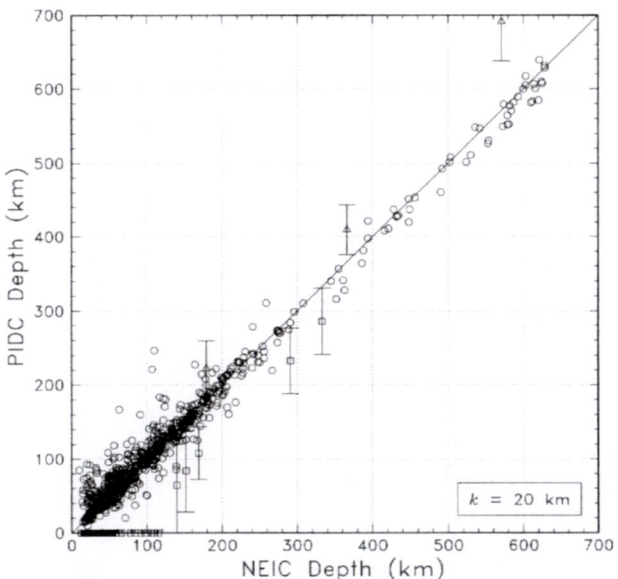

Figure 10

Comparison of PIDC free-depth solutions and NEIC depth-phase solutions for 1086 common events worldwide during 1997. Error bars ($\pm 2\sigma_D$) are shown for events with PIDC depth confidence intervals that do not contain the PDE depth estimate. All but three of the PIDC depth confidence intervals are consistent with or shallower than the NEIC depth solutions.

We also examined SEB depth-phase solutions, including a comparison of 903 common events with depth-phase solutions reported in both the SEB and PDE. The SEB and PDE depth-phase solutions generally compare quite well, as shown in Figure 11, although some of the agreement may be artificial since the PDE uses some of the same depth phase data that were used in obtaining the SEB depth estimates. We have found several events for which the analysts misidentified depth phases, leading to over-estimated depths in the SEB. For some cases, the depth-phase validation criteria (defined above) were not satisfied, either because there were less than three depth phases or because the moveout of $pP\text{-}P$ was less than 1.5 seconds. Thus, these events were not screened out as being deep because the depth-phase criteria were not satisfied. However, we found two cases with erroneous depth phases for which the depth-phase validation criteria were satisfied. Note that, in virtually all cases, the errors were made by inexperienced analysts, and that a senior analyst was able to quickly determine and correct the erroneous depth phase picks. Thus, the IDC Technical Experts recommended that senior analysts should review all depth phase solutions and that improved criteria be developed to validate depth phases.

Numerical Indication of Meeting the Event Screening Criteria

Scores are computed for each event to indicate the degree to which that event does, or does not, meet the event-screening criteria. Based on the criteria defined in

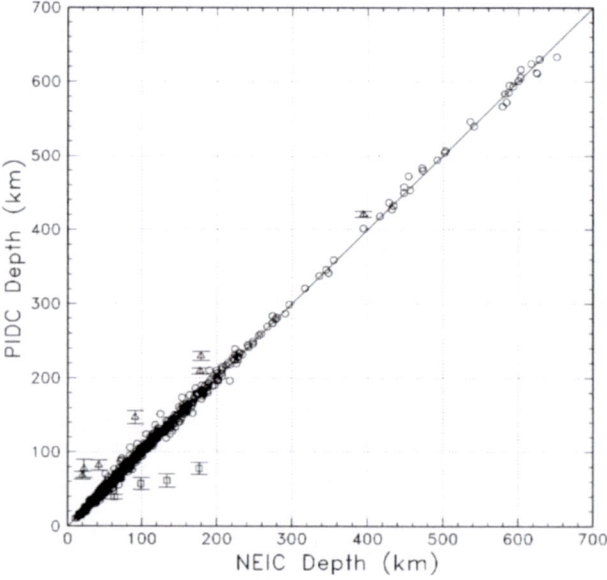

Figure 11

Comparison of PIDC and NEIC depth-phase solutions for 903 common events worldwide during 1997. Error bars ($\pm 2\sigma_D$) are shown for events with SEB depth estimates that differ significantly from the corresponding NEIC depth estimates.

Eqs. (8) and (10), the following scores are computed at the PIDC and included in the SEB for events with $m_b \geq 3.5$.

$$SCORE_{M_s:m_b} = [2.20 - (1.25\overline{m}_b - \overline{M}_s)]/2\sigma_M - 1.0; \text{ if there is an } M_s \text{ measurement}$$
$$= -999.0; \text{ otherwise} .$$

$$(12)$$

$$SCORE_{Depth} = (\hat{D} - 10.0 \text{ km})/2\sigma_D - 1.0; \text{ if the depth of the estimate is constrained}$$
$$= -999.0; \text{ otherwise} .$$

$$(13)$$

The IDC Technical Experts recommended that an event be screened out if $SCORE_{Depth}$ or $SCORE_{M_s:m_b}$ is greater than zero. The physical basis for this is that $M_s:m_b$ generally applies to shallow events. For deeper events, less long-period surface wave energy is generated, leading to $M_s:m_b$ values that may not satisfy the screening criterion. On the other hand, the depth criterion is intended to screen out deep earthquakes. Thus, the screening criteria are complementary.

Event-Screening Categories and Criteria

Table 1 summarizes the event-screening categories and associated criteria. Each event in the SEB is assigned to one of the four categories. Only events in the "Screened Out" category are excluded from the SSEB. Each event is also categorized as *offshore, onshore* or *mixed*, using the 90% location error ellipse and an onshore/offshore grid with five-minute resolution (based on ETOPO5). The fraction of offshore grid cells within or touching the 90% location error ellipse is computed and provided in the SEB and SSEB. However, this analysis is not used at this time to screen out any events.

Results of Applications

We now present results of applying the $M_s:m_b$ and depth-screening criteria to a very large sample of 41,731 events in the SEB with $m_b \geq 3.5$, between 1 January, 1996

Table 1

Summary of recommended screening categories and criteria

Screening category	Screening criteria
Not considered	$m_b < 3.5$
Insufficient data	$SCORE_{Depth} = -999.0$ and $SCORE_{M_s:m_b} = -999.0$
Not screened out	$SCORE_{Depth} \leq 0.0$ and $SCORE_{M_s:m_b} \leq 0.0$ and at least one score is greater than -999.0
Screened out	$SCORE_{Depth} > 0.0$ or $SCORE_{M_s:m_b} > 0.0$

and 30 June, 1998, and to 13 UNE's in the SEB between 1995 and 1998, four at Lop
Nor, one in India, two in Pakistan, and six at the French Polynesian test site. We also
apply the M_s:m_b screening criterion to 104 historical UNE's at the Nevada,
Semipalatinsk, Novaya Zemlya, and Lop Nor nuclear test sites, and to four peaceful
nuclear explosions (PNE's) in the Former Soviet Union.

Appropriately, all of the known explosions in the SEB have depths constrained to
the surface. Note also that M_s measurements are available in the SEB for only the
three UNE's that were conducted at the Lop Nor, Indian, and Pakistan test sites on
1996/06/08, 1998/05/11, and 1998/05/28, respectively. No PIDC M_s measurements
are available for two Lop Nor UNE's in 1995, nor for the six UNE's conducted by
France on the Mururoa and Fangataufa atolls between September 1995 and January
1996. M_s could not be measured for the 1998/05/30 Pakistan UNE due to a large
interfering earthquake and aftershock sequence in Afghanistan at that time.

M_s:m_b Screening Results

The M_s:m_b screening criterion, defined in Eq. (8), was applied to all 8608 SEB
events from 1 January, 1996 to 30 June, 1998 with $m_b \geq 3.5$ and an M_s measurement
– about 21 percent of the SEB events above m_b 3.5 during this period. (This
percentage has roughly doubled over time as PIDC capabilities to detect and measure
long-period surface waves have improved.) Included in this data set are the three
UNE's conducted at the Lop Nor, Indian, and Pakistan test sites on 1996/06/08,
1998/05/11, and 1998/05/28, respectively. The M_s:m_b criterion was also applied to
108 historical nuclear explosions, 45 at NTS, 48 at Semipalatinsk, 8 at Novaya
Zemlya, 3 at Lop Nor, and 4 PNE's in the Former Soviet Union. Note that the
magnitudes used here are based on historical m_b and M_s values that have been
adjusted for the known average biases of Eqs. (1) and (2) between NEIC and PIDC
magnitude estimates. This assumes that the average magnitude bias corrections,
inferred from SEB and NEIC earthquake data, are applicable to the magnitudes for
these explosions. The validity of this assumption is being carefully evaluated by
reprocessing waveform data for historical explosions.

Figure 12 shows M_s versus m_b for the SEB events and the 108 historical
explosions. The legend associates the various events with the marker types. All of the
explosions, including the three in the SEB, have values of $1.25\overline{m}_b - \overline{M}_s$ greater than
or equal to 2.20. Thus, none of the 111 UNE's and PNE's with available M_s:m_b data
are screened out, while 6841 of the 8608 SEB events are screened out based on the
M_s:m_b criterion. This corresponds to a screening rate of about 80 percent for events
with $m_b \geq 3.5$ and an M_s measurement. Of the SEB events that are not screened out,
most have values of $1.25\overline{m}_b - \overline{M}_s$ less than 2.20, but with some portion of their one-
sided 97.5% confidence interval greater than the threshold. These are typically events
with M_s computed for a single station, leading to relatively large uncertainty. About
two percent of the SEB events have values of $1.25\overline{m}_b - \overline{M}_s > 2.20$. Further

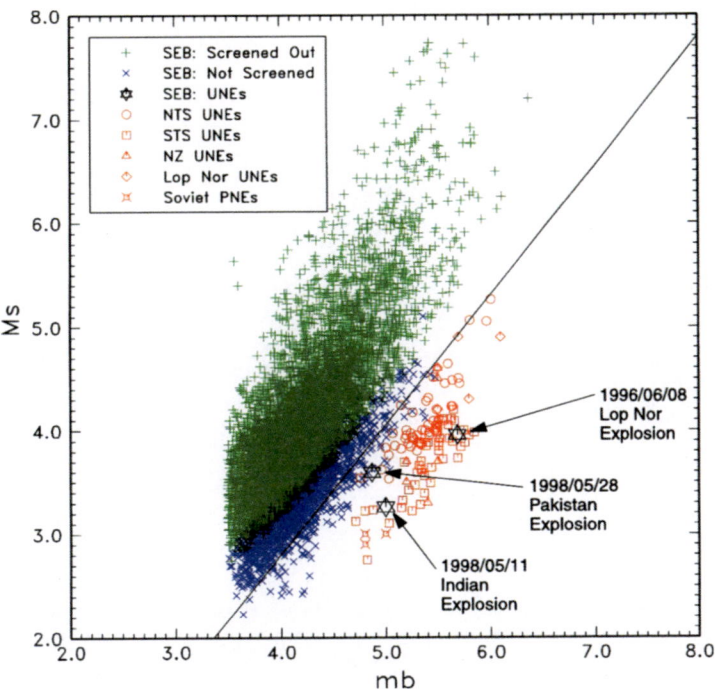

Figure 12

M_s versus m_b values for 8608 SEB events from 1 January, 1996 to 30 June, 1998 and a total of 111 underground nuclear explosions at the Nevada, Semipalatinsk, Novaya Zemlya, Lop Nor, Indian, and Pakistan test sites, and including four Soviet PNE's.

investigation is required, but it is likely that some of these events are associated with periodic data processing problems for M_s that occurred during the early stages of development. In addition, some of these events may actually be deep, although their depth estimates in the SEB are too poorly determined (or constrained to the surface by the analyst) to make this assessment. Such events, if actually deep, would be expected to have small M_s values relative to m_b.

Figure 13 shows a histogram of the M_s:m_b screening scores defined in Eq. (12). All of the available explosions have negative scores, indicating that they are not screened out. About 20 percent of the presumed earthquakes in the SEB also have negative scores, mostly due to large uncertainties. However, the three events in the SEB with the lowest, or among the lowest, M_s:m_b scores are the three known UNE's at the Lop Nor, Indian, and Pakistan test sites.

Depth Screening Results

The depth screening criterion was tested in two main ways: (1) by comparing to other reliable independent bulletins to assess the validity of the SEB depth confidence

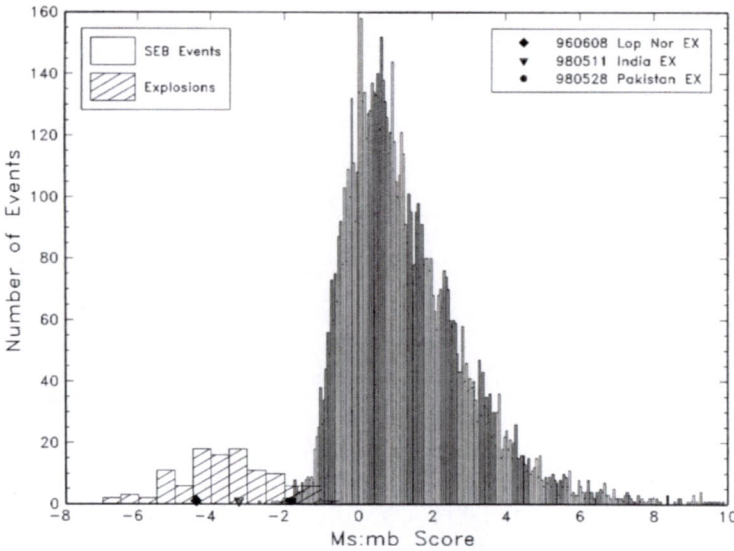

Figure 13

Histograms of $M_s:m_b$ screening scores for the same events as in Figure 12. All of the underground nuclear explosions have negative $M_s:m_b$ scores. About 80 percent of the SEB events with $M_s:m_b$ data have positive scores.

intervals (as described above); and (2) by assessing the number of SEB events that are screened out. Of the events in the SEB with $m_b \geq 3.5$ from 1 January, 1996 to 30 June, 1998, 13,926 events (about 34 percent) have depth estimates that are not constrained to the surface. Using $k = 40$ km for free-depth solutions (as recommended by the IDC Technical Experts in June 1998), about 46 percent of these events satisfy the depth screening criterion. Note that all known underground explosions in the SEB from 1995 to the present have their depths constrained to the surface. Thus, none of the known explosions in the SEB are screened out based on depth.

Overall Event-screening Performance

We now present results of applying the combined $M_s:m_b$ and depth-screening criteria. Table 2 provides representative numbers for a six-month period, 1 January to 30 June, 1998. During this period there were 8980 events in the SEB, of which 1936 (about 20 percent) were below m_b 3.5 and, hence, not considered for application of the screening criteria. The remaining numbers of events in the last three rows are divided into the location categories, defined above. Overall, about 41 percent of the events of $m_b \geq 3.5$ are screened out. It is evident from Table 2 that there are a significant number of events that are either not screened out (about 13 percent) or have insufficient data (about 46 percent) with which to apply the screening criteria.

Table 2

Screening results for seismic events in the SEB (1 Jan.–30 Jun. 1998)

Screening category	Number of events		
Total SEB events	8980		
Not considered	1936		
	Onshore	Mixed	Offshore
Insufficient data	474	1057	1708
Not screened out	185	218	499
Screened out	728	427	1748

Figure 14 depicts the cumulative screening results as functions of m_b for events during the same period. It indicates that the percentage of events that are screened out increases dramatically with increasing m_b (e.g., to about 77 percent for events above m_b 4.5). This is expected since larger events are typically recorded by more stations, improving the ability to compute surface-wave magnitudes (for relatively shallow events) and/or depth estimates with smaller uncertainty. Such events are more likely to satisfy the $M_s{:}m_b$ or depth screening criteria.

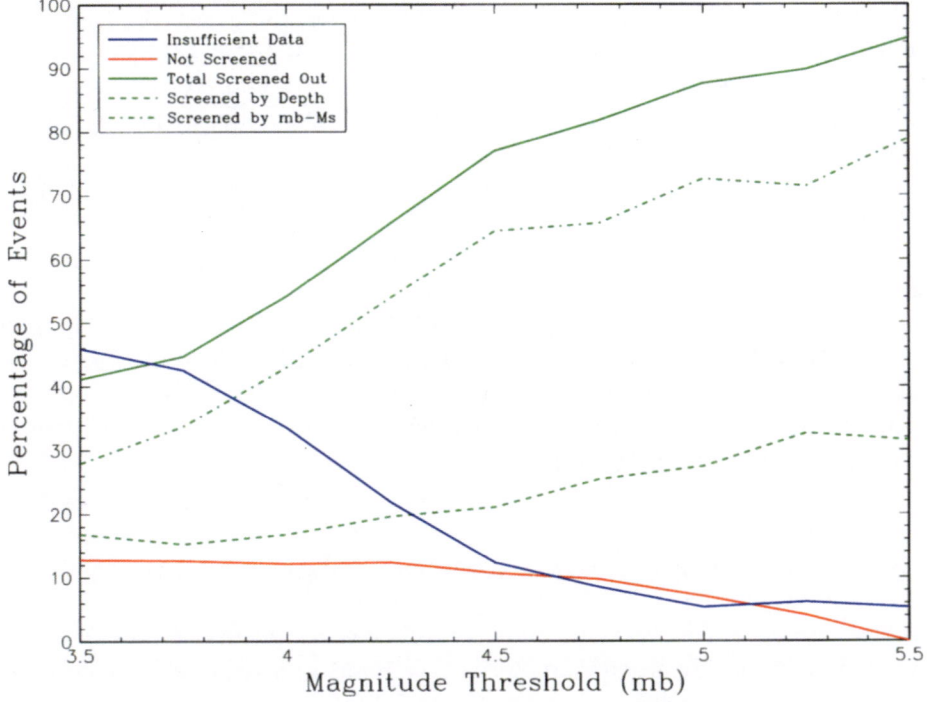

Figure 14

Percentages of SEB events during a six-month period (1 January, to 30 June, 1998) assigned to the various event-screening categories, as cumulative functions of magnitude (m_b), based on the $M_s{:}m_b$, depth, and combined screening criteria.

Conclusions

Experimental seismic event-screening criteria, based on $M_s{:}m_b$ and event depth, have been implemented and tested at the PIDC, with applications to thousands of SEB events and to historical explosions. Based on the results of testing, the criteria appear to be quite conservative. None of the available 117 UNE's at the Nevada, Semipalatinsk, Novaya Zemlya, Lop Nor, Indian, Pakistan, and French Polynesian test sites, nor four Soviet PNE's, are screened out, while about 41 percent of the SEB events above m_b 3.5 are screened out. Of 41731 SEB events above m_b 3.5 from 1 January, 1996 to 30 June, 1998, about 21 percent have M_s measurements and about 34 percent have depth estimates that are not constrained to the surface. About 80 percent of the events with $M_s{:}m_b$ data, and about 46 percent of the events with unconstrained depth estimates, are screened out. Note that the percentage of events with M_s data has increased significantly over time as PIDC capabilities to detect and measure long-period surface waves have been improved. For example, during the last six months of 1999, about 45 percent of SEB events above m_b 3.5 have M_s measurements.

Screening scores were presented that indicate numerically the degree to which an event meets, or does not meet, the $M_s{:}m_b$ and depth-screening criteria. Although a significant number of presumed earthquakes in the SEB are not screened out, due to the conservative nature of the screening criteria, the 1996/06/08 Lop Nor, the 1998/05/11 Indian, and the 1998/05/28 Pakistan UNE's have the lowest (or among the lowest) $M_s{:}m_b$ scores of all the events in the SEB.

Seismic events are also categorized as *onshore*, *offshore*, or *mixed*, using their 90% location error ellipses and an onshore/offshore grid with five-minute resolution (based on ETOPO5). However, this analysis is not used at this time to screen out any events.

As indicated in Figure 14, the percentage of events that are screened out increases dramatically with increasing m_b, as should be expected. Table 2 and Figure 14 also indicate that additional robust screening capabilities will need to be developed and tested, in order to improve the overall screening performance, particularly for events of relatively small seismic magnitude.

Discussion

There are several caveats to note regarding these results. First, the explosion data used in this study are limited to $m_b > 4.5$ and to a relatively small number of locations. Further work is needed to evaluate the robustness of the $M_s{:}m_b$ screening criterion for explosions below m_b 4.5 and at additional locations. Second, the SEB does not include a complete listing of global seismicity down to m_b 3.5, based on the current IMS network. Third, the calibration of the screening criteria may need to be adjusted over time as the IMS is completed and as additional data become available. Thus, the screening criteria and the results presented in this paper should be considered experimental and preliminary. In fact, we fully expect the explicit screening criteria and the screening performance to change over time as the IMS evolves and as enhanced or re-calibrated screening capabilities are introduced.

Significant improvements in screening performance, at fixed magnitude, are expected as the IMS network is completed, which should lead to better estimates of $M_s{:}m_b$ and/or depth for a greater percentage of events. It is also expected that many of the events in the *offshore* category can be screened out based on hydroacoustic characteristics (e.g., absences of a bubble pulse signature and high-frequency energy). Calibration of location estimates should also reduce the size of the location error ellipses, leading to fewer events in the *mixed* category, some of which will be in the *offshore* category and can be screened out by hydroacoustic criteria. Further improvements, particularly at lower magnitude, will require the use of regional seismic data.

At the third IDC Technical Experts Meeting on Event Screening, held on 21–25 May, 1999 in Vienna, Austria, recommendations were produced regarding future enhancements to the experimental screening procedures and criteria. The recommendations include: (1) The use of preliminary m_b station corrections to reduce the variance of network-averaged m_b values (station corrections for M_s are also under development); (2) reduction of the depth uncertainty term, k, from 40 km to 20 km; (3) improvements to the depth-phase validation criteria and related analyst procedures; and (4) screening criteria for offshore events based on the minimum ocean water depth within the location error ellipse, assessment of predicted hydroacoustic signal blockage, and hydroacoustic measures related to the presence or absence of a bubble pulse signature and/or high-frequency energy (e.g., above 32 Hz). They further recommended that an experimental screening approach, using regional high-frequency P- and S-wave amplitude ratios, be tested. These enhancements are expected to significantly improve the overall screening performance, while maintaining a conservative approach of not screening out any known explosions. We plan to report on the results of testing these recommendations in the future.

Acknowledgements

We would like to recognize the IDC Technical Experts on Event Screening for their contributions and recommendations regarding the experimental screening procedures and criteria presented in this paper. We also thank David Bowers for his comments on this manuscript. This work was supported by the Defense Threat Reduction Agency.

REFERENCES

BONDÁR, I. (1998), *Evaluation of PIDC Error Ellipses*, Technical Report CMR-98/23, Center for Monitoring Research, Arlington, VA.

BONDÁR, I., NORTH, R. G. and BEALL, G. (1999), *Teleseismic Slowness-azimuth Station Corrections for the International Monitoring System Seismic Network*, Bull. Seismol. Soc. Am. *89*, 989–1003.

BONDÁR, I., YANG, X., NORTH, R. G. and ROMNEY, C. (1999), *Location calibration data for CTBT Monitoring at the Prototype International Data Center*, Pure appl. geophys. Special Issue on Monitoring a Comprehensive Nuclear Test Ban Treaty: Source Location, in press.

BRATT, S. R. and BACHE, T. C. (1988), *Locating Events with a Sparse Network of Regional Arrays*, Bull. Seismol. Soc. Am. *78*, 780–798.

CTBT/PC/V/WGB/JP/6, *Case Study Using Japanese Events Inside a Very Dense Network*, submitted by the Japanese delegation to Working Group B, Preparatory Commission of the CTBT, Vienna, Austria, 1998.

CTBT/WGB-6/TL-2/10, *Report of the IDC Technical Experts Meeting on Event Screening*, 5–9 June 1998, Preparatory Commission of the CTBT, Vienna, Austria, 1998.

HE, D. (1999), personal communication.

FISK, M. D. and BONDÁR, I., *Evaluation of PIDC Depth Estimates*, Proc. IDC Technical Experts Meeting on Event Screening, 5–9 June 1998, Vienna, Austria, 1998.

GSE/CRP/243, Group of Scientific Experts, Conference Room Paper 243, Operations, Volume Two, Appendix E (available at *http://www.pidc.org*).

JEPSEN, D. and FISK, M. D. (1999), *Evaluation of PIDC Seismic Depth Estimates and Uncertainties*, Proc. IDC Technical Experts Meeting on Event Screening, 21–25 May 1999, Vienna, Austria.

JORDAN, T. H. and SVERDRUP, K. A. (1981), *Teleseismic Location Techniques and their Application to Earthquake Clusters in the South-central Pacific*, Bull. Seismol. Soc. Am. *71*, 1105–1130.

KENNETT, B. (1991), *IASPEI 1991 Seismological Tables*, Research School of Earth Sciences, Australian National University.

MARSHALL, P. D. and BASHAM, P. W. (1972), *Discrimination Between Earthquakes and Underground Explosions Employing and Improved Ms Scale*, Geophys. J. R. Astr. Soc. *28*, 431–458.

MARSHALL, P. D., SPRINGER, D. L. and RODEAN, H. C. (1979), *Magnitude Corrections for Attenuation in the Upper Mantle*, Geophys. J. R. Astr. Soc. *56*, 609–638.

MINSTER, J. B., JORDAN, T. H., MOLNAR, P. and HAINES, E. (1974) *Numerical Modeling of Instantaneous Plate Tectonics*, Geophys. J. R. Astr. Soc. *36*, 541–576.

MURPHY, J. R. (1997), *Calibration of IMS Magnitudes for Event Screening Using the M_s/m_b Criterion*, Proc. Event Screening Workshop, 4–7 November 1997, Beijing, China.

MURPHY, J. R. (1993), *Comment on 'Q for short-period P Waves: Is It Frequency-dependent?' by A. Douglas*, Geophys. J. Int. *113*, 535–540.

MURPHY, J. R., *P-wave coupling of underground explosions in various geologic media. In Identification of Seismic Sources – Earthquake or Underground Explosion* (E. S. Husebye and S. Mykkeltveit, eds.) (D. Reidel Publishing Company, Dordrecht 1981) pp. 201–205.

STEVENS, J. L. (1986), *Estimation of Scalar Moments from Explosion-generated Surface Waves*, Bull. Seismol. Soc. Am. 76, 123–151.

STEVENS, J. L. and DAY, S. M. (1985), *The Physical Basis of m_b:M_s and Variable Frequency Magnitude Methods for Earthquake/Explosion Discrimination*, J. Geophys. Res. 90, 3009–3020.

STEVENS, J. L., and MCLAUGHLIN, K. L. (1997), *Improved Methods for Regionalized Surface Wave Analysis*, MFD-TR-97-15887, Maxwell Technologies, San Diego, CA.

WGB/TL7/Rev. 2, Initial Plan for the Development of Event-Screening Criteria, Preparatory Commission of the CTBT, Vienna, Austria, 1997.

YANG, X., BONDÁR, I., MCLAUGHLIN, K., NORTH, R. G. and NAGY, W. (2000), *Path-dependent Travel-time Corrections for the International Monitoring System in North America*, Bull. Seismol. Soc. Am., Submitted.

YANG, X., BONDÁR, I., MCLAUGHLIN, K. and NORTH R. G. (1999), *Source Specific-station Corrections for Regional Phases at Fennoscandian Stations*, Pure appl. geophys. Special Issue on Monitoring a Comprehensive Nuclear-Test-Ban Treaty: Source Location, in press.

(Received June 25, 1999, revised December 11, 2000, accepted December 22, 2000)

Pure appl. geophys. 159 (2002) 889–903
0033–4553/02/040889–15 $ 1.50 + 0.20/0

⌈ Pure and Applied Geophysics

Testing for Multivariate Outliers in the Presence of Missing Data*

WAYNE A. WOODWARD,[1] STEPHEN R. SAIN,[1] H. L. GRAY,[1]
BOJUAN ZHAO,[1] and MARK D. FISK[2]

Abstract — We consider the problem of multivariate outlier testing for purposes of distinguishing seismic signals of underground nuclear events from training samples based on non-nuclear seismic events when certain data are missing. We consider the case in which the training data follow a multivariate normal distribution. Assume a potential outlier is observed on which k features of interest are measured. Assume further that the available training set of n observations on these k features is available but that some of the observations in the training data have missing features. The approach currently used in practice is to perform the outlier testing using a generalized likelihood ratio test procedure based only on the data vectors in the training data with complete data. When there is a substantial amount of missing data within the training set, use of this strategy may lead to a loss of valuable information. An alternative procedure is to incorporate all n of the data vectors in the training data using the EM algorithm to appropriately handle the missing data in the training set. Resampling methods are used to find appropriate critical regions. We use simulation results and analysis of models fit to Pg/Lg ratios for the WMQ station in China to compare these two strategies for dealing with missing data.

Key words: Outlier testing, nuclear monitoring, EM algorithm, multivariate normal, missing data.

1. Introduction

Presume we are given k observed features (P/S ratios, etc.) from a regional event of unknown source, and we wish to use these features to determine whether the event should be considered to be an outlier to the population of earthquakes in that region. Denote this observation by $\mathbf{X}_u = (X_{u1}, X_{u2}, \ldots, X_{uk})'$, where X_{uj} denotes feature j observed on the unknown event. Further, assume that there exists a training set of previous earthquakes in that region. If Π denotes the population of previous earthquakes, then the question of interest is one of deciding whether $X_u \in \Pi$ or $X_u \notin \Pi$. This question can be addressed by outlier testing. FISK *et al.* (1996) considered the problem of using a likelihood ratio test for detecting outliers from a multivariate normal (MVN) distribution fit to the training data. They applied the

---
* This research was partially supported by DTRA01-99-C-0018
[1] Southern Methodist University
[2] Mission Research Corporation

method to a variety of seismic data sets and demonstrated excellent results. TAYLOR and HARTSE (1997) also successfully applied this procedure to seismic data from the WMQ station in western China.

However, the methodology employed in these papers assumes no missing data. That is, given

$$\mathbf{X}_u = (X_{u1}, X_{u2}, X_{u3}, \ldots, X_{uk})',$$

then the assumption is that we have training data

$$\mathbf{X}_1 = (X_{11}, X_{12}, X_{13}, \ldots, X_{1k})'$$
$$\mathbf{X}_2 = (X_{21}, X_{22}, X_{23}, \ldots, X_{2k})'$$
$$\vdots$$
$$\mathbf{X}_n = (X_{n1}, X_{n2}, X_{n3}, \ldots, X_{nk})'.$$

In this case standard methodology applies. However, this problem, has commonly missing data due, for example, to poor signal-to-noise ratios at the higher frequencies. That is, the data most likely resembles the data vectors which follow:

$$\mathbf{X}_1 = (X_{11}, -, X_{13}, \ldots X_{1j}, -, X_{1,j+2}, \ldots X_{1k})'$$
$$\mathbf{X}_2 = (-, X_{22}, X_{23}, -, \ldots X_{2k})'$$
$$\vdots$$
$$\mathbf{X}_n = (X_{n1}, -, X_{n3}, \ldots X_{nk})'$$

where "$-$" denotes that the particular feature is missing for that event. Thus, to apply standard or recently developed outlier detection methodology, one must reduce the data set to a subset of the original training set denoted by

$$\mathbf{Y}_1 = (Y_{11}, Y_{12}, \ldots, Y_{1\ell})'$$
$$\mathbf{Y}_2 = (Y_{21}, Y_{22}, \ldots, Y_{2\ell})'$$
$$\vdots$$
$$\mathbf{Y}_m = (Y_{m1}, Y_{m2}, \ldots, Y_{m\ell})'$$

where $m < n$ and $\ell \leq k$. That is, it may be the case that if there is an excessive amount of missing data for some features, those features may be eliminated altogether in order to have a training sample with sufficient observations. When there is a nontrivial amount of missing data, it is clear that such a procedure will result in a substantial loss of data and associated information which leads to a loss of detection power. It should be noted that this problem increases as the dimension increases Consider, for example, the case in which 25% of the data are missing from a training sample of size $n = 50$ events. When data are missing at random, the expected number of complete cases (i.e., events for which all k features are available) is given by

$50(1 - 0.25)^k$. We give expected number of complete cases in Table 1 for $k = 2, \ldots, 6$.

The table shows that the sample size of complete vectors drops dramatically as the number of features increases. From the table we see, for example, that if we want to use four features, we would expect only about 16 of the original 50 events to have all four features recorded. Intuitively, though, it seems that there should be enough partial information in the data set to result in a substantial improvement over the use of only the 16 complete cases.

Another problem arises if there are no cases or only a very few cases in which all k of the features are observed. If the "complete-vector strategy" is being used, then some of the features may need to be deleted, i.e., $\ell < k$. This may result in some important feature variables not being used.

It should also be pointed out that we are not considering the case in which there is missing data in the outlier. In fact, in a practical situation, the features observed in the outlier determine the features to be used in the outlier test.

This paper develops a methodology which eliminates the requirement for complete data vectors in the training data, and hence allows all of the observed (non-missing) data to be used for outlier detection. This method is based on the EM algorithm (see DEMPSTER et al., 1977) and makes more optimal use of the available data than the approach of keeping only complete data vectors. In section 2 we summarize the outlier detection method for the no missing data scenario and then extend that methodology to eliminate the need for complete vectors. In Sections 3 and 4 we give empirical evidence which shows that, using the approach developed here, the detection power can be greatly improved for training sets with substantial missing data.

2. A New Outlier Testing Approach when Some Data are Missing

(a) The Modified Likelihood Ratio Test

As in the previous section, the training sample is denoted by

$$\mathbf{X}_1, \ldots, \mathbf{X}_n \in \Pi,$$

and it is considered to be a sample of size n from the population (Π) of non-nuclear events (possibly simply of earthquakes) in the region of interest. In this paper we

Table 1

Expected number of observations with complete data from a training sample of size $n = 50$ with 25% missing

Number of Features	2	3	4	5	6
Expected number of complete vectors	28	21	16	12	9

assume that Π is well modeled by a multivariate normal distribution or has been transformed to allow modeling by a multivariate normal distribution. A new observation, X_{n+1} (denoted as X_u in the previous section), is obtained, and given the training sample, we test the hypothesis

$$H_0 : \mathbf{X}_{n+1} \in \Pi$$

versus

$$H_1 : \mathbf{X}_{n+1} \notin \Pi.$$

The classical likelihood ratio test statistic is the ratio of the maximized likelihood functions under H_0 and H_1. We let $L_0(\theta) = \left(\prod_{s=1}^{n} f(\mathbf{X}_s; \theta) \right) f(\mathbf{X}_{n+1}; \theta)$ denote the likelihood function under H_0 (i.e., under the assumption that $\mathbf{X}_{n+1} \in H_0$ and where θ is an unknown vector-valued parameter associated with the distribution of $\mathbf{X}$ under H_0). We also let

$$\tilde{L}_1(\theta) = \prod_{s=1}^{n} f(\mathbf{X}_s; \theta)$$

denote the likelihood based only on the training sample $\mathbf{X}_1, \ldots, \mathbf{X}_n$ from the multivariate normal distribution. In the mixture-of-normals setting, WANG *et al.* (1997) and SAIN *et al.* (1999) used the likelihood-ratio test statistic

$$W = \frac{\sup\limits_{\theta \in \Theta} L_0(\theta)}{\sup\limits_{\theta \in \Theta} \tilde{L}_1(\theta)}. \tag{1}$$

The likelihood ratio rest used by FISK *et al.* (1996) and TAYLOR and HARTSE (1997) involves another factor in the denominator consisting of the likelihood function associated with the outlier population. However, little is known about the outlier population from which we have only one observation, X_{n+1}. FISK *et al.* (1996) and TAYLOR and HARTSE (1997) dealt with this problem by making the assumption that the covariance structure of the outlier population is the same as that of the training data, and they found critical values based on the F-distribution. In order to avoid the equal variance assumption (which does not appear to be justified), we use the test statistic in (1) which makes no assumption about the outlier population. The use of W in (1) disadvantageously does not allow for the use of the F-distribution for critical values. However, in the case considered here with missing data, it is not clear what the appropriate degrees-of-freedom would be for the F-statistic since the degrees-of-freedom are based on the number of observations. It should be noted that W in (1) was originally developed to provide an appropriate likelihood ratio test for the case in which the training data are modeled by a mixture of normals. Use of the test statistic proposed here in a bootstrap setting has been shown through simulations to have detection capabilities comparable to those of the test of FISK *et al.* (1996) and TAYLOR and HARTSE (1997) in the non-mixture case (see WANG *et al.*, 1997). We have also used it here for the sake of consistency with our recent papers (WANG *et al.*, 1997 and SAIN *et al.*, 1999) dealing with a mixture of normals (of which the normal is a special case).

It is easily seen in (1) that if $\mathbf{X}_{n+1}$ does not belong to Π, then W will tend to be small. Hence the rejection region is of the form $W \leq W_\alpha$ for some W_α picked to provide a level α test. The null distribution of W has no known closed form, thus we use a bootstrap procedure (see EFRON and TIBSHIRANI, 1993) to approximate it. Specifically,

Step 1: Given the training sample $\mathbf{X}_1, \ldots, \mathbf{X}_n$ and potential outlier $\mathbf{X}_{n+1}$, calculate W based on (1).

Step 2: For each integer $b, b = 1, \ldots, B$, draw a sample of size n from the training data. This could be done nonparametrically by sampling n times with replacement from the training sample or parametrically by generating n observations from the distribution fit to the training data. Additionally, an $(n+1)$st observation is also drawn from the population of the training sample, or its estimated distribution, since we are approximating the distribution of W when H_0 is true. For each b, we use the resampled data to compute the statistic in (1). This test statistic is denoted by W_b^*.

Step 3: Define W_α to be the (100α)th percentile of the W_b^*'s. Specifically, if $\alpha = j/(B+1)$, then W_α is the j-th smallest value of $\{W_b^*\}_{b=1}^B$ (see MCLACHLAN, 1987).

Step 4: Reject H_0 and conclude that the $(n+1)$st point is an outlier if $W \leq W_\alpha$.

(b) Use of the EM Algorithm to Estimate Model Parameters

The EM algorithm is a very general method for obtaining maximum likelihood estimates when some data are missing, and it is applicable to the multivariate normal case considered here. MILLER *et al.* (1993) consider the problem of applying the EM algorithm to the likelihood ratio statistic used by FISK *et al.* (1996) and TAYLOR and HARTSE (1997) when some data are missing. They give results of a small simulation study based on 10% missing data which compared the use of the EM algorithm with the very simple method of replacing missing data with means of available data. MILLER *et al.* (1994) consider the extension to the case in which some data may be categorical.

The EM algorithm may be thought of as a formal procedure for performing the following intuitively appealing approach to dealing with missing data: (a) Estimate missing values conditional on the observed values; (b) estimate the parameters using both observed data and missing value estimates; (c) using these parameter estimates, revise the estimates of the missing values. Iteration on (b) and (c) is continued until convergence is obtained. The EM algorithm thus has two steps: (I) The *expectation step (E-step)* in which the conditional expectation of the sufficient statistics is calculated given available data and the current estimates of the parameters, and (II) the *maximization step (M-step)* in which the estimated sufficient statistics calculated in the *E*-step are used to compute revised parameter estimates. When all data are available, then the sufficient statistics for the parameters of the multivariate normal model are given by

$$\mathbf{S} = \sum_{i=1}^{n} \mathbf{X}_i,$$

$$\mathbf{SS} = \sum_{i=1}^{n} \mathbf{X}_i \mathbf{X}_i'. \tag{2}$$

Parameter estimates are obtained as

$$\hat{\mu} = \frac{\mathbf{S}}{n},$$

$$\hat{\Sigma} = \frac{\mathbf{SS}}{n} - \hat{\mu}\hat{\mu}'. \tag{3}$$

When some data are missing, the *E*-step of the EM algorithm consists of finding conditional expectations of the missing data values and cross products needed for the calculation of the sufficient statistics, and the *M*-step involves simply calculating the corresponding parameter estimates using the conditional expectations of the sufficient statistics found in the *E*-step in place of the sufficient statistics themselves. Upon convergence of the EM algorithm, the parameter estimates obtained in the final iteration are called the EM estimates. It should be noted that the EM algorithm is guaranteed to increase the likelihood at each iteration, and thus convergence implies, at least, that a local maximum in the likelihood has been found. More details concerning the EM algorithm and the calculation of the conditional expectations can be found in JOHNSON and WICHERN (1998), LITTLE and RUBIN (1987) and MILLER *et al.* (1993).

(c) The Outlier Testing Algorithm

In this section we discuss an algorithm for outlier testing when some data are missing. It is important to note that we assume the outlier contains no missing data, since the data available in the outlier define the variables that are appropriate for use in the outlier testing. Thus the EM algorithm is not being used to "fill in" missing feature values in the event to be tested since this event defines the features to be used in the outlier test. The algorithm is as follows:

Step 1: Using the k features under consideration, fit a multivariate normal to the training data using the EM algorithm.

Step 2: The modified likelihood ratio statistic W, as defined in (1), is calculated for the data. The likelihood function in the denominator of W is calculated using the EM algorithm for the n events in the training sample. The numerator of W is obtained by augmenting the training sample with the outlier point and recalculating the EM estimates and associated likelihood function.

Step 3: The bootstrap is used to find the distribution of W. At each bootstrap iteration we use the starting values obtained from the n events in the training sample. Our studies have shown that parametric bootstrapping is preferable when substantial data is missing. This is easily understood by recalling that when resampling to obtain $n + 1$ events from the training data, the $(n + 1)$st event must have complete data for the features under consideration. If we used nonparametric bootstrapping, then only those training sample values with complete data would be available for resampling as the $(n + 1)$st event. Thus, if there are only a few observations in the training data with complete data, then nonparametric bootstrapping would not be desirable. Consequently, we will use parametric bootstrapping in this paper, and each of the B bootstrap samples is generated with the same pattern of missing data as is seen in the training data.

It should be noted that an assumption of the EM algorithm is that data are missing at random. In some cases in the current application there are missing data patterns, and the missing at random assumption must be questioned. For example, $\log(P_g/L_g)$ ratios at higher frequency bands are more likely to be missing than are those at lower frequencies because of attenuation effects on high frequencies, and consequently, data for a given event often have the pattern that if a $\log(P_g/L_g)$ ratio is missing at a given frequency, then it is also likely to be missing at all higher frequencies. In order to account for this, as was mentioned in Step 3 above, we always resample with the same missing data structure that was present in the original data set so that our critical regions are appropriate regardless of the underlying missing data mechanism.

3. Simulations

Consider the situation in which the event in question is measured at three feature variables which we want to simultaneously use in the outlier testing. The question is whether it is preferable to do the outlier testing: (a) Using only those data vectors in the training set for which all three features were observed, or (b) using all data vectors for which at least one of the three features variables was observed.

Table 2

Empirical detection probabilities comparing use of EM versus complete vectors only. (1000 replications)

		(a)		(b)		(c)		(d)	
		Complete vectors	EM	Complete vectors	EM	Complete vectors	EM	Complete vectors	EM
	0.00	0.944 (40) (0.061)	–	0.824 (40) (0.060)	–	0.700 (40) (0.054)	–	0.688 (40) (0.045)	–
p_m	0.38	0.805 (10) (0.065)	0.901 (0.043)	0.642 (10) (0.083)	0.784 (0.055)	0.542 (10) (0.087)	0.641 (075)	0.578 (10) (0.099)	0.610 (0.066)
	0.50	0.637 (6) (0.118)	0.876 (0.067)	0.562 (6) (0.071)	0.748 (0.072)	0.425 (6) (0.100)	0.620 (0.078)	0.446 (6) (0.094)	0.591 (0.064)
		SE = 0.016 (SE = 0.007)							

Consider the case in which a training sample of size n is obtained from the multivariate normal population with μ and covariance Σ. Further suppose that the new event comes from a population that is multivariate normal with mean μ_0 and covariance Σ_0. In Table 2 we show results for several values of μ, Σ, μ_0, and Σ_0. In the table we show the empirical powers based on 1000 replications. In all examples reported here, we use $B = 199$ in the bootstrap-based outlier test. The "complete vector" columns of the table correspond to the case in which only complete data vectors are retained and the second column corresponds to the case in which the EM algorithm is used on the complete data set consisting of all observations for which at least one of the three features is available. The rows in the table correspond to the missing data scenario. We consider a variety of probabilities of a missing feature (p_m) in order to illustrate the impact of various amounts of missing data. In the simulations we use $p_m = 0$ (i.e., there is no missing data), 0.38, and 0.5. Let the generated observations from the training sample be denoted by $x_i = (x_{i1}, x_{i2}, x_{i3})', i = 1, \ldots, n$. A random procedure is used to give each of the $x_{ij}, (i = 1, \ldots, n$ and $j = 1, 2, 3)$ a p_m probability of being declared missing and thus replaced in the data set by a missing data indicator. If, however, by using this procedure, all three features of a vector, $x_{i'}$, in the simulated training sample are declared to be missing, we then repeat the procedure of randomly assigning these individual features as missing until at least one of $x_{i'1}, x_{i'2}$, and $x_{i'}$ is not declared to be missing. Based on a given value of p_m and this procedure for assigning missing values, the expected number of complete vectors, i.e. vectors for which all of the observations are available, is given by $n_F = (n(1 - p_m)^3)/(1 - p_m^3)$ which (because of the simulation strategy mentioned here) is slightly different from the expected sample size as discussed in Table 1. See the Appendix for more details on the calculation of n_F. In the "complete vector" column we indicate in parentheses the expected number of retained vectors in each case. The "EM" columns in Table 2 correspond to the case in which all n observations, some of which may contain missing features, are analyzed using the EM algorithm. The numbers given (without parentheses) in the body of the table are empirical detection probabilities, i.e., they are the proportion of the 1000 replications for which the outlier was detected. The $p_m = 0$ entry in the "EM" column is left blank in each case since there is no missing data, and obviously, in this case the EM algorithm requires no iteration and would produce the results given in the "complete vectors" column. In the "complete vectors" columns of the table it can be seen that, as would be expected, the presence of missing data has reduced the detection probability. Also, it is seen that the detection probabilities using the EM-algorithm are higher. It can be seen that the difference in detection probabilities is particularly substantial when $p_m = 0.50$. Also given in the table in parentheses under the detection probabilities are the observed false alarm rates under each of the scenarios considered. In each case these are obtained using simulations identical to those used for the detection probabilities but for which the "outlier" to be tested in each case is generated from the same distribution as the training data. While these tend in general to be somewhat inflated over the nominal 0.05 level, especially for the smaller number of observations, it can be seen that this is much more pronounced for the complete vectors cases with small expected small sizes. It should be noted that when the expected sample size is small, occasionally the actual number of complete vectors out of a sample size of $n = 40$ will be so small that the outlier test fails because of insufficient sample size. This occurred about 15% of the time when $p_m = 0.38$ (i.e., the expected sample size is 10) and about 30% of the time when $p_m = 0.50$ (i.e., the expected sample size is 6). The detection probabilities and false alarm rates in these cases are calculated over the simulated samples for which the outlier test could be run. It should be noted that in all cases simulated, the results using the EM algorithm were obtainable.

(a) Training sample of size $n = 40$ from MVN $(\boldsymbol{\mu}, \Sigma)$ where $\boldsymbol{\mu} = (0,0,0)'$ and

$$\Sigma = \begin{pmatrix} 1 & 0 & 0 \\ 0 & 1 & 0 \\ 0 & 0 & 1 \end{pmatrix}.$$ The outlier is from a MVN $(\boldsymbol{\mu}_0, \Sigma)$ where $\boldsymbol{\mu}_0 = (2.5, 2.5, 2.5)'$.

(b) Training sample of size $n = 40$ from MVN $(\boldsymbol{\mu}, \Sigma)$ where $\boldsymbol{\mu} = (0,0,0)'$ and

$$\Sigma = \begin{pmatrix} 1 & 0.2 & 0.2 \\ 0.2 & 1 & 0.2 \\ 0.2 & 0.2 & 1 \end{pmatrix}.$$ The outlier is from a MVN $(\boldsymbol{\mu}_0, \Sigma)$ where $\boldsymbol{\mu}_0 = (2.5, 2.5, 2.5)'$.

(c) Training samples of size $n = 40$ from MVN $(\boldsymbol{\mu}, \Sigma)$ where $\boldsymbol{\mu} = (0,0,0)'$ and

$$\Sigma = \begin{pmatrix} 1 & 0.5 & 0.5 \\ 0.5 & 1 & 0.5 \\ 0.5 & 0.5 & 1 \end{pmatrix}.$$ The outlier is from a MVN $(\boldsymbol{\mu}_0, \Sigma)$ where $\boldsymbol{\mu}_0 = (2.5, 2.5, 2.5)'$.

(d) Training samples of size $n = 40$ form MVN $(\boldsymbol{\mu}, \Sigma)$ where $\boldsymbol{\mu} = (0, 0, 0)'$ and

$$\Sigma = \begin{pmatrix} 1 & -0.7 & -0.7 \\ -0.7 & 1 & 0.7 \\ -0.7 & 0.7 & 1 \end{pmatrix}. \text{ The outlier is from a MVN } (\boldsymbol{\mu}_0, \Sigma) \text{ where}$$

$$\boldsymbol{\mu}_0 = (1, 1, 1)' \text{ and } \Sigma_0 = \begin{pmatrix} 1 & 0 & 0 \\ 0 & 1 & 0 \\ 0 & 0 & 1 \end{pmatrix}.$$

4. Outlier Testing Based on WMQ Data

In this section we discuss simulation results based on the $\log(P_g/L_g)$ ratios at various frequency bands from the WMQ station in western China (see HARTSE *et al.*, 1997). The training data consists of $n = 134$ events which are primarily earthquakes. Data are also available on a few nuclear events observed at WMQ. The data have been distance corrected as discussed in HARTSE *et al.* (1997). It should be noted that there were no missing observations in the data analyzed by SAIN *et al.* (1999). Those authors analyzed the WNQ data and presented their results using a bivariate feature vector consisting of $\log(P_g/L_g)$ ratios for bands 1–2 Hz and 4–8 Hz since these features were the best pair of features for outlier detection. The bivariate case was used by those authors to allow graphical illustration of results, but simulations not shown there indicated that the best set of features for outlier detection were the three $\log(P_g/L_g)$ ratios in the frequency bands 0.5–1, 1.5–3, and 4–8 Hz. The current analysis of the WMQ data will be based on this three-component feature vector. Using these three features, a multivariate normal fit to earthquake training data has mean $\boldsymbol{\mu}_{EQ} = (-0.024, 0.016, 0.015)'$, where the features are in the order 0.5–1, 1.5–3, and 4–8 Hz respectively, and covariance

$$\Sigma_{EQ} = \begin{pmatrix} 0.0468 & 0.0268 & 0.0134 \\ 0.0268 & 0.0319 & 0.0170 \\ 0.0134 & 0.0170 & 0.0188 \end{pmatrix}.$$

The multivariate normal fit to the nuclear explosions has mean $(-0.044. 0.559, 0.637)'$ and covariance

$$\Sigma_0 = \begin{pmatrix} 0.0058 & -0.0041 & -0.0002 \\ -0.0041 & 0.0190 & 0.0038 \\ -0.0002 & 0.0038 & 0.0018 \end{pmatrix}.$$

In Figure 1 we show the three two-dimensional scatterplots and the three-dimensional scatterplot for the China data. There it can be seen that the nuclear data (indicated by triangles) is very well separated from the earthquake events (denoted by squares). It should be noted that the earthquakes in this data set occur in a wide variety of locations with good azimuthal and distance distribution and in fact

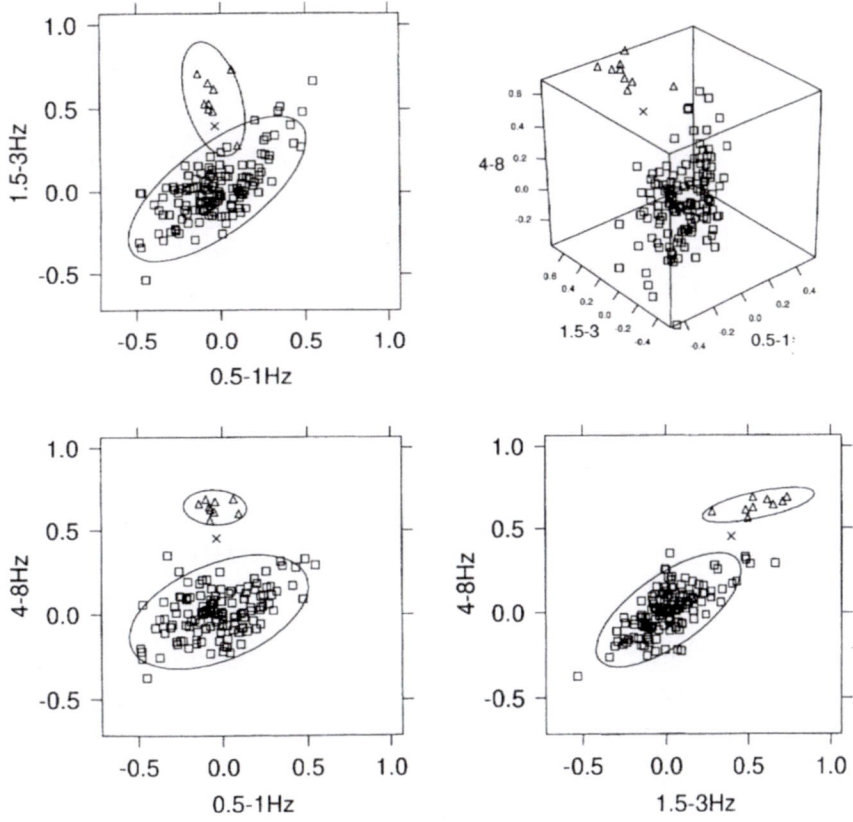

Figure 1
Two-dimensional and three-dimensional scatterplots for the China data.

some of the earthquakes are close to Lop Nor. Thus, we believe that the differences observed reflect real differences between the two source types and are not simply there because of site-path differences between the two classes. SAIN et al. (1999) mention that, in this case, the nuclear explosion population is so far removed from the population of the training data that detection is essentially assured using any reasonable approach. Thus, in order to investigate the detection capability of the outlier test when the separation between the outlier population and the training data is not so extreme, we will use the approach of Sain et al. (1999) of artificially moving the outlier population closer to the training data along a path connecting the mean of the earthquake data and the mean of the nuclear data. Specifically, in the simulations given here we consider the outlier population to be multivariate normal with mean $\mu_0 = (-0.038, 0.396, 0.450)'$ and with covariance Σ_0 given above. It is clear that $\log(P_g/L_g)$ ratio in the frequency band 4–8 Hz is the best single feature. However, it is also clear that its use in combination with one or both of the other features increases

Table 3

Empirical detection probabilities using the models fit to the WMQ data. Training samples from MVN
$(\mathbf{\mu}_{EQ}, \Sigma_{EQ})$. *Outlier from MVN* $(\mathbf{\mu}_0, \Sigma_0)$ *given in section 3*

		$n = 50$		$n = 100$		$n = 150$	
		Complete vectors	EM	Complete vectors	EM	Complete vectors	EM
	0.00	0.958 (50) (0.041)	–	0.964 (100) (0.057)	–	0.968 (150) (0.051)	–
p_m	0.38	0.764 (13) (0.080)	0.891 (0.064)	0.892 (25) (0.058)	0.956 (0.051)	0.934 (38) (0.052)	0.967 (0.051)
	0.50	0.610 (7) (0.089)	0.854 (0.070)	0.813 (14) (0.078)	0.937 (0.063)	0.894 (21) (0.061)	0.949 (0.055)
		SE = 0.016 (SE = 0.007)					

the separation between the two populations. The " × " indicates the position of the
new mean of the outlier population $\mathbf{\mu}_0$ (or its two-dimensional version) used in the
simulations.

In Table 3 we report the results of a simulation study based on the multivariate
normal distribution fit to the earthquake training data, with the outlier population
being the multivariate normal distribution with mean and covariance given by $\mathbf{\mu}_0$ and
Σ_0 above. As in Table 2 we give the empirical detection probabilities based on 1000
replications with $B = 199$ in the bootstrap-based outlier test. The table is constructed
in the same fashion as Table 2 except that we consider three training sample sizes,
$n = 50, 100$ and 150. We again consider the cases in which p_m takes on the values 0,
0.38, and 0.5.

In Table 3 we again see that in every case, the use of the EM algorithm improves
the detection probability over the use of only the complete vectors. As would be
expected, this improvement is more apparent when the percent missing is higher and
when the original sample size is smaller. Of particular interest is the fact that for
$n = 100$ and $n = 150$, the use of the EM algorithm with $p_m = 0.38$ and $p_m = 0.50$
yields detection probabilities very similar to those given in the first row of the table
for the case in which no data are missing. False alarm rates are shown, and these are
seen to be inflated in the "complete vector" cases when $p_m = 0.38$ and $p_m = 0.50$ for
$n = 50$ and when $p_m = 0.50$ for $n = 100$. In these simulations the test failed because of
insufficient sample size in some of the "complete vector" simulations. Of the 1000
repetitions, this happened 126 times when $n = 50$ and $p_m = 0.50$, 3 times when $n = 50$
and $p_m = 0.38$, and 1 time when $n = 100$ and $p_m = 0.50$. As with the simulations
reported in Table 2, in all cases simulated the results using the EM algorithm were
obtainable.

5. Concluding Remarks

In this paper we have shown that use of the EM algorithm can provide a substantial gain in detection probability over the use of only the complete data vectors. This gain is seen to be substantial when the use of only complete vectors results in a sample size that is quite small. Some algorithms currently in use for outlier testing based on only complete data, delete a feature or use another feature or set of features when complete vector sample sizes become too small. However, if in so doing, a good feature is replaced by an inferior one, then detection capabilities may suffer dramatically. For example, in the WMQ data, the high frequency P_g/L_g ratios are better discriminators than those ratios at low frequencies. Although the data we used as our baseline data set had no missing data, as was mentioned previously it is the case in practice that the high frequency ratios are often not observed due to attenuation effects on high frequencies. However, it is clear that elimination of high frequency ratios from the analysis, because of sample size considerations, would probably be a very poor decision. Use of the EM algorithm allows the retention of good feature variables even when the data on these features are sparse.

Another comment regards the observed significance levels (i.e., the false alarm rates) of outlier tests based on (1). WANG et al. (1997) note that the modified likelihood testing procedure given in Section 2(b) has observed significance levels (false alarm rates) that are slightly higher than nominal levels. In the case in Table 1 these typically ranged 0.06–0.07 when the nominal level was 0.05. However, the exception to this was in the complete data case (i.e., the EM algorithm is not used) with very small sample size, e.g., $n_F = 6$. In these cases, the significance levels were inflated to about 0.09. Our simulations show that the use of the EM algorithm helps control this false alarm rate over that which would be obtained using a very small number of complete vectors.

WANG et al. (1997) and SAIN et al. (1999) have considered the problem of modeling the training data as a mixture of normals due to the existence of more than one event type in the training data or to deal with non-normality of the training data. We are currently investigating the use of the EM algorithm to deal with missing data in the mixture setting.

Appendix

Consider the case of k feature variables, each of which has $p_m^{(i)} \times 100\%$ chance of being missing, $i = 1, \ldots, k$. That is, we assume that the features are missing at random although there may be differing missing value probabilities for different features. For example, in many regions the P_g/L_g ratio at high frequencies is more likely to be missing than the P_g/L_g ratio at low frequencies. The algorithm used in our simulation studies for assigning missing values to an observed k-component feature vector $(x_1, \ldots, x_k)'$ is as follows:

(a)   Generate k independent uniform $(0, 1)$ deviates, $U_1, \ldots, U_k$.

(b)   If $U_i < p_m^{(i)}$ we set feature x_i to the missing value code.

(c)   If at least one of the features in this vector is declared not missing, we use the resulting vector (containing possibly some missing values) as the observation vector. If all features in this resulting vector are declared to be missing, then we restore the original values $(x_1, \ldots, x_k)'$ to the vector observation and repeat (a) and (b) until the resulting vector is not all missing.

Since we are comparing the use of the EM algorithm with that of using only those vectors for which all features are observed for a given set of $p_m^{(i)}, \ldots, p_m^{(k)}$, it is informative to know the probability that a given vector will have all features observed. The probability that all features are present in a k-feature vector is given by the probability that all features are present on the first application of (a) and (b), plus the probability that all features are missing on the first application of (a) and (b) but that all are present on the second application, and so forth. Specifically this probability is

$$
(1 - p_m^{(1)}) \cdots (1 - p_m^{(k)}) + p_m^{(1)} \cdots p_m^{(k)} (1 - p_m^{(1)}) \cdots (1 - p_m^{(k)})
$$
$$
+ (p_m^{(1)} \cdots p_m^{(k)})^2 (1 - p_m^{(1)}) \cdots (1 - p_m^{(k)}) + \cdots
$$
$$
= \frac{(1 - p_m^{(1)}) \cdots (1 - p_m^{(k)})}{1 - p_m^{(1)} \cdots p_m^{(k)}}.
$$

Therefore, the expected number of complete data vectors in a sample of size n is $n_c = n(1 - p_m^{(1)}) \cdots (1 - p_m^{(k)})/(1 - p_m^{(1)} \cdots p_m^{(k)})$. For example, if $n = 40$ and $p_m^{(1)} = p_m^{(2)} = p_m^{(3)} = 0.5$, then the expected number of complete data vectors is $n_c = 40(1 - 0.5)^3/(1 - 0.5^3) = 6$. Note that this is slightly different from the formula used to calculate the values in Table 1. In that case, we were calculating the expected number of complete cases based on an unconditional calculation. However, in the Appendix the formula for the expected number of complete cases is based on our simulation procedure that involves resampling if all observations in a simulated sample vector are missing.

References

DEMPSTER, A. P., LAIRD, N. M., and RUBIN, D. B. (1977), *Maximum Likelihood Estimation from Incomplete Data via the EM Algorithm (with discussion)*, J. Roy. Statist. Soc. B39, 1–38.

EFRON, B. and TIBSHIRANI, R. J. *An Introduction to the Bootstrap.* (Chapman and Hall, New York 1993).

FISK, M. D., GRAY, H. L., and McCARTOR, G (1996), *Regional Event Discrimination Without Transporting Thresholds*, Bull. Seismol. Soc. Am. *86*, 1545–1558.

HARTSE, H. E., TAYLOR, S. R., PHILLIPS, W. S., and RANDALL, G. E. (1997), *A Preliminary Study of Regional Seismic Discrimination in Central Asia with Emphasis on Western China*, Bull. Seismol. Soc. Am. *87*, 551–568.

JOHNSON, R. A. and WICHERN, D. W., *Applied Multivariate Statistical Analysis*, Fourth Edition (Upper Saddle River, New Jersey: Prentice Hall 1998).

LITTLE, R. J. A. and RUBIN, D. B., *Statistical Analysis with Missing Data* (John Wiley and Sons, Inc. New York 1987).

MCLACHLAN, G. J. (1987), *On Bootstrapping the Likelihood Ratio Test Statistic for the Number of Components in a Normal Mixture*, Appl. Statist. *36*, 318–324.

MILLER, J. W., GRAY, H. L., and WOODWARD, W. A. (1993), *Discriminant Analysis and Outlier Testing when Data are Missing*, Phillips Laboratory Technical Report, ARPA F29601-91-K-DB25.

MILLER, J. W., WOODWARD, W. A., GRAY, H. L., FISK, M. A., and MCCARTOR, (1994), *A Hypothesis-testing Approach to Discriminant Analysis with Mixed Categorical and Continuous Variables when Data are Missing*, Technical Report No. SMU/DS/TR-273, Department of Statistical Science, Southern Methodist University.

SAIN, S. R., GRAY, H. L., WOODWORD, W. A., and FISK, M. D. (1999), *Outlier Detection when Training Data are Unlabled*, Bull. Seismol. Soc. Am. *89*, 294–304.

TAYLOR, S. R. and HARTSE, H. E. (1997), *An Evaluation of Generalized Likelihood Ratio Outlier Detection to Identification of Seismic Events in Western China*, Bull. Seismol. Soc. Am. *87*, 824–831.

WANG, S., WOODWARD, W. A., GRAY, H. L., WIECHECKI, S., and SAIN, S. R. (1997), *A New test for Outlier Detection from a Multivariate Mixture Distribution*, J. Computat. and Graph. Statist. *6*, 285–299.

(Received June 25, 1999, revised December 12, 2000, accepted December 20, 2000)

To access this journal online:
http://www.birkhauser.ch